WILDLAND FIREFIGHTING PRACTICES

WILDLAND FIREFIGHTING PRACTICES

Joseph D. Lowe

Africa • Australia • Canada • Denmark • Japan • Mexico • New Zealand • Philippines •
Puerto Rico • Singapore • Spain • United Kingdom • United States

NOTICE TO THE READER

Publisher does not warrant or guarantee any of the products described herein or perform any independent analysis in connection with any of the product information contained herein. Publisher does not assume, and expressly disclaims, any obligation to obtain and include information other than that provided to it by the manufacturer.

The reader is expressly warned to consider and adopt all safety precautions that might be indicated by the activities described herein and to avoid all potential hazards. By following the instructions contained herein, the reader willingly assumes all risks in connection with such instructions.

The publisher makes no representations or warranties of any kind, including but not limited to, the warranties of fitness for particular purpose or merchantability, nor are any such representations implied with respect to the material set forth herein, and the publisher takes no responsibility with respect to such material. The publisher shall not be liable for any special, consequential, or exemplary damages resulting, in whole or in part, from the readers' use of, or reliance upon, this material.

Cover photo: the Laguna Beach Fire, courtesy of Battalion Chief Pete Lawrence, Oceanside, Calif.

Delmar Staff

Business Unit Director: Alar Elken:
Executive Editor: Sandy Clark
Acquisitions Editor: Mark Huth
Developmental Editor: Jeanne Mesick
Editorial Assistant: Dawn Daugherty
Executive Marketing Manager: Maura Theriault
Channel Manager: Mona Caron
Marketing Coordinator: Kasey Young
Executive Production Manager: Mary Ellen Black
Project editor: Barbara L. Diaz
Art director: Rachel Baker

Printed in the United States of America
4 5 6 7 8 9 10 XXX 05

For more information, contact Delmar at 3 Columbia Circle, PO Box 15015, Albany, New York 12212-5015; or find us on the World Wide Web at http://www.delmar.com

Asia
Thomson Learning
60 Albert Street, #15-01
Albert Complex
Singapore 189969

Australia/New Zealand
Nelson/Thomson Learning
102 Dodds Street
South Melbourne, Victoria 3205
Australia

Canada
Nelson/Thomson Learning
1120 Birchmont Road
Scarborough, Ontario
Canada M1K 5G4

International Headquarters
Thomson Learning
International Division
290 Harbor Drive 2nd Floor
Stamford, CT 06902-7477
USA

Japan
Thomson Learning
Palaceside Building 5F
1-1-1 Hitotsubashi, Chiyoda-ku
Tokyo 100 0003
Japan

Latin America
Thomson Learning
Seneca, 53
Colonia Polanco
11560 Mexico D. F. Mexico

Spain
Thomson Learning
Calle Magallanes, 25
28015-Madrid
Espana

UK/Europe/Middle East
Thomson Learning
Berkshire House
168-173 High Holborn
London
WC1V7AA United Kingdom

Thomas Nelson & Sons Ltd.
Nelson House
Mayfield Road
Walton-on-Thames
KT 12 5PL United Kingdom

Library of Congress Cataloging-in-Publication Data

Lowe, Joseph.
Wildland firefighting practices / Joseph Lowe.
p. cm.
Includes index.
ISBN 0-7668-0147-0 (alk. paper)
1. Wildfires—Prevention and control. I. Title.
SD421 .L69 2000
634.9'618—dc21

00-060322

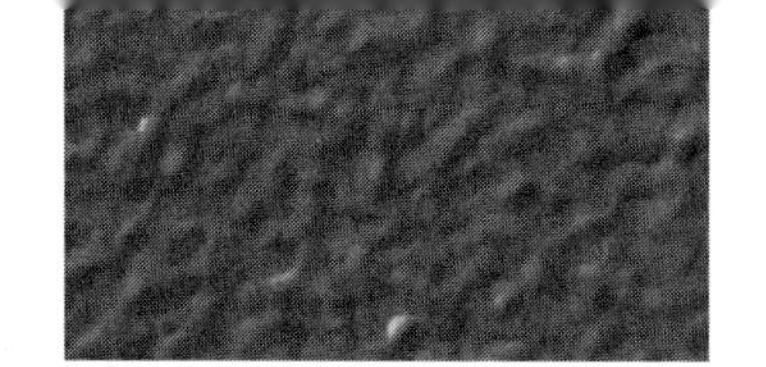

Contents

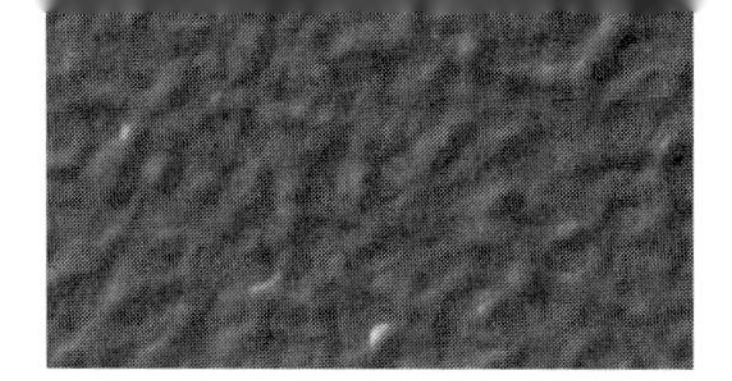

Preface

Wildland firefighting requires stamina, fortitude, awareness, and an understanding of the factors that drive the fire. This book represents a complete treatise on the subject.

ABOUT THE BOOK

This text was written for both the new wildland firefighter as well as those who are already involved in wildland fire suppression. For those already involved in fighting wildland fires, it can serve as a reference text to be used during the department's annual refresher training classes.

This book covers all aspects of wildland firefighting and introduces new advances in technology for wildland fire suppression, such as fire-blocking gels and the use of GPS. The book was written to provide actual how-to knowledge of those basic skills needed by the wildland firefighter. Numerous pictures and illustrations demonstrate or further support what is written.

The latest information about the use of a fire shelter, a new deployment technique, and the human factors a wildland firefighter needs to watch out for are also included.

HOW TO USE THIS TEXTBOOK

The text is divided into thirteen chapters. It starts with an overview of how wildland fires are fought and then covers each subject area in depth. Some chapters, such as the chapter on wildland fire safety and fire behavior, are more important than others.

A list at the beginning of each chapter defines the learning objectives for the chapter. First, skim over the chapter and make note of all key points and safety concerns, which are identified in the margin. This will give you a sense of what the chapter is about. Then go back through and read it for understanding. When you have completed reading the chapter, take the review test at the end of the chapter. If a question is missed, go back through the chapter and review the portion that applies to the missed question. An extensive glossary has been included to help you with any term that you do not understand.

ABOUT THE AUTHOR

Joseph Lowe is a seasoned Battalion Chief who has spent most of his career in wildland fire suppression. He works for the San Bernardino County (California) Fire Department, the largest county in the United States. Chief Lowe had twenty years of experience with the Orange County Fire Authority, including several years on the department's numerous Type III Engines, in the development and training of a handcrew academy, and as burn boss for the

department. He was Fire Management Officer for Airborne Fire Attack, which operates several PBY water-scooping air tankers in southern California.

He has completed all the National Wildfire Coordinating Group (NWCG) fire behavior classes through S490, "Advanced Fire Behavior Calculations," as well as RX 340, "Fire Effects." Fire Effects is a NWCG class that deals with how fire affects the forest ecosystem.

Battalion Chief Lowe's incident command system qualifications are as follows:

- Safety Officer
- Situation Unit Leader
- Field Observer
- Helispot Manager
- Air Support Group Supervisor
- Strike Team Leader
- Division Supervisor
- Check-In Recorder

Lowe and his wife own and operate Wildland Fire Consultants and Seminars, which produces the *Wildland Essentials* video series. These videos have been purchased by thousands of fire agencies throughout the United States and Canada. Wildland Fire Consultants and Seminars' office number is 760-942-4684.

ACKNOWLEDGMENTS

I am a product of all those I have fought fire with and those mentors who have invested in me. They helped give me the knowledge to write this book.

I want to thank the Orange County Fire Authority for sending me to all the best fire schools that are available. I applaud those within the organization who have worked in its brush stations, on handcrews, and who operate dozers, as they share the same passion for wildland fire suppression as I do. A special thanks to Battalion Chief Marc Hawkins, Crew Foreman Stan Sutton, Superintendent Ole Bardwell, Captain Rick Reeder, Captain Kirk Summers, and former USDA Forest Service FMO and current Assistant Chief of the Governor's Office of Emergency Services Gary Glotfelty. They all have that same passion.

A special thanks to the National Interagency Fire Center and National Wildfire Coordinating Group, who helped with some of the pictures and their support to all wildland firefighters through classes, research, and educational materials they have provided. I also want to thank the Florida Department of Forestry's Assistant Chief Jim Karels, District Chief Doug Voltolina, and Safety Officer Gene Madden for their support. They furnished pictures and advice on the Florida fire problem.

Thanks to Christie Baumgart with Word Factory for much of the word processing.

I would also like to thank my loving wife, Wendy, for her editing and extensive typing, but more importantly for her belief in me and support in helping me write this book.

And thank you to Delmar, who offered me the opportunity to put all my teaching knowledge into a book that will reach readers worldwide.

Finally, both Delmar and I would like to thank the following reviewers who generously offered their time, expertise, and talent. Our gratitude is extended to:

Chris Miller, L.A.F.D. (Retired)
Morongo Fire Department
Morongo, CA

Tom Romaine
Department of Natural Resources—Forestry
Lewiston, MN

Tim Irwin
Mayer Fire District
Mayer, AZ

John Hogmire
South Trail Fire Department
Fort Myers, FL

Don Perry
Clarkston Fire Department
Clarkston, WA

Wayne Bemis
Kings River Community College
Reedley, CA

Thomas Welle
Red Rocks Community College
Lakewood, CO

John Craney
CA Department of Forestry and Fire Protection
Sacramento, CA

Grant Estell
Unity College
Unity, ME

George Cooper
Monticello, FL

With all of us working together, we can make a difference in the suppression of wildfires. I hope to see some of you in the future on some smoky ridge.

Ground and Air Resources Used on a Wildland Fire

Learning Objectives

Upon completion of this chapter, you should be able to:

- List the types of resources that are available to fight a wildland fire.
- Explain the use of each of these resources.
- Identify the differences found in each of the apparatus and aircraft groupings.

INTRODUCTION

Wildland fires are dynamic events that require a variety of resources and staffing levels. Before you can develop an overall plan for fighting a wildland fire, you must understand the types of resources available and their capabilities. Some engines are capable of operating off road, some are not. Handcrews can operate in areas where bulldozers cannot, removing the fuels with handtools in the path of the advancing fire. Aircraft are a great initial attack tool; however, they must be used efficiently and wisely because of the costs involved. Larger fires require the use of management teams and overhead personnel. This chapter acquaints you with the different resources typically found on a wildland incident.

Resources are classified (typed) in general categories: engines, water tenders, bulldozers, tractor plows, helicopters, and air tankers. Classifying resources enables incident commanders to order the appropriate resources to match the incident objectives.

Different systems exist for classifying resources. For instance, a Type I engine in Riverside County, California, would be called a Class A engine in Palm Beach County, Florida. Know the system in use for your area. The classification system used in this book is based on the National Wildfire Coordinating Group's standards, found in Appendix A of the *Fireline Handbook*. The *Fireline Handbook* is intended to serve as a field reference guide for wildland fire agencies using the incident command system in the control of wildland forest and range fires.

ENGINES

■ Note
Type I engines are typical urban fire apparatus.

A wide variety of types of engines can be found on almost any brush or interface fire. Type I engines (Figure 1-1) are typical urban fire apparatus. They must meet the following minimum standards:

- Have a pump capacity of 1,000 gallons per minute
- Have a tank capacity of at least 400 gallons
- Carry 1,200 feet of 2½-inch hose
- Carry 400 feet of 1½-inch hose
- Carry 200 feet of 1-inch hose
- Be able to produce a 500-gallon-per-minute fire stream
- Carry 20 feet of extension ladders
- Have a minimum of four personnel

■ Note
Type II engines should have the capability to mobile pump.

mobile pump
a method whereby a fire engine pumps a hose line while moving

These apparatus typically are not used off the roadway and are used primarily for structure protection. This does not preclude their use along the side of a roadway to supply water to wildland hose lays.

Type II engines (Figure 1-2) can be used for structure protection or wildland fire suppression. They are versatile and can be used to respond off road on a wildland fire. Type II engines should have the capability to **mobile pump.** They

Figure 1-1 *Type I engine.*

Figure 1-2 *Type II engine.*

generally have a smaller wheelbase than Type I engines and must meet the following minimum standards:

- Have a pump capacity of 500 gallons per minute
- Have a water tank size of 400 gallons
- Carry 1,000 feet of 2½-inch hose
- Carry 500 feet of 1½-inch hose
- Have 300 feet of 1-inch hose as minimum complement
- Carry 20 feet of extension ladders
- Have a minimum of three personnel

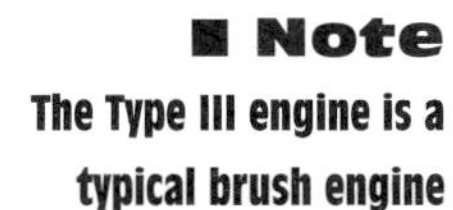

Note
The Type III engine is a typical brush engine used in wildland areas.

The Type III engine (Figure 1-3) is a typical brush engine used in wildland areas. It is a short wheelbase, high clearance vehicle that can be used off road. It has the ability to mobile pump (pump and roll) and does not carry larger diameter hose complements. Type III engines must meet the following minimum standards:

- Have a pump capacity of 120 gallons per minute
- Have a water capacity of 500 gallons
- Carry 1,000 feet of 1½-inch hose
- Carry 800 feet of 1-inch hose
- Have at least three personnel

Figure 1-3 *Type III engine.*

Figure 1-4 *Marine Corps base Camp Pendleton Type VI brush engine. It has a 50-gallon per minute pump, holds 340 gallons of water, and carries a crew of two.*

Table 1-1 *Minimum standards for Type IV, V, VI, and VII engines.*

	IV	V	VI	VII
Pump capacity (gallons per minute)	70	50	50	20
Tank capacity (gallons)	750	500	200	125
1½-inch hose (feet)	300	300	300	200
1-inch hose (feet)	300	300	300	200
Personnel minimum	3	3	2	2

Note
Type IV, V, VI, and VII engines are small pickup/patrol vehicles.

Type IV, V, VI, and VII engines (Figure 1-4) are small pickup or patrol vehicles found in many rural areas. They must meet the minimum standards shown in Table 1-1.

Note
Bulldozers are a great fireline resource that can clear fuels rapidly in advance of the fire.

BULLDOZERS

Bulldozers (Figure 1-5) are a great fireline resource that can clear fuels rapidly in advance of the fire. However, their progress can be slowed in rocky soils or areas of dense timber. Some conditions, such as steep slopes and heavy fuels, even limit

Figure 1-5 *United States Forest Service bulldozer.*

their use. Use of bulldozers must be managed closely as they can damage fragile soils and leave scars that remain for decades. With proper management of this resource, the damage caused by their use can be minimized. Three different types of bulldozers, rated by size, are found on firelines:

- Type I bulldozer is a heavy bulldozer (e.g., D7, D8, D9, or equivalent).
- Type II bulldozer is a medium bulldozer (e.g., D5, D6, or equivalent).
- Type III bulldozer is a light bulldozer (e.g., D4 or equivalent).

A complete discussion on how to use this resource effectively is found in Chapter 9.

TRACTOR PLOWS

■ Note
Tractor plows are best used in level-to-rolling topography with soils that have a minimum of rock.

Tractor plows (Figure 1-6) are effective fireline tools used to build a control line. They are used in the flat woods and coastal plains of the South, the Southeast, and Florida. They are also used in the Midwest, the Great Lakes states, and the Northeast. Tractor plows are best used in level-to-rolling topography with soils that have a minimum of rock. Rocky soils and increased slopes are a problem for plows. Plows are pulled by a bulldozer or a four-wheeled vehicle. The *Fireline Handbook* lists six different tractor plow types.

Figure 1-6 *Tractor plow. Photo courtesy Florida Division of Forestry.*

HANDCREWS

■ **Note**
Handcrews construct handlines in areas where it is too steep to use mechanized equipment such as a bulldozer.

Handcrews (Figure 1-7) can be found on any major wildland fire and are used primarily to construct firelines. Handcrews construct handlines in areas where it is too steep to use mechanized equipment, such as a bulldozer.

Two types of handcrews are found on firelines. Type I crews have the most training and no operational restrictions. They are best suited for hot line assignments. The minimum crew size for a federal Type I crew is 18 to 20. A Type II crew receives less training and has some fireline restrictions. The minimum size for a federal Type II crew is also 18 to 20.

Handcrews are an effective resource and can work directly on the fire's edge constructing handline, if the fire intensity level is not too great. If the fire intensity level is too great, handline is constructed at a distance from the fire's edge and the intervening fuels are usually burned out. Handcrews can also be used to assist or put in hose lays. Handcrews are usually the most experienced at firing operations of any of the fire personnel found on the fireline. Other uses of handcrews might be mop-up and logistical support functions.

■ **Note**
Fixed-wing aircraft are generally used to drop long-term retardant.

AIRCRAFT

Two broad categories of aircraft are found on a wildland incident: fixed-wing air tankers and helicopters. Fixed-wing aircraft (Figure 1-8) are generally used to

Figure 1-7 *Handcrew.*

Figure 1-8 *Two fixed-wing air tankers drop Class A foam in tandem. Photo courtesy Air Tractor Inc.*

drop long-term retardant, which slows the spread of the fire, just ahead of the flame front. They can also be used to drop water or short-term retardant.

Four classified types of aircraft are used today. The four types, classified by the minimum gallons of water or retardant they carry, are:

- Type I air tanker: minimum 3,000 gallons of water or retardant
- Type II air tanker: minimum 1,800 gallons of water or retardant
- Type III air tanker: minimum 600 gallons of water or retardant
- Type IV air tanker: minimum 100 gallons of water or retardant

Fixed-wing aircraft can deliver large volumes of retardant or fire suppressants very quickly on a wildland fire. They work well as an initial attack tool, keeping the fires small and easy to extinguish. Some of the disadvantages to their use include long load and return times to retardant bases, and unsuitability for steep slopes, deep canyons, and heavy timber canopy with a surface fire. Chapter 9 further discusses the advantages and limitations of aircraft.

Helicopters (Figure 1-9) usually drop water directly on the flaming fire front or pick up spot fires. The advantages of helicopters are (1) they can work in terrain that is too difficult for safe and accurate air tanker drops and (2) they have a shorter turnaround time between water drops as they do not have to return to an air attack base to refill. They can also be used to fly logistical missions on the fire.

Figure 1-9 *Type I helicopter.*

■ Note
Helicopters are classified by the number of seats, weight capacity, and water or retardant capacity.

Helicopters are classified by the number of seats, weight capacity, and water or retardant capacity. The helicopter types and their specifications are:

- Type I helicopter
  - no less than 16 seats
  - 5,000 pounds carded weight capacity
  - 700 gallons of retardant or water capacity
- Type II helicopter
  - 10 to 15 seats
  - 2,500–4,999 pounds carded weight capacity
  - 300–699 gallons of retardant or water capacity
- Type III helicopter
  - 5 to 9 seats
  - 1,200–2,499 pounds carded weight capacity
  - 100–299 gallons of retardant or water capacity
- Type IV helicopter
  - 3 to 4 seats
  - 600–1,199 pounds carded weight capacity
  - 75–99 gallons of retardant or water capacity

overhead
personnel assigned to supervisory positions, including Incident Commander, Command Staff, General Staff, Branch Directors, Supervisors, Unit Leaders, Managers, and staff

incident command system (ICS)
standardized on-scene emergency management concept specifically designed to allow users to adopt an integrated organizational structure equal to the complexity and demands of single or multiple incidents without being hindered by jurisdictional boundaries

FIRE MANAGEMENT

The fire size and complexity dictates how management of the fire is established. Smaller fires may require only an incident commander, while larger fires require management teams and additional **overhead** personnel.

Chapter 13 discusses how a fire transitions from the initial attack phase to an extended attack incident. The **Incident Command System (ICS),** an all-risk management system used on wildland fires, and its five key elements—Command, Operations, Logistics, Planning, and Finance/Administration—are discussed in depth in Chapter 13.

Summary

Seven different types of engines are found on a wildland fire. Type I engines are not generally used off road; they are primarily used to provide structure protection. Type II engines can be used to fight wildland fires or structure fires. They may be capable of mobile pumping. The short wheelbase and high ground clearance of the Type II make it a versatile engine for use in the wildland interface. Type III engines are larger brush engines and are used throughout the United States. Type III engines carry no large diameter hose lines. In the East, Midwest, and South, Type IV through VII engines are in common use.

Bulldozers are a great resource that can be used to clear wildland fuels. There are three different types of bulldozers found on a wildland fire, and they are rated by size.

Tractor plows are pulled by a bulldozer or a four-wheeled vehicle and used to construct control lines. They are best suited to fairly level topography.

Handcrews are used to construct control lines in areas that are too steep for mechanized equipment. Two types of handcrews are found on firelines.

On any large wildland fire you will find aircraft being used. Fixed-wing aircraft are generally used to drop long-term retardant in advance of the flaming front. Helicopters drop water or Class A foam directly on the fire. They can also be used to shuttle crews or supplies to the fire.

Fixed-wing aircraft are classified by the amount of water or retardant they carry. Helicopters are classified by the number of seats, weight capacity, and water or Class A foam they can deliver.

Many resources are available for use on wildland fires, each with a different purpose and use. It is important that you are acquainted with each type if you are going to be involved in the suppression of wildland fires.

Review Questions

1. What is the minimum amount of 2½-inch hose carried on a Type I engine?
2. What is the minimum pump capacity for a Type II engine?
3. Do Type III engines carry large diameter hoses?
4. What type of topography is best suited for a tractor plow?
5. What is the minimum size for a Type II crew?
6. What is the minimum gallonage of water or retardant carried on a Type I air tanker?
7. How are helicopters classified?
8. What is the minimum number of seats found on a Type I helicopter?

References

National Wildfire Coordinating Group, *Fireline Handbook* (Boise, ID: National Interagency Fire Center, 1998).

National Wildfire Coordinating Group, *S270 Basic Air Operations* (Boise, ID: National Interagency Fire Center, 1991).

Wildland Fire Behavior

Learning Objectives

This chapter is one of the most important in this book. You must understand wildland fire behavior because all tactical decisions and safety concerns are based on an understanding of these principles.

Upon completion of this chapter, you should be able to:

- Describe how wildfire spreads.
- List the types of wildland fuels and how they burn.
- Identify wildland fuel characteristics.
- Explain the effect topography has on the fire.
- Explain how weather factors influence a fire.
- Identify extreme fire behavior watchouts.

INTRODUCTION

A wildland fire spreads in one of three ways or a combination of the three. Fuels provide the heat energy on a wildland fire, and this transfer of heat energy spreads the fire. Topography affects the direction of the fire and how fast it spreads. Weather factors cause some of the most dramatic changes on a wildland fire. They can result in changes to the **rate of spread** and the direction of the fire and they can cause the fire to increase in intensity. Each of these factors is explained in detail in this chapter.

■ **Note**
Fuels provide the heat energy on a wildland fire.

rate of spread
the relative activity of a fire in extending its horizontal dimensions

WILDLAND FUELS

A wildland fuel can be any **organic material,** either living or dead, that can ignite and burn. Fuels can be loosely grouped into three levels.

organic material
fraction of the soil that includes plant and animal residues at various stages of decomposition, cells and tissues of soil organisms, and substances synthesized by the soil population

Types

Ground Fuels Combustible material lying on or beneath the ground, **ground fuels** (Figure 2-1) include root systems, deep duff, rotting buried logs, and other woody fuels in various states of decomposition. Fires in ground fuels are slow moving and do not play a major role in fire spread. They become important in line construc-

ground fuels
all combustible materials below the surface litter that normally support a glowing combustion without flame

Figure 2-1 *Ground fuels are combustible material lying on or beneath the ground.*

■ Note
Topography affects the direction the fire goes and how fast it spreads.

■ Note
Weather factors cause some of the most dramatic changes on a wildland fire.

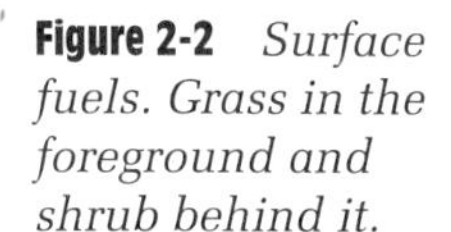

Figure 2-2 *Surface fuels. Grass in the foreground and shrub behind it.*

tion and mop-up operations. Root systems have carried fire across control lines, and smoldering fires below ground have been known to hold fire through the winter. In the Great Lakes states, Southeast, and Alaska, organic soils, especially during drought periods, play an important part in fire spread. Consider this especially when constructing fire line and mopping up fires.

surface fuels
fuels lying on or near the surface of the ground

aerial fuels
standing and supported live and dead combustibles not in direct contact with the ground, such as foliage, branches, stems, bark, and vines

crown fires
a fire that advances from top to top of trees or shrubs, more or less dependent on a surface fire

Surface Fuels The major cause of wildland fire spread, **surface fuels** (Figure 2-2) include such materials as grass and brush. These fuels act as kindling for heavier fuels. Most fires start and spread in surface fuels. Surface fuels include ground litter, such as leaves, needles, grasses, and brush up to 6 feet in height.

Aerial Fuels Fuels that are greater than 6 feet in height are considered **aerial fuels** (Figure 2-3). Trees or tall brush fall into this category. The main concern with these fuels is the crown or canopy closure, that is, how close the fuels are to each other. Timber stands that have open canopies as shown in Figure 2-4 usually have faster spreading surface fires with individual torching of trees. Closed canopy stands (Figure 2-5) greater than 6 feet in height, occurring in tall brush or trees, present the greatest opportunity for running **crown fires.** Crown fires are the fastest spreading of all wildland fires and burn with the greatest intensity.

Crown fires are not confined just to trees; they occur in the crowns of other aerial fuels as well. Examples of brush models where this could occur include

Figure 2-3 *A Florida Division of Forestry tractor plow working in aerial fuels. Photo courtesy Florida Division of Forestry.*

Figure 2-4 *An open stand of pines in Florida's Everglades National Park.*

Figure 2-5 *Closed stands of timber in Plumas National Forest, Northern California.*

oakbrush, as found in the southern Rocky Mountains; an old age class chaparral bed found in Southern California; or palmetto and gallberry as found in the coastal plains of the Southeast. Access your region and identify fuels that have this potential. Do not be caught by surprise.

Characteristics

Fuel characteristics determine how fast a fire travels (rate of spread) and how intensely it burns.

Horizontal Continuity Horizontal continuity refers to the way that wildland fuels are distributed at various levels. Continuity of the fuel bed determines where the fire will spread and whether it will travel along the surface fuels, the aerial fuels, or both.

Horizontal distribution of fuel refers to patchy or continuous fuel beds. Patchy fuel beds (Figure 2-6) are areas of widely distributed fuels. It requires a strong wind or steep slope for a fire to travel through these fuel beds. Continuous fuel beds (Figure 2-7) provide ample fuels to burn and the opportunity for the fire to spread rapidly, especially when large concentrations of continuous fire fuels are present. Horizontal continuity also applies to aerial fuels such as the treetops.

vertical fuel arrangement
fuels above ground and their vertical continuity, which influences fire reaching various levels of vegetation strata

Vertical Arrangement An important concept, **vertical fuel arrangement** must be evaluated. It has to do with the different heights of the fuels that are present. Vertical fuel arrangement should also be part of an evaluation. This factor allows a fire to reach different fuels, such as fires moving in surface fuels reaching the aerial fuels. The term *ladder fuels* (Figure 2-8) defines this stepping effect. Fires usually

Figure 2-6 *Patchy fuels. Note the areas void of vegetation.*

Figure 2-7 *Continuous fuels. All the fuels are close together with no areas void of fuel.*

Figure 2-8 *Note the presence of ladder fuels that would allow the fire to reach the treetops. Ocala National Forest, Florida.*

start in light fuels, such as grasses, then step up into the brush. Once the brush starts burning, the potential exists for the fire to move into the aerial fuels.

compactness spacing between fuel particles

Compactness The spacing between fuel particles can be defined as **compactness.** The more loosely compacted (Figure 2-9) the fuel bed, the more easily oxygen can move in and around the fuels. Usually the rate of spread is greater as a result of an increased oxygen supply. Tightly compacted (Figure 2-10) fuels usually have slower rates of spread because oxygen cannot move freely around the fuels.

Live to Dead Fuel Ratio The amount of dead fuels (Figure 2-11) present in a fuel bed is important when evaluating fire potential. This concept is called the *live to dead ratio*. The dead fuels increase as the fuel bed matures. Overmature fuel complexes, frost kill, prolonged drought, insect damage, previous fire damage, disease, wind, and snow all increase the amount of dead fuels present in a fuel bed. Dead fuels carry the fire and heat the live fuels to their ignition point. The more dead fuels present, the greater the chance of ignition of the living fuels.

Figure 2-9 *This bed of leaves is an example of a loosely compacted fuel bed.*

Figure 2-10 *Matted grass is an example of a tightly compacted fuel. Air can not move freely around the fuels.*

Figure 2-11 *Note how much of the shrub is dead in relation to how much is still living.*

fuel moisture
the amount of water in a fuel

live fuel moisture
ratio of the amount of water to the amount of dry plant material in living plants

dead fuel moisture
the amount of water contained in dead plant tissue

Fuel Moisture A significant player in the spread of a wildland fire, **fuel moisture** is an important factor in how easily a fuel ignites and burns. In general, when fuel moistures are high, a fire will probably burn poorly. If the fuel moistures are low, fires start easily, spread, and burn rapidly.

There are two types of fuel moistures: **live fuel moisture** and **dead fuel moisture.** Live fuel moisture is the moisture found in living plants. Live vegetation has a much higher moisture content than dead fuels. Live fuels are more affected by seasonal changes than daily changes in weather. Fuel moistures are highest when foliage is fresh and the plant is in its growing cycle. They are usually at their lowest point in late summer or autumn.

Further examination of the types of living fuels reveals either woody plants or herbaceous plants. Herbaceous plants are soft and do not develop woody tissue. Herbaceous annuals sprout from seed each year. Cheatgrass (Figure 2-12) is a good example of this type of herbaceous fuel. Before a fire can carry in this fuel type, one-third of the grass must be dead. These grasses usually cure in early summer and are responsible for much of the fire spread in other fuel types. Perennial

Figure 2-12
Cheatgrass.

herbaceous vegetation sprouts each year from its base. Some perennial grasses never cure. Woody vegetation (Figure 2-13) reacts to seasonal changes, and, normally, live fuel moistures decrease as the growing season comes to an end. Of most concern are the small twigs, leaves, and needles that would be consumed by the **flaming front.**

flaming front
that zone of a moving fire in which the com bustion is primarily flaming

Dead fuel moisture is the moisture found in dead fuels such as annual grasses. Fine dead fuels react rapidly to precipitation or changes in the relative humidity and temperature. If it rains, they reach their saturation point quickly. When humidity levels are high, there is a net gain in fine fuel moistures. When humidity levels are low, these fuels give up moisture to the air.

How long it takes for this equalization of fuel moisture with the air is called the *timelag principle*. This principle uses stock fuel diameters as its reference. They are as follows:

equilibrium moisture content
moisture content that a fuel particle attains if exposed for an infinite period in an environment of specified constant temperature and humidity

- 1-hour fuels are less than ¼ inch in diameter and reach **equilibrium moisture content** in 1 hour.
- 10-hour fuels are ¼ inch to 1 inch in diameter and reach equilibrium moisture content in 10 hours.
- 100-hour fuels are 1 inch to 3 inches in diameter and reach equilibrium moisture content in 100 hours.
- 1000-hour fuels are 3 inches to 8 inches in diameter and reach equilibrium moisture content in 1,000 hours.

Figure 2-13
Manzanita is an example of woody vegetation.

When a fuel type approaches 60 to 70% of its equilibrium moisture content with the surrounding air, the equilibrium change is said to be complete. Firefighters are most concerned with fine dead fuels that are less than ¼ inch in diameter. These 1-hour fuels are most susceptible to a change in relative humidity and are the prime carriers of a wildland fire.

surface area to volume ratio
the ratio between the surface area of an object, such as a fuel particle, to its volume

The size and shape of wildland fuels are also an important consideration to evaluate. Smaller fuels have greater **surface area to volume ratio.** Smaller fuels ignite easier, as less heat is required to drive off the moisture in the fuels and raise them to their ignition temperature.

Think of lighting a fireplace. Small fuels are needed to ignite the larger pieces of wood. Larger fuels have smaller surface-to-volume ratio and are harder to ignite and sustain combustion.

Fuel size and surface area examples are shown in Figure 2-14.

Many wildland fuels contain chemicals and minerals that can enhance or slow combustion. Volatile substances, such as oils, resins, wax, and pitch, are present in many wildland fuels throughout the world. They are responsible for many

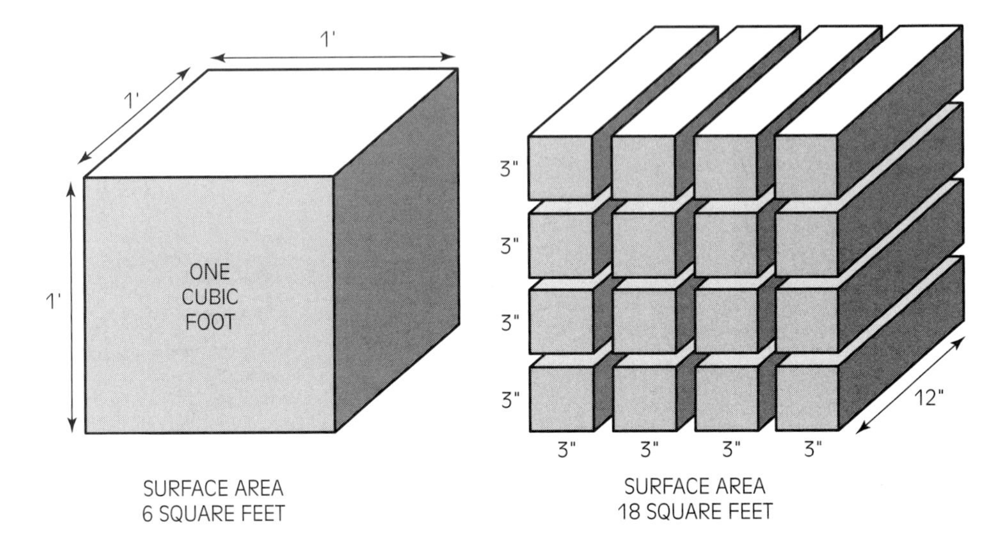

Figure 2-14 *Surface area to volume ratio. All fuel pictured here, as blocks, has a volume of 1 cubic foot. However, because of their shape and size, the surface area varies greatly.*

high intensity, rapidly spreading fires. Fires that are high in mineral content retard the spread and intensity of the fire.

Fuel Temperature Fuel temperature also plays a big part in fuel moisture levels. Each day the radiant energy of the sun heats the Earth's surface and the fuels close to it. As the ground surface and heated fuels become warmer, they begin to warm the air close to the ground. This warming of the air causes a reduction of humidity and wind and an increase in air temperature.

The first slope to be heated by the sun's energy faces east. The direction the slope faces in relation to the cardinal compass points is called **aspect.** Fuel types also vary by aspect (Figure 2-15). Each aspect receives its peak solar influence at a certain time of the day. Eastern aspects (east-facing slopes) peak at around 0900 hours, southern aspects (south-facing slopes) peak at around 1200 hours, western aspects peak in the afternoon around 1500 hours. Northern aspects also peak around 1200 hours; however, they have the highest fuel moistures and the lowest daytime temperatures. Times may vary due to seasonal changes. The fine dead fuels in shaded areas have higher moisture contents than those found in direct sunlight (Figure 2-16).

aspect
cardinal direction toward which a slope faces

When present, smoke, haze, and clouds interfere with the radiant energy from the sun, thus fuels on the ground surface are cooler. Changes in elevation also affect fuel moistures. The higher one goes on a mountainside, the cooler it gets. Cooler temperature means higher relative humidity and higher fuel moistures. Each 1,000-foot change in elevation represents a 3½°F temperature change (Figure 2-17).

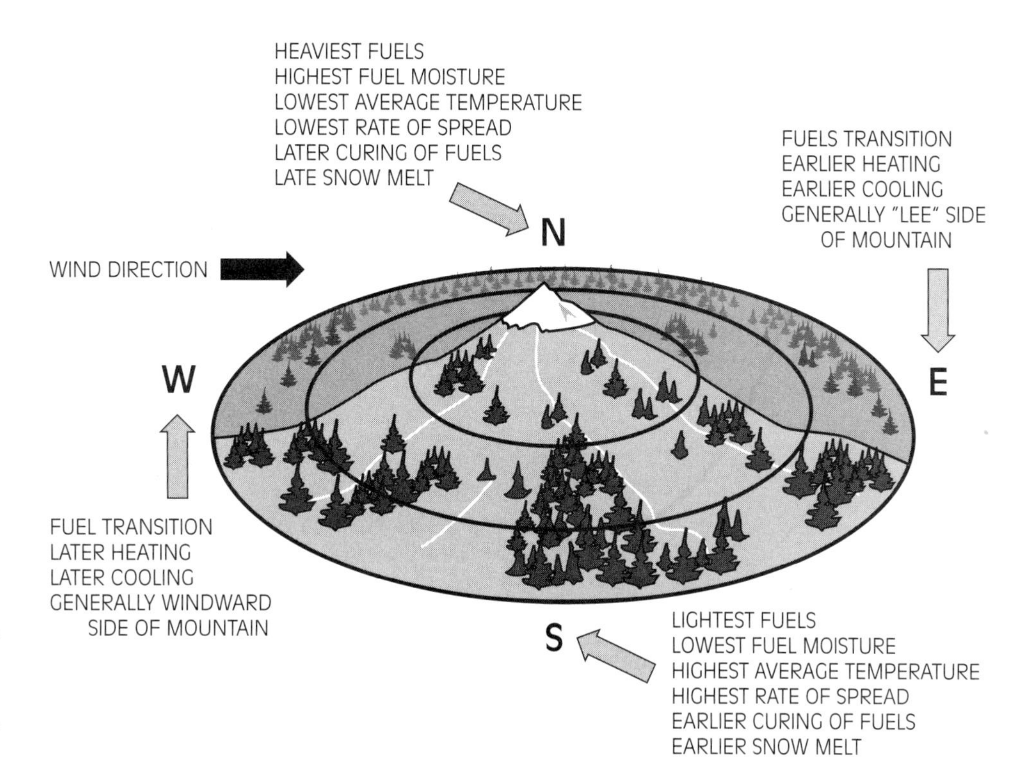

Figure 2-15 *Fuel types, fuel temperature, and moisture, related to aspect.*

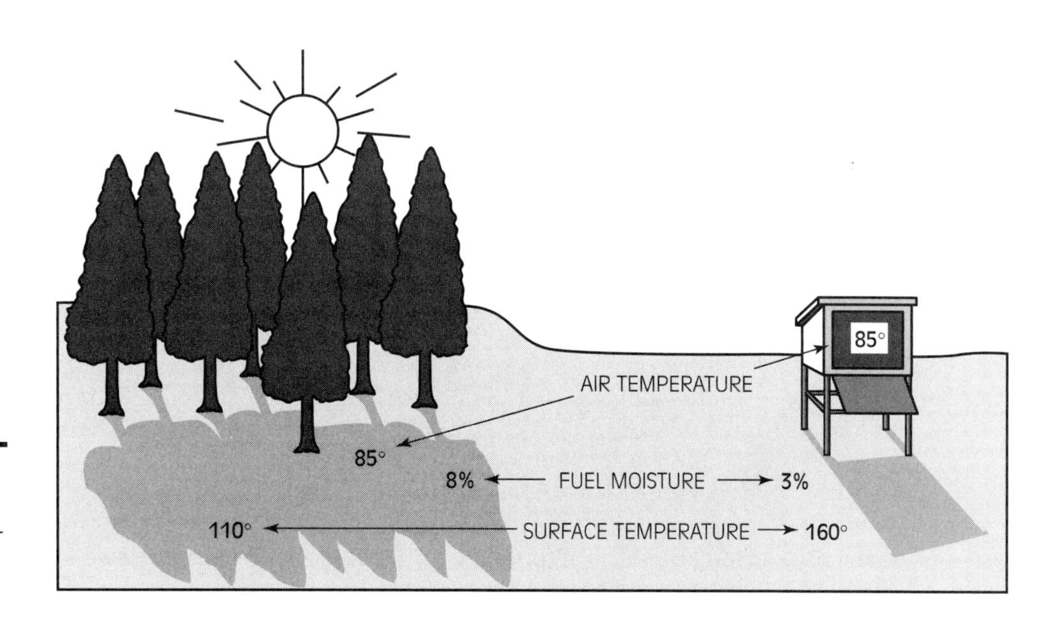

Figure 2-16 *From the diagram, note the effect that shading has on fuel temperature.*

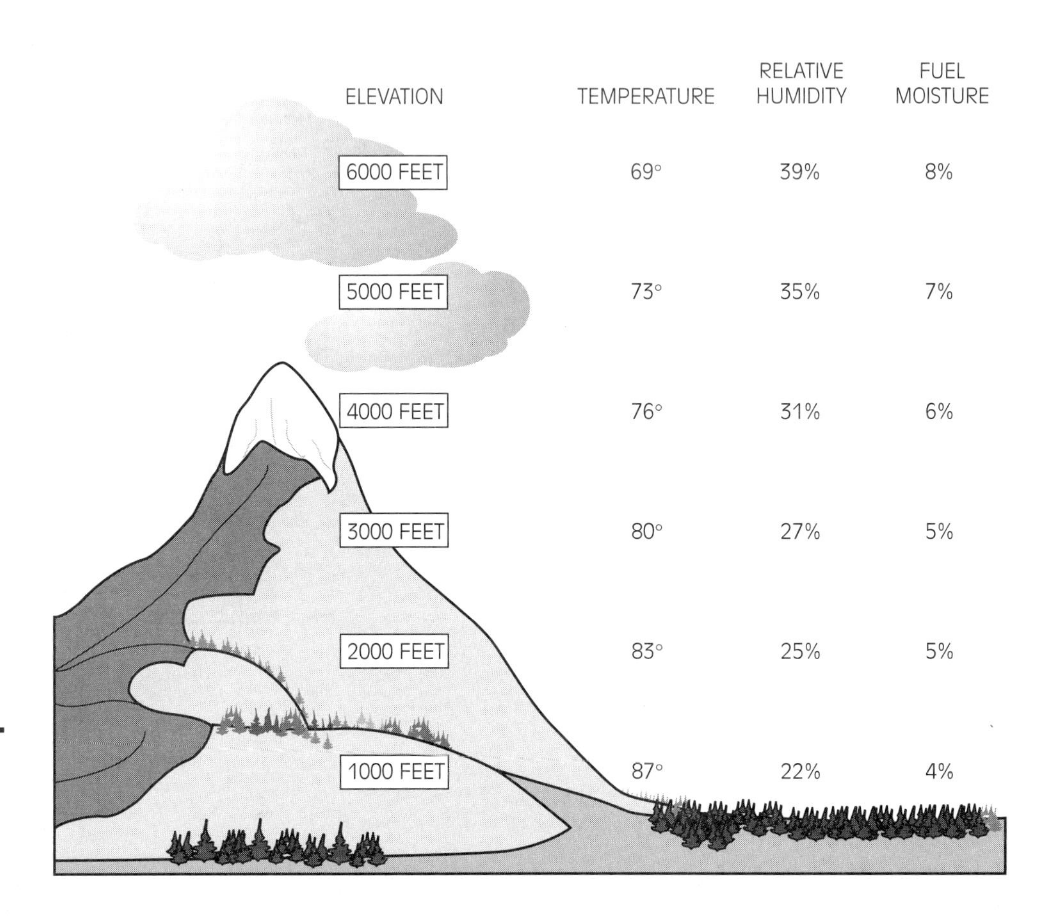

Figure 2-17 *How elevation affects temperature and fuel moisture—daytime example.*

Slope also has an effect on fuel moisture; surfaces perpendicular to the sun absorb more of the solar energy. South and west slopes receive considerably more of the incoming solar radiation than north or east. Thus, they have higher temperatures and lower fuel moistures. The angle of the sun's radiation changes throughout the day and with the seasons.

Wind can also affect fuel moisture levels. During calm periods, the air that is saturated with water vapor has a chance to act on the fine fuels. Fuels reach equilibrium moisture content with the air at a faster rate when it is windy. Stronger winds during the day may prevent surface temperatures from rising and thus actually cool the fuels. This cooler air raises the relative humidity. Night winds can cause a turbulent mixing of air, which may prevent surface temperatures from reaching the dew point. A wind that causes fuels to dry rapidly is a **foehn wind.** These winds occur when heavy, stable air pushes across a mountain range and descends rapidly on the lee side of the mountain range, the side of the mountain range away from the wind. This rapid descent compresses the air and makes it hotter and dryer, in turn reducing the fuel moistures.

foehn wind
warm, dry, and strong general wind that flows down into the valleys when stable, high pressure air is forced across and then down the lee slopes of a mountain range

shrub
a woody perennial plant differing from a perennial herb by its persistent and woody stem and from a tree by its low stature and habit of branching from the base

brush
stands of vegetation dominated by shrubby, woody plants or low growing trees, usually undesirable for livestock or timber management

■ Note
Grasses are the primary carrier of wildland fires.

■ Note
Fire behavior fuel models 1, 2, and 3 are grass models.

■ Note
Fuel models 4, 5, and 7 contain both dead and live fuels.

Classifications

The *Fireline Handbook* lists thirteen fire behavior model descriptions. These thirteen fuel models are aids used to estimate and calculate fire behavior. For purposes of this book, all thirteen are included in four major categories.

Grass Models Grasses (Figure 2-18) are the primary carrier of wildland fires. The depth varies between 1 and 2½ feet. Fires in grasses burn rapidly and the fuel is usually entirely consumed. Wind greatly affects the spread of fire in grass models. These fine fuels also respond quickly to changes in relative humidity. Intensity levels are usually low when compared to that of other fuel models. Grass is found in most areas; however, it is more prevalent in desert and range areas. Fire behavior fuel models 1, 2, and 3 are grass models as defined in Appendix B of the *Fireline Handbook*.

Shrub Models Fuel bed depths from 2 to 6 feet are found with **shrub** or **brush** models (Figure 2-19). A good example of this type of fuel is mixed chaparral fuels found in the Southwest. Most of these fuels are less than 1 inch in diameter with small leaves. Fuel models 4, 5, and 7 contain both dead and live fuels. Some of the fuels in this grouping contain volatile compounds that allow the fuel to burn in

Figure 2-18 *Firing operation in a grass model.*

Figure 2-19 *Fire in a shrub model.*

■ **Note**
Shrub fuel models are 4, 5, 6, and 7.

higher fuel moistures. A good example of this is Model 7 fuels, such as palmetto and gallberry, that are found in Florida. Once a fire progresses into these fuel beds, the rate of spread generally slows. Fire intensity levels are higher than those found in grass models. Shrub fuel models are 4, 5, 6, and 7 as defined by the Fire Behavior Fuel Model descriptions found in Appendix B of the *Fireline Handbook*.

litter
the top layer of forest floor

Timber Litter Models Timber **litter** models (Figure 2-20) are composed largely of leaves, mixed litter, and occasionally, twigs and large branch wood. The fuel bed compactness can range from loose to tight. Fires generally run through the surface and ground fuel (Figure 2-21); however, when the fire finds a large concentration of dead fuel, crowning out and torching are possible. Timber litter models are 8, 9, and 10 in the *Fireline Handbook* Fire Behavior Fuel Model descriptions.

■ **Note**
Timber litter models are 8, 9, and 10.

slash
debris left after a logging operation

Logging Slash Models Logging **slash** (Figure 2-22) is basically the debris left after a logging operation. Slash can also be produced by thinning and pruning operations. Fires in these fuel models tend to have moderate to higher rates of spread. They also tend to produce moderate to higher intensity fires. Fires burning in areas where many trees have been blown down, such as along the East Coast after the passage of a hurricane, will burn like Models 11, 12, and 13. Models 11, 12, and 13 represent the logging slash models and are also found in the *Fireline Handbook* Fire Behavior Fuel Model descriptions.

■ **Note**
Models 11, 12, and 13 represent the logging slash models.

Each fuel model burns at a different rate of spread and produces varying fireline intensities. Evaluate your area and identify the different fuel types. Predict

Figure 2-20 *A timber litter fuel model.*

Figure 2-21 *A surface fire burning in a stand of timber.*

Figure 2-22
Firefighters walking through logging slash preparing for a controlled burn.

what type of fire intensities and rates of spread you would expect to see. From this evaluation, you will be better prepared to make appropriate tactical decisions should a wildland fire occur.

Note
Topography represents the physical features of the land.

TOPOGRAPHY

Topography represents the physical features of the land in a given region. Topography can be divided into four categories: slope, aspect, elevation, and terrain features that aid in fire spread.

Slope

spotting
behavior of a fire producing sparks or embers that are carried by the wind and that start new fires beyond the zone of direct ignition by the main fire

firebrands
any sources of heat capable of igniting wildland fires

Fires on steep slopes spread more rapidly because the flames are closer to the fuels, thus preheating upslope fuels (Figure 2-23). The upslope fuels are then more susceptible to **spotting** from aerial **firebrands.** As the fire burns upslope, the products of combustion move in an upslope direction and oxygen rushes in at the fire's base to feed the growing fire. As the slope factor increases, the rate of spread and flame length also increase. Fires on steep slopes burn in wedge-shaped patterns.

On steep slopes you also need to watch for burning, rolling material as it may roll across the control line and start a fire below it, potentially trapping you between two fires.

Where the fire starts on the slope is also an important factor when considering the rate of spread. Fires that start close to the base of the slope have more avail-

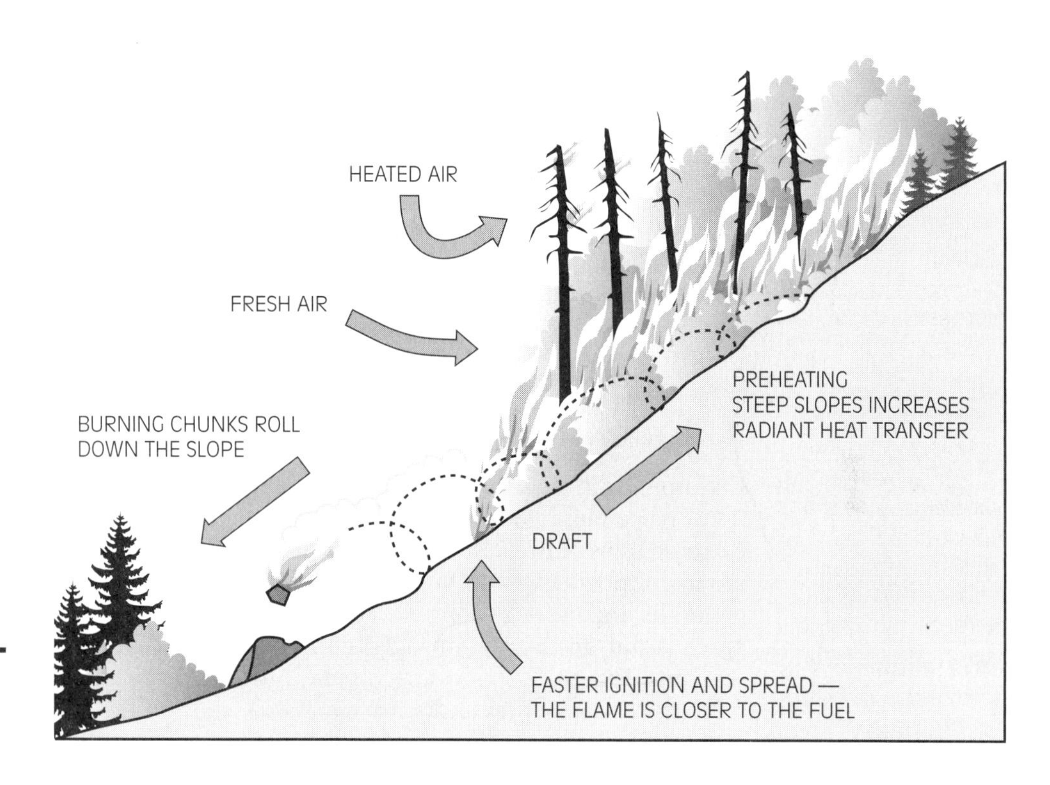

Figure 2-23
Preheating upslope fuels.

able fuel to burn and preheat upslope fuels. Fires on the slope are also influenced by the normal upslope daytime winds. Fires starting near the top of a slope generally do not have as much available fuel to burn and therefore do not generally grow large in size.

Aspect

■ Note
Aspect is the direction a slope faces.

Aspect is an important factor to consider when evaluating fire behavior. Aspect is the direction a slope faces in relation to a cardinal compass point. Each aspect heats at different times of the day. This heating raises the fuel temperature, reduces the humidity (see previous discussion under Fuel Temperature), and starts the upslope winds. In addition, each of these aspects has different types of fuels. A good example of these types of fuel variations occurs on a southern aspect, which has lighter type fuels. Fires on these slopes usually become large fires quickly. Northern aspects, however, have the heaviest fuels, highest fuel moisture, and the lowest average temperature of all the aspects.

Elevation

mean sea level
average height of the surface of the sea for all stages of the tide over a 19-year period

Elevation above **mean sea level** has an effect on the fuel types and their ability to burn. There are also climate changes, especially at higher elevations such as in a

chimney
a terrain feature that has a channeling effect on the convective energy of the fire

saddles
low topography points between two high topography points

eddy
a circular-like flow of a fluid drawing its energy from a flow of much larger scale and brought about by pressure irregularities, as on the downwind side of a solid obstacle

narrow canyons
steep canyon walls that are close together, possibly creating problems for fire crews if fire from one canyon wall jumps to the opposite canyon wall

area ignition
ignition of several individual fires in an area that cause the main body of the fire to produce a hot, fast-spreading condition

box canyons
steep-sided, dead-end canyons

mountain range (Figure 2-24). On the lower slopes, grass models are the predominate fuel types. As the elevation increases, shrub models start developing. Further up the slope, timber models develop. Above the tree line, found at very high elevations, there is an absence of fuels as a result of the extreme weather factors found there. Besides a difference in fuel types, there is also a decline in temperature as elevation increases. The accepted figure for temperature change is a 3½°F decrease per each 1,000-foot increase in elevation. It is important to remember that as the temperature decreases there is a corresponding change in the relative humidity. The cooler the temperature, the higher the relative humidity. Higher relative humidity in fine dead fuels makes these fuels harder to ignite.

Terrain Features That Aid in Fire Spread

Certain terrain features cause the fire to spread at a greater rate. A **chimney** (Figure 2-25) has a channeling effect on the convective energy of the fire. In a chimney, extreme rates of fire spread and spotting are likely to occur. **Saddles** (Figure 2-26) are low topography points between two high topography points. The wind speed generally increases as the wind passes through these constricted areas and then spreads out on the lee side. An **eddy** generally occurs on the lee side. **Narrow canyons** (Figure 2-27) are a problem for fire crews because the steep canyon walls are close to each other and a fire can easily start on the opposite canyon wall as a result of radiant heat or spotting. This proximity can also cause **area ignition,** where a large canyon can explode in seconds. **Box canyons** (Figure 2-28) are canyons with three steep walls that create very strong upslope winds. These

Figure 2-24 *Climate changes as the elevation changes. Fuel types change as a result.*

■ Note
As the temperature decreases, there is a corresponding change in the relative humidity.

Figure 2-25 *Two chimneys run to the top of the mountain. The sun was low on the horizon. Note the shading in each of the chimneys. One side will have hot fuels and the other cold fuels.*

Figure 2-26 *Saddle.*

Figure 2-27 *A narrow canyon area near homes. The canyon walls are close together, and fire can easily start on the opposite canyon wall.*

Figure 2-28 *Box canyon.*

upslope winds create rapid fire spread conditions. As the convective column rises, air is drawn up from the canyon bottom. Box canyons are areas where extreme fire can occur.

WEATHER

Weather factors are the greatest concern on a fireline as they have the greatest effect on a wildland fire. The following weather factors are the primary elements that cause change on a wildland fire.

Wind

Wind drives the fire, gives it direction, causes spotting, preheats fuels ahead of the fire, and brings a fresh supply of oxygen to a fire. Winds are responsible for a rapid **rate of spread** on wildland fires. Wind direction is defined by the direction from which the wind is coming; an east wind blows from the east. There are two types of winds: general winds and local winds.

rate of spread
relative activity of a fire in extending its horizontal dimensions

General Winds Produced by the broad scale pressure gradients associated with high and low pressure differences, **general winds** or **gradient winds** are large-scale winds that are the result of temperature variations. The areas of high and low pressure are always trying to equalize each other. In the United States, these pressure cells typically move from west to east. Air in a high pressure cell moves outward and in a clockwise direction. In low pressure cells, the air moves toward the center in a counterclockwise direction (Figure 2-29).

general winds
large-scale winds caused by high- and low-pressure systems, generally influenced and modified in the lower atmosphere by terrain

gradient winds
winds that flow parallel to pressure isobars or contours; they occur at a height of about 1,500 feet above mean terrain height

Foehn Winds A special type of general wind associated with mountainous areas that causes great concern on firelines is called a foehn wind. Local names are given to foehn winds. They are called Chinook winds in the Rocky Mountains, Santa Ana winds in Southern California, north winds in Northern California, and east winds in Washington and Oregon. Foehn winds also occur in other parts of the country, for example, on the eastern slopes of the Appalachian Mountains. Foehn winds are

Figure 2-29 *A high pressure cell moves in a clockwise direction, while a low pressure cell moves in a counterclockwise direction.*

down-flowing winds that are usually warmer and drier as a result of compression as the air descends the leeward side of a mountain range. If these winds combine with the nightly local down-slope winds they can become quite strong. These winds usually occur from September to April; they do not occur frequently during summer months.

The first type of foehn wind is the result of moist Pacific air being forced upward and over a mountain range. We find this occurring in the Sierra-Cascade range or the Rocky Mountain range. As the air is forced upward on the windward side, it reaches its condensation level. When it lifts further, clouds are produced along with precipitation. Upon passing over the mountain range, precipitation is lost and the air then starts to descend and warm. As it descends the leeward side of the mountain range, it is warmed by compression at a rate of 5.5°F per 1,000 feet of fall. As this wind pushes through the mountain passes or other constricted topographical features, it gains speed. When it arrives in the lowlands it is a strong, gusty, and dry wind. A good example of this type of wind is a Chinook wind that occurs on the eastern side of the Rocky Mountains in the fall or winter.

The second type of foehn wind occurs when a cold, dry, stagnated high pressure air mass is centered over the Great Basin located in Nevada. The air mass is separated by a mountain barrier from the low pressure center located on the opposite side of the barrier. Since the mountains block the flow of surface air, the air flow must come from aloft. This air flow from the high pressure mass flows toward the area of lower pressure. On the leeward side of the mountain range, the air on the surface is forced out of the way by pressure gradients and is replaced by heavy air flowing from aloft and downslope. These downslope winds are also influenced by **subsidence.** Winds like these are typified by the Santa Ana winds that occur in Southern California during the fall and winter. Santa Ana winds (Figure 2-30) cause many of the rapidly spreading, large wildland fires common in Southern California. Surface wind speeds are commonly in the 40 to 60 mile per hour range. In a severe Santa Ana condition, winds can reach as high as 90 miles per hour. These winds can last for days before they gradually weaken and stop.

subsidence
downward or sinking motion of air in the atmosphere

Local Winds Smaller scale winds that occur daily, **local winds** are the result of local temperature differences and are most influenced by terrain factors. A good example of local winds would be the upslope winds that form as the result of the heating of Earth's surface. Upslope winds are covered below in the section on slope winds.

local winds
winds generated over a comparatively small area by local terrain and weather

Local winds play an important role in fire behavior and must be accounted for when making tactical decisions on firelines. These winds occur daily. Local winds include three types—sea and land breezes, slope winds, and valley winds—as described in the following list:

- *Sea Breezes.* Sea breezes develop daily along coastal areas as the result of the differential heating that occurs locally each day. Each day the land mass warms and the ocean remains cooler than the land mass. The cooler air from the

Figure 2-30 *The Laguna Beach fire. More than 400 homes were lost during this Santa Ana wind-driven fire.*

ocean moves onshore to replace the heated air rising off the warmer land mass (Figure 2-31). The result is wind. This sea breeze (wind) begins between mid-morning and early afternoon. These times vary due to seasonal changes and location. Typical wind speeds are approximately 10 to 20 miles per hour. They can be stronger in different regions of the country. For example, winds along the California, Oregon, and Washington coasts can attain speeds of 20 to 30 miles per hour. Similar winds also occur on the Eastern shoreline.

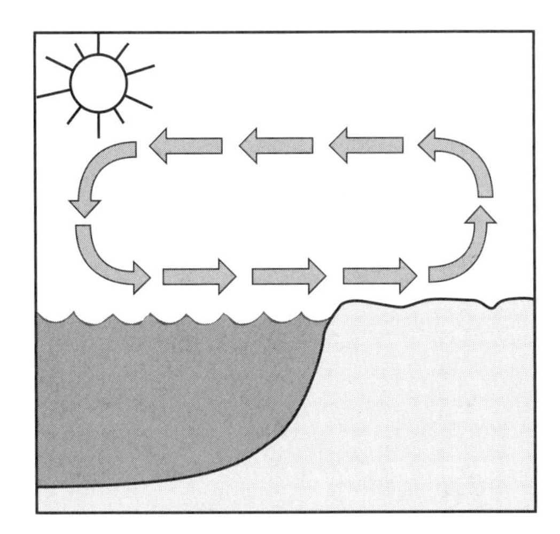

Figure 2-31 *Sea breeze development.*

• *Land Breezes.* Land breezes are the result of the land mass cooling more quickly than the water surfaces at night (Figure 2-32). This difference in air pressure causes the air to flow from the land to the water. These land breezes develop approximately 2 to 3 hours after sunset. The nighttime land breeze is not as strong as the daytime sea breeze. Wind speeds are typically between 3 and 10 miles per hour.

■ **Note**
Slope winds affect fire by increasing the rate of spread.

effect of slope
the effect topography has on the rate of fire spread—fire travels 16 times faster uphill than downhill

valley winds
daily winds that flow up-valley during the day and down-valley at night caused by differences in temperature between air in the valley and air at the same elevation over the adjacent plain or larger valley

pressure gradients
the difference in atmospheric pressure between two points on a weather map

• *Slope Winds.* Upslope winds (Figure 2-33) occur during the day as a result of the heating of the land surfaces. The air in the valleys becomes warmer than the air on the mountain top. This differential causes the movement of air upslope. Upslope winds start on eastern aspects as a result of the sun's exposure on the land mass. Maximum upslope winds occur about midafternoon. Upslope winds are stronger than downslope winds and range between 3 and 8 miles per hour.

Downslope winds (Figure 2-34) occur at night and are weak superficial winds that blow downslope. These winds are usually no stronger than 2 to 5 miles per hour. Gravity is the principal force behind these winds. Downslope winds continue during the night until the slope warms again and the upslope flow reoccurs. Calm periods do exist in the evening and early morning hours when these winds are changing direction.

Slope winds affect fire by increasing the rate of spread. Should the downslope wind become strong enough to overcome the **effect of slope,** a reversal in the direction of the fire could occur.

Always ask the local area residents about the wind conditions that occur in their area. The information will help you develop better technical plans and enable you to operate safely on the fireline.

• *Valley Winds.* Daily (diurnal) winds that flow up-valley during the day and down-valley at night (Figure 2-35), **Valley winds** are the result of local **pressure gradients** caused by differences in temperature between the air in the valley and

Figure 2-32 *Land breeze development.*

Figure 2-33 *Upslope winds.*

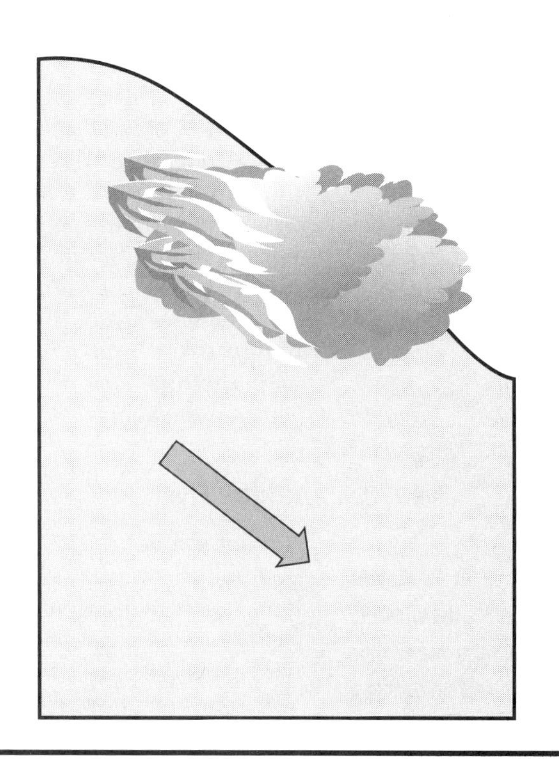

Figure 2-34 *Downslope winds.*

Figure 2-35 *Diurnal valley winds.*

the air at the same elevation over an adjacent plain or a larger valley. Upslope winds start within minutes after the sun strikes a mountain slope. The up-valley winds do not start until the whole mass of air within the valley becomes warmed. That is why up-valley winds do not start until late morning or early evening. Common up-valley wind speeds range between 10 and 15 miles per hour.

The transition to down-valley winds starts gradually. First to appear are light 2 to 5 mile per-hour downslope winds in the canyon bottoms. This takes place early in the night. The time depends on the size of the valley or canyon, on factors favoring cooling, and the establishment of a temperature differential. The downslope winds gradually deepen during the early night and become down-valley winds. Down-valley wind speeds are in the 5 to 10 mile per hour range. These winds continue through the night and diminish after sunrise.

front
the boundary between two air masses of differing atmospheric properties

Weather Fronts

A **front** is a boundary between two air masses. Most increased wind activity is in this boundary area. The passage of a cold front causes fire crews the most problem. Pressure gradients are tight in cold fronts, and strong upper winds are more easily mixed down to the surface as unstable air, which results in an increase in wind speed as the front approaches and strong gusty winds as the front passes. Warm fronts are weaker systems than cold fronts and are not as great of a concern to firefighters.

■ Note
Expect the winds to increase and change direction during a frontal passage.

Expect the winds to increase and change direction during a frontal passage. Heighten your awareness, think safety, and anticipate needed changes in tactics.

■ Note
Downdrafts can push the fire in many directions and cause it to jump already constructed control lines.

Thunderstorms

Two characteristics make thunderstorms a great concern to firefighters on firelines: (1) They are often accompanied by cloud to ground lightning, which often starts wildland fires, and (2) they are accompanied by strong **downdraft winds,** which spread out upon meeting the ground and produce erratic gusting winds. Downdrafts can push the fire in many directions and cause it to jump already constructed control lines.

downdraft winds
winds associated with a thunderstorm

Thunderstorms start as cumulus clouds, however only a few cumulus clouds develop into thunderstorms. If there is instability of the air mass only in the lower atmosphere, and there is stability aloft, then the convectional activity essential for a thunderstorm is not there. However, if the air is conditionally unstable through a deep layer, well beyond the freezing level, then cumulonimbus clouds (Figure 2-36) can develop.

thermal lifting
air being lifted aloft by local heating of the land mass

orographic lifting
air being forced aloft by a slope, hill, or mountain range

The triggering mechanism necessary to release the instability is usually one of four forms of lifting: thermal, orographic, frontal, and convergence. **Thermal lifting** (Figure 2-37) is the result of strong heating of the air near the ground. The heated, moisture-laden air rises high enough in the atmosphere to form cumulus clouds. **Orographic lifting** occurs in mountainous areas where the heated moist air

Figure 2-36 *Cumulonimbus clouds from an aircraft.*

frontal lifting air being forced up the slope of a warm or cold front

convergence horizontal air currents merging together or approaching a single point

is forced upslope and is a result of the presence of a slope. Thermal lifting and orographic lifting often work together in mountainous areas. **Frontal lifting** occurs when a moving cooler air mass pushes its way under a warmer air mass and lifts it. Lifting by **convergence** occurs when there is more horizontal air movement into an area than movement of air out. When this happens, the air is forced upward. It always occurs around a low pressure system. Convergence is present in the aforementioned lifting systems but can also occur independently.

Typically, thunderstorms have three stages of development and decay: the cumulus, mature, and dissipating stages.

The cumulus stage starts with a rising column of moist air. The column of rising moist air then develops into a cumulus cloud that expands vertically. The cloud has a cauliflower-like appearance. During this stage the cloud has strong indrafts (air drawn inward and upward) (Figure 2-38). If such a cloud passes over a wildland fire, the convective energy from the fire may join with the updraft of the cloud and both may reinforce each other. This in turn may strengthen the inflow at Earth's surface, causing an increase in surface winds and fire activity.

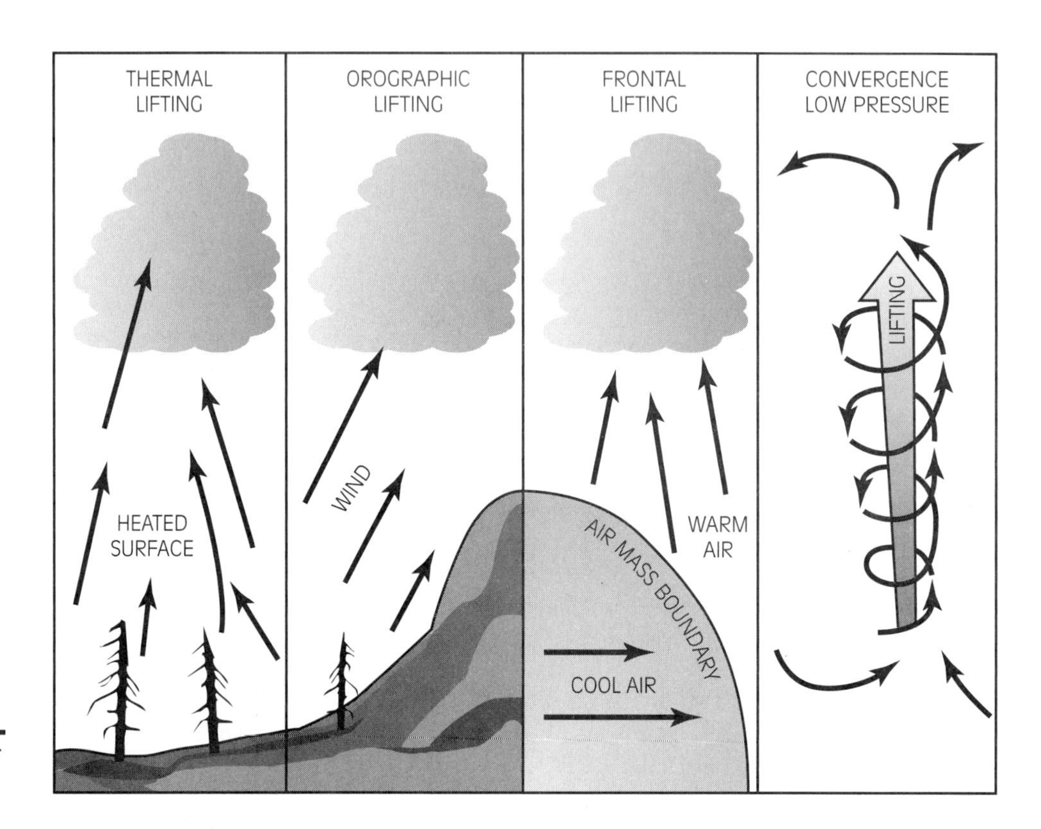

Figure 2-37 *Types of lifting.*

Safety **Downdrafts that reach the ground result in cooler gusty surface winds that can increase the fire's activity or can cause it to change direction or jump control lines.**

virga
precipitation falling out of a cloud but evaporating before reaching the ground

The mature stage, as exhibited in Figure 2-39, causes the most concern on firelines because of the strong downdrafts that are present in the cumulonimbus clouds. Downdrafts are the result of the moist air condensing into rain droplets or ice particles. These rain drops and ice particles have grown to such an extent that they can no longer be supported by the updraft. As they fall, they create frictional drag, which in turn initiates the downdraft winds. There is a downdraft in part of the cloud and an updraft in the remainder. The downdraft winds are strongest near the bottom of the cloud. Downdrafts that reach the ground result in cooler, gusty surface winds that can increase the fire's activity or can cause it to change direction or jump control lines. Sometimes we see streaks of rain falling out of the cloud but not reaching the ground. This phenomenon is called **virga** and can be found in high-level thunderstorms. These dry thunderstorms are a problem as they often produce lightning-started wildland fires.

Thunderstorms generally travel in the direction of the winds aloft. You can note their direction of travel by the direction their anvil-shaped tops are pointing; however, to observe this you must be a considerable distance from the actual thunderstorm activity.

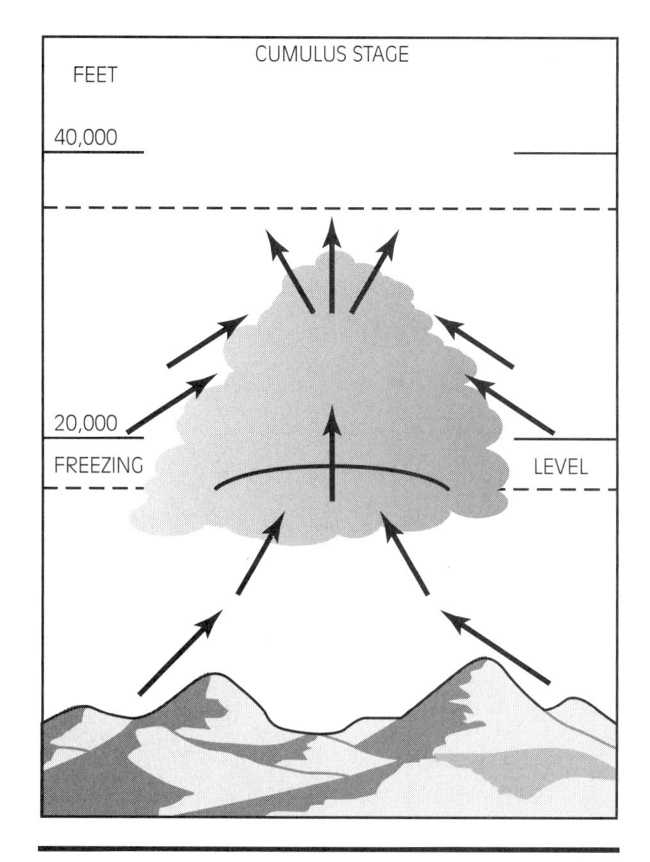

Figure 2-38 *Cumulus stage.*

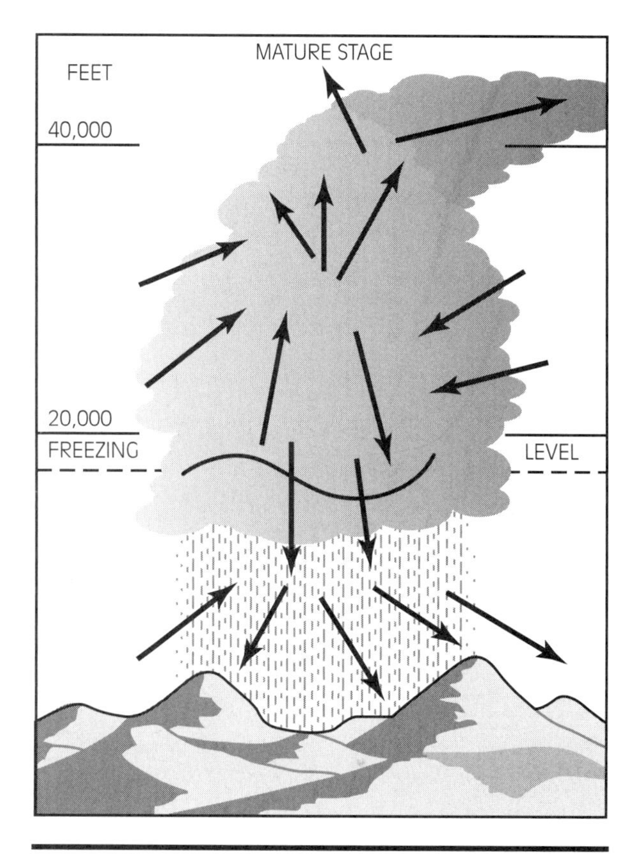

Figure 2-39 *Mature stage of a thunderstorm.*

Thunderstorms end in the dissipating stage (Figure 2-40). At that time, the downdrafts continue to develop and spread both vertically and horizontally. Once this happens, the entire thunderstorm becomes an area of downdrafts. The storm cell now enters the dissipating stage. In this stage, the updrafts, which are the source of moisture and energy, end, and the cell's growth and activity is cut off. The downdraft then gradually weakens, the rain ceases, and the clouds start to dissipate.

Air Mass Stability

atmospheric stability the degree to which vertical motion in the atmosphere is enhanced or suppressed

The stability of the air mass, which has to do with the resistance of air to vertical motion, can greatly affect a wildland fire. **Atmospheric stability** may either encourage or suppress vertical motion, which in turns affects a fire.

If there is an unstable air mass (Figure 2-41), the fire will develop a well-defined smoke column that rises vertically to great heights. This strong vertical

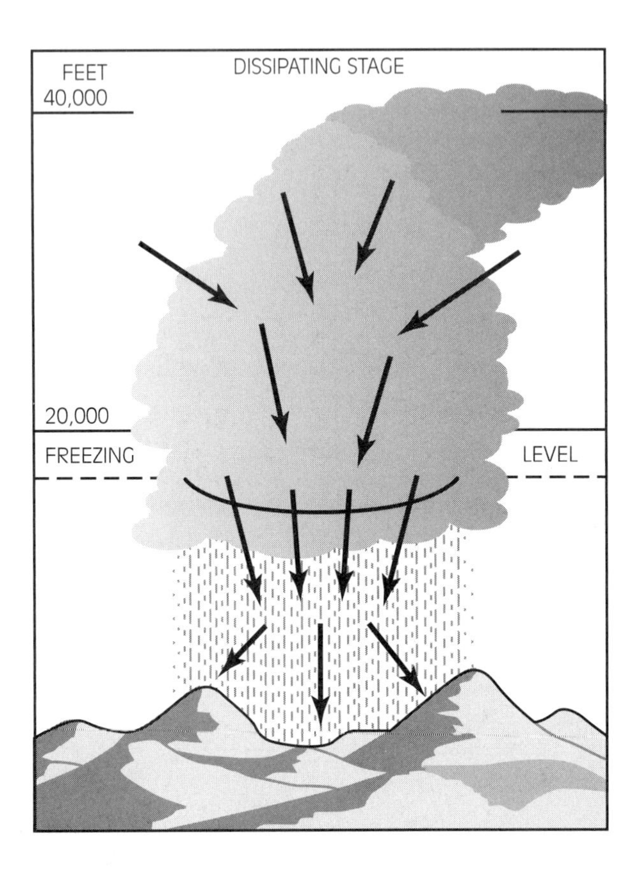

Figure 2-40
Dissipating stage of a thunderstorm.

movement of converted energy creates a strong need for oxygen to be indrafted into the fire's base. Wildland fires burn hotter and with more intensity when the air mass is unstable. A stable air mass suppresses the convective energy of the fire. After a limited rise, a smoke column drifts apart (Figure 2-42). It will appear as if it has been capped off. Fireline intensities are not as great because of the air's resistance to vertical lifting. Visual indicators that are present on the fireline give indications as to the stability of the air mass (Figure 2-43). The following lists show characteristics of stable and unstable air masses:

Unstable Air Mass Indication

- Clouds grow vertically and smoke rises to great heights
- Development of cumulus clouds
- Gusty winds
- Good visibility
- Dust devils or fire whirls

Figure 2-41 *Unstable air mass indicator—a well-defined smoke column. Photo courtesy of The National Interagency Fire Center.*

Figure 2-42 *Stable air mass indicators. Note the smoke column drifts apart after limited rise. Photo courtesy The National Interagency Fire Center.*

Stable Air Mass Indicators

- Clouds in layers
- Stratus clouds
- A smoke plume that drifts apart after a limited rise
- Poor visibility, due to haze or smoke
- Steady winds
- Fog

Use these indicators to give you additional clues as to the fireline intensities to expect.

VISIBLE INDICATORS OF A STABLE ATMOSPHERE.

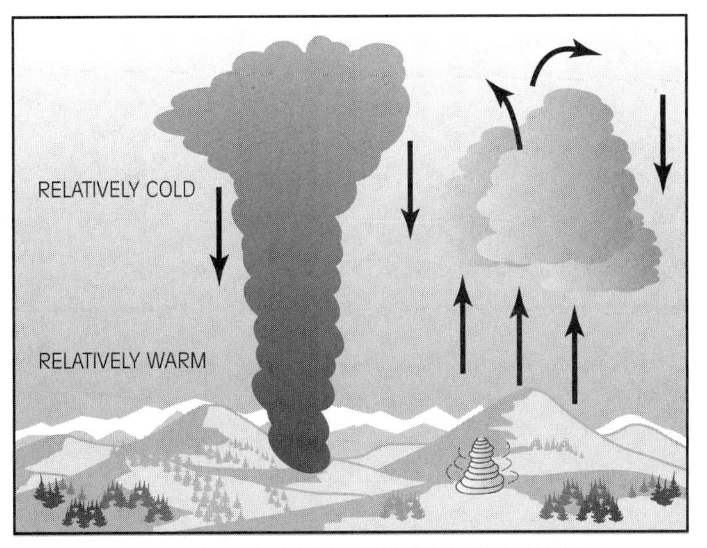

VISIBLE INDICATORS OF AN UNSTABLE ATMOSPHERE.

Figure 2-43 *Visual indicators of both stable and unstable atmospheres.*

inversion
a layer in the atmosphere that acts like a lid or cap over a fire

Inversion Layers

An **inversion** layer acts like a lid or cap over a fire. The smoke column rises and then flattens out and spreads horizontally. The fire will not burn as rapidly if a stable air mass is present.

■ Note
When the inversion layer breaks or dissipates, expect to see a change in convective column and an increase in fire activity.

When the inversion layer breaks or dissipates, expect to see a change in the convective column and an increase in fire activity. As the inversion dissipates or breaks, unstable conditions develop.

There are three types of inversion layers that can affect a fire: night inversion, marine inversion, and subsidence inversion.

night inversion
a surface layer of air in which the temperature decreases when coming in contact with the Earth's surface at night

Night inversion or radiation inversion (Figure 2-44) is the most common type of inversion and is found predominantly in mountainous terrain and inland valleys. Air is cooled at night as it comes in contact with Earth's surface. At night Earth loses its heat through radiation and the air, in contact with the ground, cools and becomes dense. This cold, dense air readily flows downslope and gathers in air pockets and valleys. Radiation or night inversions deepen as the night progresses. This creates a condition of cool, heavier air below warm air in the valley floor. In mountain areas, topography plays a decided role in both the formation and intensity of night inversions.

Conditions usually start to reverse after sunrise. As the sun rises, surface heating takes place and begins to warm the cold air. Expansion of the air takes place, and the inversion top may rise slightly. Heating starts destroying the inversion along the slopes, and upslope winds start to develop. As this continued heating and mixing of the air continues, the inversion layer continues to dissipate until final dissipation is complete. When this occurs, unstable conditions develop on the fireline and fire activity increases.

Figure 2-44 *A surface inversion clearing just after sunrise.*

marine inversions
warm season inversions occurring mainly on the West Coast of the United States in which cool, moist air from the ocean spreads over the low-lying land, varying in depth from a few hundred feet to several thousand feet, causing fog or stratus clouds

subsidence inversion
an inversion caused by subsiding air, often resulting in very limited atmospheric mixing conditions

Marine inversions are common warm season inversions that happen predominantly on the West Coast of the United States. In this type of inversion, the cool, moist air from the ocean spreads over the low-lying land. The cool air mass is capped off by a layer of much warmer, drier, and relatively unstable air. The layer of cool air can vary in depth from a few hundred feet to several thousand feet. A deep marine layer can spread over coastal mountains into the inland valleys of the central basin. Marine inversions are strongest at night and dissipate as the day goes on. If the cold air is shallow, fog will form; however, if it is deep, stratus clouds are likely to form.

Subsidence inversion is caused by a warm high pressure in the upper atmosphere. The sinking air in the higher pressure area warms and dries as it descends. This subsiding air may reach the ground surface with only a slight change in moisture. The air is also warmed and dried adiabatically as it sinks. Mountain tops are affected first even though coastal slopes may be under the influence of a massive humid layer. Subsiding air is also responsible for foehn winds, which contribute to some of the most significant weather to be found anywhere.

Thermal Belt

A thermal belt (Figure 2-45) is any area in a mountainous region that typically experiences the least changes in temperature on a daily basis. Here we find the highest daily minimum temperatures and the lowest nighttime fuel moistures.

Figure 2-45 *Thermal belt.*

Thermal belts are found in mountainous areas below the main ridges, typically the middle third of a slope. The location of a thermal belt also varies from night to night in these areas. At night, air comes in contact with the upper slopes and cools. This dense, cool air flows downhill into the low-lying mountain valleys. An inversion layer then develops above the pool of cool air. Where the inversion layer comes in contact with the mountain slopes, we see an area known as a thermal belt. This area contains relatively warmer air.

At night, wildland fires remain quite active in the thermal belt. Identify and plan appropriate tactics for these areas.

Relative Humidity

■ Note
If the relative humidity levels are low, then fine forest fuels increase in flammability and the fire danger increases.

Relative humidity is a ratio of the amount of moisture in a volume of air to the total amount that volume can hold at a given temperature and atmospheric pressure (Figure 2-46). Changes in relative humidity have a significant effect on fine fuel moisture. High relative humidity causes these fuels to burn slowly. Fires are also harder to ignite. However, if the relative humidity levels are low, then the flammability of fine forest fuels increases. Rates of spread are also greater with low humidity levels.

Time of day also plays an important part in relative humidity levels. In the early morning hours, temperatures are lowest when the relative humidity levels are higher. As the day progresses, temperatures start to rise and relative humidity levels fall. Usually in the late afternoon temperatures peak and relative humidity levels are at the minimum. Therefore it is important to check humidity levels periodically; instrument readings will indicate how fire fuels will burn. These fire fuels carry the fire.

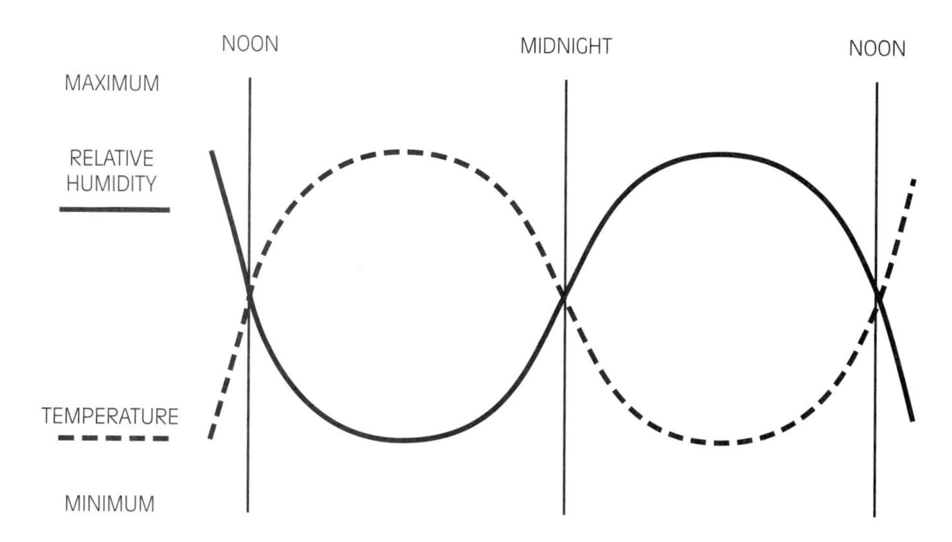

Figure 2-46 *Temperature/relative humidity chart.*

Figure 2-47 *A large firewhirl.*

Vortices

A vortex (Figure 2-47) is a whirling mass of air with a low pressure area in its center that tends to draw fire and objects, such as firebrands, into the vortex action. The vortex action becomes strongest at the center.

There are two types of vortices: vertical and horizontal. Vertical vortices (firewhirl or dust devil) are triggered in one of three ways:

1. Thermally driven vortices are triggered by intense heating of the land mass or intense burning of portions of the wildland fire.
2. Convective column vortices are produced by unequal convective activity in parts of the convective column.
3. Wake type vortices are triggered by physical obstructions to wind movement, such as ridge tops, convective columns, and trees. They occur on the lee side of the physical obstruction.

Horizontal vortices are a phenomenon associated with extreme burning conditions. They are further divided into two types. One occurs on the surface along the flanks of wildland fires, and the other occurs in the convective column and can affect air operations on the fire. Horizontal vortices tend to form more readily over flat or gentle terrain and at low-to-moderate windspeeds.

Vertical vortices aid in fire spread through heat and mass transfer as the vortex (firewhirl) moves along the surface, lifting firebrands into the air. The firewhirl action concentrates the localized wind.

Horizontal vortices are difficult to predict and, therefore, can make a safe flank of the fire a dangerous place.

EXTREME FIRE BEHAVIOR WATCHOUTS

If several of these indicators exist simultaneously, then expect that a high intensity fire is likely to occur. Always observe the indicators; they will help you predict a change in the fire's behavior early on.

Fuels

- Unusually low fuel moistures (in the fine dead fuels and live fuels)
- Large amounts of fine, dead fuels on slopes, especially if the fuels are continuous
- Dried crown foliage as the result of a previous surface fire
- Large concentrations of snags
- A large portion of the fuel bed containing dead fuels (frost kill, bug kill, drought conditions)
- Many ladder fuels present
- Fuel bed containing fuels with highly flammable oils

Topography

- Aspects with increasing fuel temperatures
- Steep slopes
- Narrow canyons
- Box canyons
- Chimneys
- Saddles
- Thermal belts; they can burn with intensity at night
- Tops of mountains; fires burning on slopes, under the influence of local winds, may come under the influence of gradient winds at the ridge top

Weather

- Low relative humidity
- Thunderstorm activity in the area
- Approaching cold fronts

- Foehn winds
- Inversion layers
- Battling winds or sudden calm, which may indicate a wind shift
- Lenticular clouds, indicating strong wind aloft; these winds may become surface winds on the lee side of the mountain range
- High winds aloft, as indicated by high, fast-moving clouds that are blowing opposite the direction of the surface winds
- High temperatures in the morning

Indicators of an Unstable Air Mass

- Good visibility
- Well-defined smoke column
- Firewhirls or dust devils
- Frequent spot fires
- Cumulus clouds
- Gusty winds

Summary

A wildland fire spreads as a result of three factors: fuel, topography, and weather. Fire spread is usually caused by a combination of all three. In flat terrain, topography is not considered as a force that drives the fire.

The element of fuel, when consumed by the combustion process, releases heat energy allowing the fire to spread. There are thirteen Fire Behavior Fuel Models, and in this chapter they were grouped into four broad classes.

Grass models (1, 2, and 3) are the primary carriers of a wildland fire. Wind really effects the spread of fire in these fuels. Grasses, depending on the time of year, burn rapidly, and the fuel is usually entirely consumed. These fine fuels also respond quickly to changes in relative humidity.

Shrub models (4, 5, 6, and 7) usually burn with greater intensity than grass models. Fuel models 4, 5, and 7 contain both dead and live fuel moistures. Generally speaking, once the fire moves into a shrub model, the fire's progress slows and the intensity levels rise, compared to grass models.

Timber litter models (8, 9, and 10) are composed largely of leaves, mixed litter, occasional twigs, and large branch wood. The litter is naturally occurring and not the result of a logging operation. Fires generally run through the surface and ground fuels. However, when the fire finds a large concentration of dead fuel, crowning or torching are possible.

Logging slash models (11, 12, and 13) are the result of logging operations. Fires in these fuel models tend to have moderate to higher rates of spread. They also tend to produce moderate to higher intensity fires. Areas of large, blown-down trees will also burn like logging slash models.

Topography represents the physical features of the land. Slope, aspect, elevation, and certain terrain features aid in fire spread. They are all elements of topography.

Weather factors are the primary elements that cause changes in a wildland fire. Wind brings a fresh supply of oxygen to the fire, gives the fire direction, and causes increased spread rates. Weather fronts bring with them increased wind spreads and changes in wind direction. Thunderstorms are accompanied by strong downdraft winds that spread out upon meeting the ground. The strong erratic winds produced by a thunderstorm often cause the fire to jump control lines.

Stability of the air mass has to do with the resistance of air to vertical motion. Atmospheric stability may either encourage or suppress vertical motion, which in turn affects fire behavior.

To thoroughly grasp fireline safety concepts, you need to have a good knowledge of wildland fire behavior. You must understand wildland fire behavior before responding to wildland fires, because tactical decisions are based on an understanding of these precepts. Review this chapter until you know its contents. This will be your foundation for a successful career as a wildland firefighter.

Review Questions

1. Explain briefly how a wildland fire spreads.
2. What is a surface fuel?
3. What does the fuel characteristic "compactness" mean?
4. In the Fire Behavior Fuel Modeling system, what fuel model numbers define the grass group?
5. Why is logging slash different from timber litter models?
6. Why do fires spread on slopes faster than on flat ground?
7. What is a foehn wind?
8. Define the term *atmospheric stability*.

References

Campbell, Doug, *The Campbell Prediction System,* A Wildland Fire Prediction System (Ojai, CA, 1991).

National Wildfire Coordinating Group, *The Fire Environment Concept* (Boise, ID: National Interagency Fire Center, 1972).

National Wildfire Coordinating Group, *Fireline Handbook 3* (Boise, ID: National Interagency Fire Center, 1998).

National Wildfire Coordinating Group, *S 190 Introduction to Wildland Fire Behavior* (Boise, ID: National Interagency Fire Center, 1983).

National Wildfire Coordinating Group, *S 290 Intermediate Wildland Fire Behavior* (Boise, ID: National Interagency Fire Center, 1994).

National Wildfire Coordinating Group, *S 390 Introduction to Wildland Fire Behavior Calculations* (Boise, ID: National Interagency Fire Center, 1994).

National Wildfire Coordinating Group, *S 490 Advanced Wildland Fire Behavior Calculations* (Boise, ID: National Interagency Fire Center, 1993).

Safety on Wildland Fires

Learning Objectives

Upon completion of this chapter, you should be able to:

- Use personal protective equipment properly.
- Use a fire shelter correctly and understand its value.
- Know what to do when trapped by fire in a vehicle or building.
- Identify wildland fire behavior watchouts.
- Explain the proper safety procedures when working around snags.
- Describe the proper safety procedures for working near a bulldozer.
- Describe the proper safety procedures for working near aircraft.
- Explain the additional wildland back country hazards.
- Explain the term *LCES* and understand its value.
- Explain and understand the FIRE ORDERS and 18 situations that shout "watch out."

INTRODUCTION

This chapter is one of the most important in this book. Personal safety and the safety of the crew should be the prime considerations when operating at a wildland fire. Safe operation at a wildland fire can only occur with a thorough and complete understanding of the concepts presented in this chapter and the understanding of fire behavior.

! Safety Personal protective clothing, the first line of defense against a wildland fire, must be worn at all times on a wildland fire.

PERSONAL PROTECTIVE EQUIPMENT

Personal protective clothing, the first line of defense against a wildland fire, must be worn at all times on a fireline. It should meet the requirements of the National Fire Protection Association (NFPA) Standard 1977, "Standards on Protective Clothing and Equipment for Wildland Fire Fighting."

NFPA 1977

Standards on Protective Clothing and Equipment for Wildland Fire Fighting

Personal protective clothing (Figure 3-1) includes:

- Helmet and shroud
- Goggles
- A Nomex® or fire-resistive cotton shirt
- A long-sleeved, cotton T-shirt
- Nomex or fire-resistive cotton pants

Figure 3-1 *Firefighters in personal protective clothing. These firefighters are members of a California Department of Forestry and Fire Protection helitack crew.*

- Gloves
- Boots
- Web gear (a belt and harness used to carry the fire shelter and canteens), fire shelter, and canteens
- Cotton undergarments
- Cotton or wool socks

Note
Personal protective clothing should be inspected after each assignment and replaced if holes or tears exist.

Nomex is a fire-resistive fabric that reflects 43% of the radiant heat of the fire. The fiber absorbs an additional 39% of the heat energy. The remaining 18% is transmitted to the skin. It would be helpful to wear an additional long-sleeved cotton (Figure 3-2) T-shirt under the Nomex. This layering of clothing will help cut down on the additional 18% that would normally be transmitted directly to the skin.

Personal protective clothing should be inspected after each assignment and replaced if holes or tears exist. It should also be kept clean and checked after each use.

Safety
Always wear a helmet with a shroud while working on the fireline.

Always wear a helmet with a shroud (Figure 3-3) while working on firelines. A helmet protects the head from falling objects and heat. Three types of helmets can be worn on a wildland fire: a single-brimmed helmet (Figure 3-4), a wide-brimmed helmet (Figure 3-5), and a structure helmet. The structure helmet (Figure 3-6) is usually worn by municipal fire crews involved in **wildland/urban interface** fire protection. The helmet shroud (Figure 3-7) protects the neck and face. It should always be closed properly in order to protect the face.

wildland/urban interface line, area, or zone where structures and other human development meet or intermingle with undeveloped wildland or vegetative fuels

Figure 3-2 *Always wear a long-sleeved cotton T-shirt under your Nomex shirt.*

Figure 3-3 *Helmet shroud.*

Figure 3-4 *Single-brimmed helmet.*

Figure 3-5 *Wide-brimmed helmet.*

Figure 3-6 *Structure helmet.*

Figure 3-7 *Note how the helmet shroud protects the face and neck.*

Figure 3-8 *Handcrew firefighter receives his handtool and prepares for a fireline assignment. Note he is wearing his goggles.*

Figure 3-9 *Always wear gloves on the fireline. These gloves are made of leather and have fire-resistive wristlets.*

! Safety Goggles should always be worn to protect the eyes.

! Safety Gloves protect the hands.

■ Note The Nomex shirt should have a high collar to protect the neck area.

Goggles (Figure 3-8) should always be worn to protect the eyes while working on a wildland fire. The lenses should be made of plastic or industrial safety glass so they do not shatter if the goggles are struck.

Gloves (Figure 3-9) protect the hands and should be made of leather. The gloves should have a fire-resistive wristlet or pull closure to keep embers from getting inside the glove.

The Nomex shirt (Figure 3-10) should have a high collar to protect the neck area. It should also have closures at the wrist area to keep embers from entering the sleeve.

Trousers should have a closure at the bottom to keep the pant leg tight around the boot. Boots should be made of leather and fit well. They should not have a steel toe or shank, as both hold the fire's heat. They should fit well and have at least an eight-inch-high top. Many companies make custom boots such as Nicks in Spokane, Washington, that ensure a proper fit. Also, anyone with special fitting problems should consider a custom-made pair of boots. The benefits will far outweigh the cost.

Safety
A wet bandanna produces high-humidity air that can damage the lungs.

Cotton bandannas can be worn in areas where it is smoky, however, they do not offer any protection against carbon monoxide. They should be worn dry. A wet bandanna produces high-humidity air that can damage the lungs more quickly at lower temperatures than a dry bandanna.

FIRE SHELTER

Safety
The fire shelter is the most important piece of fire safety equipment a firefighter owns. Never enter a fire area without one.

The fire shelter (Figure 3-11) is the most important piece of fire safety equipment a firefighter owns. Never enter a fire area without one.

Note
A fire shelter comes in a case that should be worn on the web gear while in the fire area.

A fire shelter comes in a case that should be worn on the web gear while in the fire area (Figure 3-12). It should be worn on the web gear in a way that allows it to be removed from the case easily. It should never be left on the engine. The fire shelter case should have a Velcro closure and quick pull strap (Figure 3-13). The shelter is in its plastic enclosure inside the case. It has two red pull tabs that work like a zipper. When the shelter has been removed from the plastic, it is ready for use and can be deployed.

Note
A shelter reflects 95% of the radiant heat of the fire.

A shelter can save a firefighter's life. Do not be afraid to use it if it is necessary. It works primarily by reflecting radiant heat; in fact, it reflects 95% of the radiant heat of the fire.

Safety
A shelter should be deployed in an area that is clear of fuels or, if this is not possible, a light fuel type.

It should be deployed in an area that is clear of fuels or in a light fuel type as a last resort. The shelter can not tolerate sustained heat from high intensity,

Figure 3-10 *A firefighter in full protective clothing. Note he is wearing a Nomex shirt.*

Figure 3-11 *A fire shelter in its case.*

Figure 3-12 *Note that all three firefighters are wearing web gear and a fire shelter. The web gear is the dark color strapping that can be seen on the firefighters.*

heavy fuels. It will melt at approximately 1200°F. Fortunately, forest fire temperatures are approximately 1100°F.

There are two ways to deploy the fire shelter: static deployment, which is done in place, and dynamic deployment, done on the run. The dynamic fire shelter deployment was developed as a result of the South Canyon Fire in Colorado. Fourteen people died trying to escape that fire. Therefore, many fire agencies looked at and developed fire shelter techniques to be performed on the run, which is why it is called "dynamic deployment."

The following describes a dynamic deployment technique. It is done when the fire is close at hand when firefighters are caught between the safe zone and the flaming front.

First, discard handtools (Figure 3-14) and start moving in the direction of the safe zone. Then unsnap the web gear (Figure 3-15) and remove the shelter from its carrying case (Figure 3-16).

While picking up speed and starting to run, slip out of the web gear (Figure 3-17) and let it fall to the ground. Once this is done, hold the shelter with both hands and remove it from its plastic case. The shelter is now ready to deploy (Figure 3-18).

Figure 3-13 *Fire shelter in its case.*

Figure 3-14 *Discard handtools.*

Figure 3-15 *Unsnap the web gear.*

Figure 3-16 *Remove the shelter from its carrying case.*

Figure 3-17 *Slip out of the web gear.*

Figure 3-18 *Remove the shelter from its plastic case and hold it with both hands. It is now ready to deploy.*

There will be no time to clear an area, so look for a place to deploy that is clear of vegetation or has low intensity fuels, such as a short grass model. When a spot is located, hit the ground face up and deploy the fire shelter while lying on the ground (Figure 3-19). The optimal survival zone is within one foot of the ground.

Use arms and legs to enter the shelter (Figure 3-20). Once inside the shelter, make sure the shelter has air space (Figure 3-21) away from the body, and roll over

Figure 3-19 *Hit the ground face up and deploy the fire shelter while lying on the ground.*

Figure 3-20 *Use arms and legs to enter the shelter.*

on your stomach. The face should be covered and airway protected (Figure 3-22). Remember, always use the wind to the best advantage. Place feet toward the fire if the wind permits.

Now, let's look at a static deployment technique. This is the technique most firefighters are probably familiar with. It is done in place and not on the run.

Figure 3-21 *Once inside the shelter, make sure the shelter has air space and then roll over.*

Figure 3-22 *Once you have rolled over, cover your face and airway.*

First, find an area that is as clear of fuels as possible and away from heavy fuel concentrations. Then discard all flammable items such as fusees, fuel bottles, and chain saw (Figure 3-23). Once this is done, time permitting, continue to clear as large an area as possible (Figure 3-24), then discard handtools (Figure 3-25).

Figure 3-23 *Discard all flammable items.*

Figure 3-24 *Clear at least a 4 foot × 8 foot area. Make it larger if time permits.*

Remove the shelter from its case (Figure 3-26), remove the plastic enclosure (Figure 3-27), and shake the shelter out (Figure 3-28). Now, open the shelter (Figure 3-29). When this has been completed, with your back toward the shelter, place a leg through the hold-down strap (Figure 3-30) to keep it from blowing away. Immediately continue to open the rest of the shelter (Figure 3-31) and then put the other leg into the hold-down straps (Figure 3-32). Now place the arms into the upper hold-down straps and fall to the ground. The shelter should now resemble a pup tent (Figure 3-33). Once on the ground, directly seal all the flaps and make an air space for insulation. The face should be covered and airway protected. Keep calm. Place feet toward the fire.

If the fire is close and a decision must be made on how to deploy, remember that it is better to be on the ground before the flaming front arrives than standing up and trying to get into the fire shelter.

Safety Advisory—Fire Shelters

Fire shelter training stresses the importance of deploying in an area where there will be no direct flame contact. Recent tests by the USDA Forest Service Technology &

Figure 3-25 *Discard any handtools.*

Figure 3-26 *Remove the shelter from its case.*

Figure 3-27 *Remove the plastic case.*

Figure 3-28 *Shake the shelter out.*

Figure 3-29 *Open the shelter.*

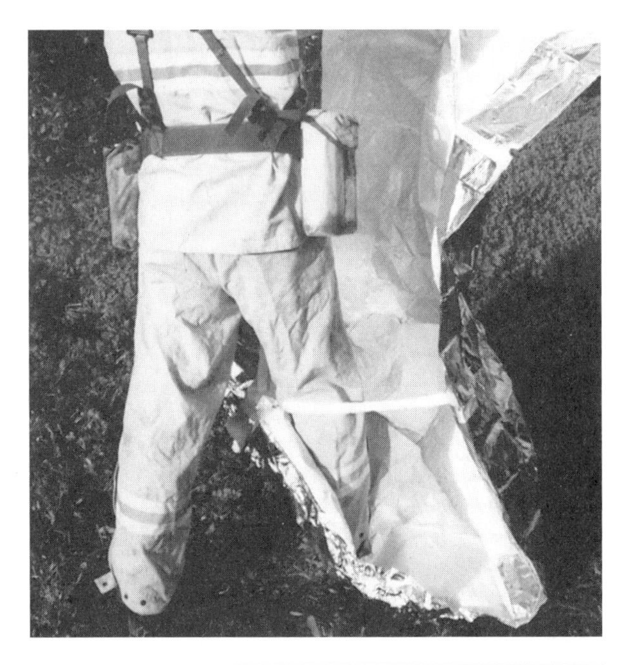

Figure 3-30 *Place a leg through the hold-down strap.*

Figure 3-31 *Continue to open the shelter.*

Figure 3-32 *Now place your second leg in the shelter.*

Development Program, though, have shown that avoiding flame contact is far more important than was understood before. Improvements in video recording equipment have shown that direct flame contact can cause fire shelter materials to break down; glue used in the shelter material produces gas and can fill the shelter with smoke. The gas and smoke are flammable, and if flame enters the shelter through a pinhole or under the edge, the gasses can ignite and cause a flashover. The shelter material continues to burn after the flashover, and shelter damage ranges from small holes in the aluminum outer layer to large holes in the aluminum and fiberglass cloth to total destruction of the shelter. Similar damage can occur from flames in grass and ground litter or from intense flame fronts.

The limiting factor on the shelter's durability appears to be off-gassing and ignition of the adhesive. The precise temperature or heat flux that causes this is not yet known, but recent test results indicate that the temperature is much lower

Figure 3-33 *After you fall to the ground, the shelter should look like a pup tent.*

than previously thought. The ignition inside the shelter causes rapid delamination and flaking of the aluminum foil but when the heat load outside is reduced, the shelter material stops burning in a matter of seconds. If the intense external heat load continues, off-gassing and combustion also continue.

Firefighters have survived entrapments in shelters with areas of delamination and missing foil. They may have experienced fire within their shelters. However, we still believe that the conditions inside the shelter are safer than those outside the shelter, but flame contact reduces the protection offered by the shelter.

Fire shelters have limitations. In light of these new findings, it is critical that firefighters recognize the importance of deploying shelters in an area that is as large as possible and as free of fuels as possible. Suggested deployment sites include gravel or paved roads, areas cleared by bulldozers to mineral soil, or black areas with no residual fuel. Remember that grasses, fine fuels, ground litter, and firefighter equipment such as fusees can ignite and burn the shelter. These precautions are in addition to those listed in *Your Fire Shelter—Beyond the Basics* (USDA Forest Service, 1996 edition), particularly those related to situations that could expose the shelter to flames or convective heat, such as chimneys, steep slopes, draws.

First and foremost, remember that all firefighting tactics must be selected to ensure firefighter safety at all times. Escape routes and safety zones must be known by all and must be continually reevaluated. The fire shelter is your last resort; it is not a guarantee of safety. If the fire shelter must be deployed, it is extremely important to deploy in an area where flames will not contact the shelter.

The above information was just sent out by the National Interagency Fire Center as a safety advisory on fire shelters.

Shelter Review Points

Here are some points to remember when deploying the fire shelter:

- Never deploy the fire shelter under power lines.
- Do not deploy the fire shelter under a snag or widow maker.
- Deploy in an area that is as free of fuels as possible. Avoid areas where the shelter could sustain direct flame contact.
- Watch terrain features: avoid deploying in saddles or chimneys.
- Always protect airways.
- Deploy in a group if possible. This action results in mutual shielding, and supervisors have better control of their crews.
- Never exit the fire shelter until the flame front has passed.
- Do not come out of the shelter until you get an all clear from the supervisor.
- If you are alone, wait for the shelter to cool.
- If time permits, clear as large an area as possible.
- Discard all flammable items such as fusees, fuel bottles, and chain saws.
- Always check the fire shelter seal once in the shelter and on the ground. Blowing embers should not enter the shelter.
- Keep the shelter away from the body by maintaining an air space inside the shelter.
- Communicate often with other crew members while in the shelter.
- Limit movement while inside the shelter.

■ Note It may be necessary to use a vehicle or building for an area of safe refuge.

Never discard the fire shelter after it has been used until replaced with a new one. It may be needed while leaving the fire area. Once in camp or at the station, get a new one immediately.

A shelter greatly improves survival chances, so never hesitate to deploy it should it become necessary. Waiting could mean the difference between life and death.

■ Note Sometimes it is not simple to decide whether it is better to be in a vehicle, a building, or a fire shelter.

VEHICLE AND BUILDING ENTRAPMENTS

It may be necessary to use a vehicle or building for an area of safe refuge. In all cases evaluate whether it is better to be in a vehicle, a building, or a fire shelter. Sometimes this is not a simple decision.

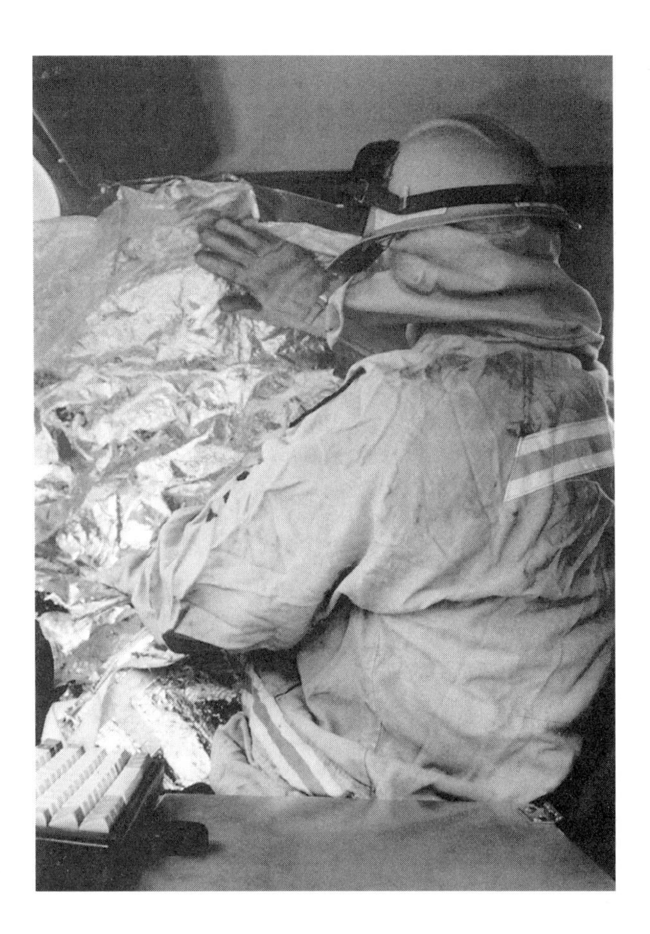

Figure 3-34 *Placing a fire shelter against the window.*

First, let us examine a vehicle as an area of safe refuge. The 1996 edition of *Your Fire Shelter—Beyond the Basics* (USDA Forest Service) states in its text the following: "There are no known instances where some people have deployed shelters just outside a vehicle while others stayed inside. Without such a comparison, recommendations are tentative."*

A vehicle can be used as a safe zone in two ways. The first circumstance is if you are driving through an area, and the fire is suddenly upon you. In this case, stay in the vehicle and place the shelter against the window (Figure 3-34). Doing so helps reflect the fire's radiant heat. If the engine has a breathing apparatus inside the cab, then wear it. If for some reason the fire shelter is not in the cab, then get to the floor of the vehicle and cover up with a brush jacket or turnout coat, if available.

*USDA Forest Service, Technology and Development Program, *Your Fire Shelter—Beyond the Basics.* (Missoula, MT) 22.

If time permits, move the vehicle away from the top of the slope, away from heavy fuel types, and out of saddles or chimneys.

Do not panic, the flaming front will pass soon. In light fuels, it will pass in about 30 seconds. However, in heavier fuels, the flame and intense heat will pass in 2 to 5 minutes.

If for some reason it becomes necessary to leave the enclosed cab, then wrap up in the fire shelter (Figure 3-35) and get out on the side away from the main body of fire. Once outside, take a quick look for a place to deploy and get to the ground quickly. Then deploy the shelter as in a dynamic deployment.

The second circumstance in which the vehicle can be used as safe zone occurs when you are in your vehicle on a wide road and it is not possible to drive out of the area. However, there is enough time, before being impacted by the fire, to safely exit the cab. In this situation, it may be beneficial to get out of the vehicle on the side away from the advancing fire and deploy the fire shelter (Figure 3-36). The vehicle will help shield some of the intensity of the main fire. Leave enough time to enter your fire shelter before the flame front arrives or do not leave the cab. If in doubt, stay in the cab and use the previously mentioned method. Always evaluate the proximity to heavier type fuels when deploying the shelter.

Figure 3-35
Firefighter exiting the cab with the fire shelter around him.

Figure 3-36
Firefighter opening the fire shelter and getting ready to deploy.

There is an increasing number of wildland/urban interface fires, so we now discuss using a house as an area of safe refuge.

First, evaluate whether the house is built well and is not too close to large concentrations of heavy fuels. If the answer is yes to both questions, then it is safe to use as a temporary barrier against the fire.

■ Note
If you plan to use the house as an area of safe refuge, all doors and windows should be closed.

If you plan to use the house as an area of safe refuge, all doors and windows should be closed. The door should be unlocked and left closed. Immediately turn off the propane tank. This should be done well in advance of the main fire. If time permits before the flaming front arrives, clear away any concentrations of heavy fuels.

Just before the flaming front arrives, enter the house with a fire shelter and web gear on. Take a shovel as it may be needed as a suppression tool later in the event the engine is burned when the flame front arrives. Also take a breathing apparatus and a portable fire extinguisher inside if the engine is so equipped.

! Safety
Should it become necessary to exit the house, wrap up in the fire shelter.

Once inside, check again to see that all doors and windows are closed. Immediately move the furniture away from the windows. Remove any flammable curtains. Then find a place near a closed doorway, on the side away from the fire, and sit down. Remember to stay low and be prepared to exit the house if necessary.

Should it become necessary to exit the house, wrap up in the fire shelter (Figure 3-37) and open the door slowly. Once outside, stay low and find a deployment area. Once in this area, immediately fall to the ground and enter the fire shelter.

Figure 3-37 *Should you have to leave the house, wrap yourself in your fire shelter and stay low.*

Remember that a well-built house will afford protection before heavy fire involvement occurs. The flaming front usually will have passed before the house burns to the ground.

FIRE BEHAVIOR WATCHOUTS

Understanding wildland fire behavior is the foundation upon which everything else is built. Without this foundation, poor tactical decisions can be made that will put others at risk.

Certain terrain features, first mentioned in Chapter Two such as chimneys, saddles, narrow canyons, box canyons, and midslope roads, are now discussed at length. They are cause for concern because they enhance fire spread.

Safety A chimney is a terrain feature that channels the convective energy of the fire.

Chimney

A chimney (Figure 3-38) is a steep drainage on the side of a mountain. This terrain feature channels all the convective energy of the fire and is responsible for rapid rates of fire spread.

Safety A fire pushes through a saddle at a rapid rate during uphill fire runs.

Saddle

A saddle (Figure 3-39) is a low topography point between two high points in a mountain range. Saddles are points of least resistance for winds and convective energy. A fire pushes through this low point at a rapid rate during uphill fire runs.

Safety The problem with a narrow canyon is that the two canyon walls are so close together that spotting can easily occur on the opposite canyon wall and trap the firefighter between two bodies of fire.

Narrow Canyon

A narrow canyon (Figure 3-40) is a terrain feature that requires great care. The problem with a narrow canyon is that the two canyon walls are so close together that spotting can easily occur on the opposite canyon wall and trap the firefighter between two bodies of fire. Area ignition can also easily occur in a narrow canyon.

Figure 3-38 *This mountainside has seven chimneys.*

Figure 3-39 *A saddle.*

Figure 3-40 *Narrow canyons are cause for concern. The canyon walls are so close together that spotting can easily occur on the opposite canyon wall.*

Note
Always have plenty of lookouts and maintain heightened awareness when working in a box canyon.

Box Canyon

Box canyons (Figure 3-41) are areas where intense updrafts occur. The steep canyon walls help force the convective column aloft. In addition, with this terrain feature there is usually only one way in and one way out. Always have plenty of lookouts and maintain heightened awareness when working in a box canyon.

Safety
The danger when working on a midslope road is the potential for becoming trapped between two bodies of fire.

Midslope Road

When working on a midslope road (Figure 3-42), always remember to watch the fuels on the unburned side. The problem with midslope roads is that there will be active fire above or below the location, and there is a danger of spotting occurring in the unburned fuel bed. The danger is the potential for becoming trapped between two bodies of fire. If spotting occurs in the unburned fuels, plan on leaving the area.

Figure 3-41 *Box canyon.*

Figure 3-42 *Midslope road.*

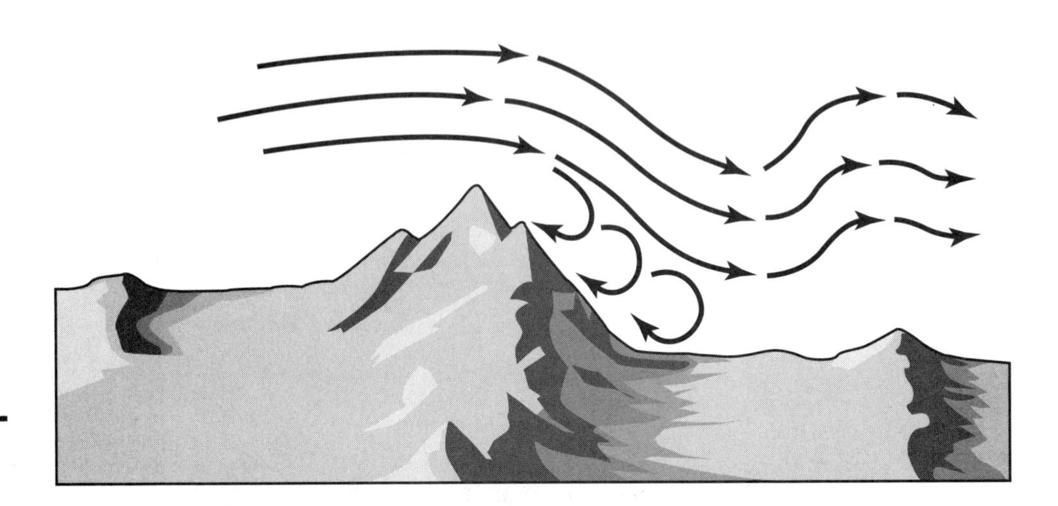

Figure 3-43 *Eddy effect.*

Safety The danger of the eddy effect is that spotting can occur below on the lee side of the ridge opposite the main fire.

Eddy Effect

As wind is blowing across a mountaintop, a rolling effect, called the *eddy effect* (Figure 3-43), can take place on the lee side (side away from the wind) of the mountain. The danger is that spotting can occur below on the lee side of the ridge opposite the main fire. If working near the top of the ridge, on the lee side, the firefighter may become trapped by the advancing flame front, which is driven by the effect of slope, wind, and the fuels. If personnel are available, it is good to place a lookout on both sides of the ridge.

Safety The danger of a thunderhead to firefighters is the downdrafts and erratic winds that occur when the mature cells pass over the fire.

Thunderhead

A thunderhead is a violent local storm produced by cumulonimbus clouds (Figure 3-44). The danger to firefighters is the downdrafts and erratic winds that occur when the mature cells pass over the fire. These winds will drive the fire in many directions. Thunderstorms may also be accompanied by lightning. If lightning is occurring, follow these rules:

- Stay out of dry creek beds.
- Do not use radios or telephones.
- Put down all tools and remove caulk boots.
- Sit or lie down, if in open country.
- Avoid grouping together.
- Do not handle flammable materials in open containers.
- Use the vehicle as a shelter whenever possible.
- Turn off machinery and electric motors.

Figure 3-44
Thunderheads forming over a mountain range.

- Where there is no shelter, avoid high objects, such as lone trees. If only isolated trees are nearby, the best protection is to crouch in the open, keeping a distance of twice the height of the tree away. Keep away from wire fences, telephone lines, and electrically conductive elevated objects.
- Avoid ridge tops, hilltops, wide-open spaces, ledges, rock outcroppings, and exposed shelters.
- Advise crew that if they feel an electrical charge—if their hair stands on end or their skin tingles—lightning may be about to strike. Drop to the ground immediately.

Adjustments to tactics and heightened awareness become necessary as thunderheads occur.

Frontal Passage

A front is a boundary between two air masses. As a front approaches, expect the wind to increase. Winds are strongest during the passage of the front and as it passes, the winds start to change direction. Winds blow counterclockwise in a low pressure area and clockwise in a higher pressure area. For this reason it is easy to see why a wind shift takes place with the passage of a front.

Remember to adjust tactical plans in accordance with the wind shift.

SAFETY WATCHOUTS

The safety watchouts in this section deal with accountability of incident personnel and working around equipment and hazards that are found on firelines. Safety watchouts are necessary for the safety of personnel and operations at a wildland fire.

Accountability of Incident Personnel

demobilization
release of resources from an incident in strict accordance with a detailed plan approved by the incident commander

Accounting for incident personnel is a major safety function at the wildland fire. The locations of all known incident resources must be tracked from check-in at the fire to **demobilization.** Without a system that tracks personnel, safety issues start to appear. Crews may become lost if isolated on the fireline. This could be a problem when they need to be fed or rested. Without food and rest, fatigue sets in and safety problems start to occur. Not knowing the whereabouts of incident resources can also set up burnover situations.

On a fire in the West, a helitack crew had been dropped off on a ridge top close to the fire's edge. The crew's location was unknown by the engine company personnel who were initiating a firing operation at the bottom of the hill. Fortunately, the helitack crew was close to the safe zone when the fire advanced uphill. This is a good example of an accountability issue that could have entrapped firefighters.

Accountability begins with proper check-in. Resources are sent either directly to the incident (immediate need resources) or used at another time (planned need resources). Check-in takes place at the following locations:

- Staging area
- Division/Group Supervisor (immediate need resources that are part of a strike team)
- Base or camp
- Helibase
- Incident command post—resource unit

Once resources have been properly checked in, they are tracked with the use of t-cards, a tag system, or some type of duty list. These systems are used to account for all personnel assigned to the fire until they are released. While personnel are on a fireline, supervisors are responsible for keeping track of their subordinates at all times.

Aircraft Safety

Aircraft are usually deployed on any large wildland fire. Firefighters working with aircraft in close proximity could potentially be in the air drop target area. Therefore it is important to know what to do if this is about to happen. If this potential exists, then consider moving out of the area if time permits. If time does not

Figure 3-45 *Firefighter preparing to be dropped on.*

permit and a drop is about to occur, then evaluate the situation in relation to overhead objects. Stay out from under large, older trees as the treetop or limbs may break due to the force of the retardant drop.

Find a place to lie down behind a large stationary object, such as a large rock. Then lie down on your stomach in the direction of the oncoming drop. Leave helmet and goggles on with face to the ground. Put arms out, with any handtool away from the body and feet apart for additional stability. Prepare for the impact of the retardant or water (Figure 3-45).

After the drop has taken place, get up slowly and make sure all crew members are unhurt. Retardant is slippery so move out of the area slowly. Clean handtools so they may be used.

Bulldozer Safety

Bulldozers (Figure 3-46) are used on any sizable wildland fire, because they are extremely effective at removing flammable vegetation in the fire's path. They are to be respected, as these large pieces of equipment are capable of dislodging huge rocks, which can roll downhill and injure fire crews. In addition, the operator may have poor visibility due to dust, smoke, and working with a helmet on.

Here is a safety checklist to use when working around a bulldozer.

- If working below a bulldozer, then watch for large rolling rocks that are displaced. It is best not to work below, if given a choice.

Figure 3-46
Bulldozer.

- When working around a bulldozer, keep a 200-foot safety radius around it. A bulldozer can change direction very quickly.
- Do not get right in front of or behind a bulldozer. The operator has poor visibility due to the blade and the tractor protective cage.
- Never sleep or sit down close to or under a bulldozer.
- Never get on or off a moving bulldozer.
- Always remain in full view of the operator.
- During night operations, always have a good headlamp or reflective vest on.
- Use hand signals for direction and safety (Figure 3-47). The operator cannot hear verbal orders.
- Do not ride on a bulldozer. The operator is the only one who has a proper seat.
- Never approach the operator unless he knows your proximity and if you are asked to do so.
- When requested to do so by the operator, approach from a 45° angle only. The operator has a better view from this angle.
- If climbing up to talk with the operator, then make sure the blade is down and use the handholds.

Figure 3-47 *Bulldozer/tractor hand signals.*

Snags

Note Snags and widow makers are the number two killers of firefighters working on firelines.

Generally speaking, a snag is a dead or dying tree that is still standing (Figure 3-48). A widow maker is a broken tree branch that is caught in the tree or other trees, waiting to fall (Figure 3-49). Snags and widow makers are the number two killer of firefighters working on firelines.

If working in an area with numerous burning trees, expect snags and widow makers. Long after the passage of the flaming front, trees continue to smolder and may finally burn through and fall, especially if a fire has burned through in previous years and has left deeply scarred trees that have little wood left to support them (Figure 3-50). In addition, the root systems may burn out and leave stump holes (Figure 3-51), which can also cause the tree to be unstable. If aircraft are making drops in the area or bulldozers are working nearby, then watch for additional instability of trees with structural defects.

Figure 3-48 *Standing dead tree—a snag.*

Figure 3-49 *A widow maker is a branch waiting to fall. Note the one in the center of the picture. This picture was taken in Maryland.*

Figure 3-50 *Snags identified by flagging.*

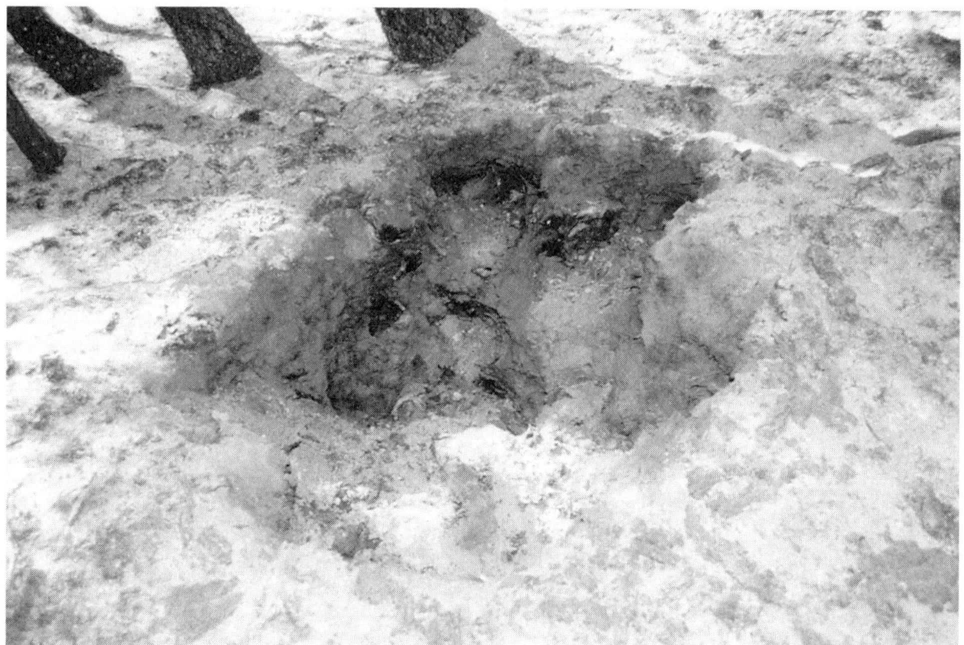

Figure 3-51 *A stump hole.*

Always identify the snags in the work area by flagging them with yellow and black flagging tape. This should be done before mop-up crews enter the burned area and start their operations.

Give snags a large safety radius and stay clear of them until they are brought down by a saw team. It is also important to post lookouts and account for and communicate with all fire crews working within the hazard area before falling a snag. Trees larger than 20 inches in diameter at breast height (DBH) should be brought down by a qualified faller (Figure 3-52). Trees less than 20 inches DBH can be felled by specially trained personnel.

Here is a list of snag watchouts to help maintain safety when working in a forested area:

- Increased wind in an area with numerous burning trees.
- Leaning dead trees.
- An area that is experiencing air drops from fixed-wing aircraft or helicopters.
- Bulldozers working in the area.
- Mop-up operations that cut root systems.
- Using strong hose streams that wash away supporting root systems.

Figure 3-52 *A faller brings down a large tree in northern California.*

Liquid Petroleum Gas Tanks

In many rural areas in the United States the primary source of cooking and heating fuel is liquid petroleum gas (LPG). This LPG is usually stored in pressurized gas tanks located on the property next to the house. Note their location before the fire arrives and clear flammable fuels around them. Turn the tank's connection to the house off.

■ Note
Keep LPG tanks cool and avoid direct flame impingement.

Keep LPG tanks cool and avoid direct flame impingement. If direct flame impingement on the LPG tank is evident, move out of the area (Figure 3-53). It will take fire water flow that is probably not available to cool the tank, including small barbecue LPG tanks, properly.

Narrow Bridges

■ Note
Narrow bridges in rural areas can become a trap for a fire engine.

Rural areas usually have narrow bridges over streams and creeks. Some are able to support the weight of a fire apparatus and others are not. They can become a trap for a fire engine. Some agencies have them denoted on preplans of the area. Use this information if it is available. If not, evaluate the bridges carefully if there are any questions as to their weight limitations. Get out and look before crossing.

Figure 3-53 *A propane tank that is venting as a result of direct flame impingement—a dangerous situation.*

Power Lines

Safety Do not park under power lines or deploy the fire shelter under power lines.

Watch overhead at all times. Do not park under power lines or deploy the fire shelter under power lines. During the heat of a wildland fire, these wires may burn through and fall. They present a dangerous electrical hazard.

Upon arrival at a wildland fire if the wires are down, cordon off the area and notify fire department dispatch of their location.

Hazardous Materials

While providing structure protection during a wildland fire, always check the garage or outbuildings for hazardous or flammable materials. Also check the brush area near the outbuilding for old tires, polyvinyl chloride (PVC) piping, or other hazardous materials.

If these items are found, try to keep the fire away from them. If this is not possible, then either evacuate the area or stay upwind of any smoke. Consult a hazardous material guide and follow the recommendations found there. If there are any doubts, then move back a safe distance and deny entry to anyone.

TAKING CARE OF YOURSELF

Safety also involves taking care of yourself before and during a wildland fire. It is important that you understand the symptoms associated with heat stress, carbon monoxide poisoning, and fatigue. Any of these conditions can debilitate a firefighter working on a fireline.

Situational Awareness

On a wildland fire, stay aware of the fire's current situation at all times by thinking globally. Be aware of weather changes. Know the predicted fire behavior, the location of the safe zone, the special hazards to be faced, and the location of fellow crew members and supervisor.

Always observe the current fire behavior, noting the fire intensity, the rate of spread, and the flame lengths. Know the direction the fire is spreading and the proximity of the flaming front. In drought years, it is critical to maintain situational awareness. Severe conditions can develop rapidly.

Heat Stress

Wildland firefighting is done in conditions that subject the firefighters to heat stress injury. Make sure to start drinking fluids before your work assignment on the fireline. Once on the fireline, consume at least one quart of water each hour.

Sports drinks, also called carbohydrates/electrolyte replacement drinks, can be used as a water replacement. However, only 25% to 50% of the total water intake should be from a sports drink.

When firefighters are mobilized from a temperate climate area to an area experiencing high temperature, they will not be able to work as many hours as they are used to working. Firefighters need to monitor their condition and make sure their work assignments do not compromise their health and safety.

Remember to set a responsible pace, take frequent breaks, and drink fluids, even if not thirsty. It is also extremely important to know the symptoms of heat stress and monitor yourself and coworkers.

Heat stress disorders are divided into four categories: *heat cramps, heat exhaustion, dehydration exhaustion,* and *heat stroke.*

Heat cramps are painful muscle cramps that may be caused by lack of fitness or failure to replace salt lost in perspiration. The victim should be treated by resting in the shade and drinking a carbohydrate/electrolyte replacement drink, lightly salted water, lemonade, or tomato juice.

Heat exhaustion is caused by a failure to hydrate properly. The symptoms include general weakness, an unstable gait or extreme fatigue, wet clammy skin, headache, nausea, and possible collapse. The victim needs to be moved to the shade and given carbohydrate/electrolyte replacement drinks. The victim should also lie down with the feet elevated slightly.

Dehydration exhaustion is another heat stress disorder caused by a failure to replace water in the system. The symptoms are weight loss and excessive fatigue. Treatment includes increasing fluid intake and resting until body weight is restored.

Heat stroke is a true medical emergency that is caused by a total collapse of the body's temperature-regulating mechanism. Call for emergency medical assistance at once, as heat stroke is a life-threatening emergency. Immediately start cooling the victim, either by immersing in cold water or soaking clothing with cool water, and fanning to promote cooling.

The victim presents with the following signs and symptoms:

- Hot, often dry skin
- A high body temperature (106°F or higher)
- Mental confusion
- Delirium
- Loss of consciousness
- Convulsions

Carbon Monoxide Poisoning

Carbon monoxide poisoning is the result of the combustion of forest fuels or the burning of gas and oil in an internal combustion engine. In a wildland fire, carbon monoxide can coexist with smoke. It becomes a problem when firefighters are exposed to heavy concentrations of carbon monoxide for an extended period of time. This invisible, odorless, and colorless gas is absorbed by the body. The carbon monoxide molecules attach to the hemoglobin molecules in the blood and displace oxygen needed by the body. Carbon monoxide is absorbed by the body at a rapid rate for the first hour of exposure, after which the rate drops slightly for the next 4 to 8 hours.

The symptoms of moderate carbon monoxide levels in the blood are a possible headache, nausea, and increased fatigue. The behavioral symptoms include increased impairment of alertness, vision discrimination, problems with physical coordination, and errors in judgment of time. As the level of carbon monoxide increases, drowsiness occurs, along with a headache, fatigue, nausea, and dizziness. Very high concentrations in the blood stream can cause convulsions and respiratory difficulty.

The treatment for carbon monoxide poisoning is to remove the firefighter from the work site to a carbon monoxide-free area and rest the victim. It may also be necessary to administer oxygen to the victim. It is important to remember that it will take about 8 hours in an uncontaminated environment to purge the carbon monoxide from the body.

Fatigue

Fatigue affects firefighters by reducing their alertness and situational awareness. It causes errors in judgment and makes people physiologically vulnerable.

Fatigue of fire crews happens in many ways. First, incident personnel are often worked too many consecutive hours, on too many consecutive days. Often those same fire crews are then moved to yet another fire.

Firefighters often pay little attention to their nutritional needs and do not hydrate properly. Both of these factors can cause fatigue, as the energy requirements of incident personnel are high. Make sure you eat right and properly hydrate.

Lack of rest is a major factor that contributes to fatigue. Incident personnel may not receive adequate rest. Incident base camps are usually busy places with a lot of noise. Portable generators usually provide base camp lighting. For security reasons, the areas are well lit, which can make sleep difficult. Night shift personnel are particularly vulnerable to the effects of chronic sleep deprivation, because they must sleep during the day.

Chronic sleep deprivation affects a firefighter's mood, reduces cognitive ability, and reduces motor skills. Get quality sleep when off shift. In incident base camps, find a dark, quiet area to set up a tent or sleeping bag. Get as much sleep as possible.

Transporting firefighters long distances also contributes to fatigue. Firefighters on large incidents are often flown across country and then placed immediately on the fireline. Similarly, incident personnel often are requested at night or very early in the morning. Sometimes, personnel are on the road all night, and upon arrival at the incident, they are immediately assigned to the fireline.

When checking in, be honest about a need for rest. If exhausted, do not be afraid to speak up. Let the supervisor know. Remember, this is a serious safety problem that needs to be addressed.

Physical Fitness

Being in good physical condition is an important factor in survival at a wildland fire. According to a recent study done by the USDA Forest Services Technology and Development Program on wildland fire fatalities in the United States between 1990 and 1998, heart attacks accounted for 21% of the deaths on wildland fires.

Put in place a good exercise program before and during the fire session. This program should not only include strength-building exercises but aerobic exercise to elevate the heart rate. Check with a physician before entering into any type of strenuous exercise program.

At some handcrew training programs, new recruits hike 3 miles in full gear with a handtool each day before class. They also do pushups as a crew to add upper body strength. Exercise programs that simulate actual fireline work conditions are best.

Many physical conditioning programs are available. The important thing is start one and stay fit. Also, watch food intake and weight. Being overweight can tax the heart.

Attitude

Attitude is an important aspect for a firefighter. A positive attitude can make the difference between success and failure in difficult situations. It can mean survival when you are fatigued, stressed, or in mental overload.

A firefighter also needs to find the balance between a lack of confidence and overconfidence or arrogance. Either attitude means trouble on a wildland fire.

Do not act impulsively. Stop and use previous experience and technical skills as a base to draw from in the decision-making process. Remember to use the acronym STOP. It stands for

S—Slow down

T—Think

O—Observe

P—Predict

Remember it and use it. Learn from any mistakes and know human beings are fallible. All human beings make mistakes from time to time, however, the lesson is what is important. Work always at improving technical skill levels, evaluating and learning from experiences, and keeping egos in check.

FIRE ORDERS AND 18 SITUATIONS THAT SHOUT "WATCH OUT!"

Look over the Fire Orders and 18 situations that shout "watch out" and analyze each.

Fire Orders

F—Fight Fire Aggressively, but Provide for Safety First Safety is always the first consideration on a wildland fire. It takes precedence over all other actions. Fight fire aggressively; however, always analyze fire behavior factors before taking hasty action. Fireline intensities may be too great for direct attack action.

I—Initiate All Actions Based on Current and Expected Fire Behavior Understand wildland fire behavior in order to meet your tactical objectives and work safely on the fireline. From this foundation, current fire behavior factors can be analyzed and objectives developed based on behavior expectations.

R—Recognize Current Weather Conditions and Obtain Forecasts One of the biggest influences on wildland fires is the weather factor. Wind brings in a fresh supply of oxygen to the fire, contributes to the rate of spread, and gives the fire direction. It is necessary to track the weather to have successful outcomes on a wildland fire.

E—Ensure Instructions Are Given and Understood If fire crews or firefighters do not understand the task at hand, then they cannot contribute effectively to the plan of attack. Once instructions are given, make sure they are understood. When laying out a tactical objective for a fire crew, check with them periodically to see where they are, the progress they are making, and if they have any support needs.

O—Obtain Current Information on the Fire Status A firefighter needs to know what the fire is doing, where it is going, and what kind of fireline intensities it is generating before a sound plan can be developed.

R—Remain in Communication with Crew Members, Supervisor, and Adjoining Forces If firefighters do not communicate with each other, then the system starts to break down. Tactical objectives are not met. Injuries happen to fire crews. A good example of this occurs when one crew is involved in a firing operation and an adjoining fire crew knows nothing about it. Yet the firing operation will severely affect that adjoining force.

D—Determine Safety Zones and Escape Routes One of the first things to do when getting to the fire is to establish escape routes and safe zones. Also, remember that as the fire progresses, both the escape route and safe zone may change. Escape routes and safe zones need to be easily accessible to all, including your slowest firefighter. Keep in mind the condition of the crew when planning escape routes. Is the crew fresh, or have they been on the line for a long time? Evaluate all these factors when planning escape routes and safe zones.

E—Establish Lookouts in Potentially Hazardous Situations Lookouts should be experienced and competent crew members. They need to be alert and always stay focused on their jobs. They need to see that a potential problem is developing and sound the alarm early, thereby enabling your crew to start toward the safe zone early or to change fireline tactics.

R—Retain Control at All Times Maintain control of your resources at all times. Make sure the instructions you give are clear, concise, and understood. Know the locations of those assigned to you and stay in communication with them. Be aware of what the fire is doing at all times.

S—Stay Alert, Keep Calm, Think Clearly, Act Decisively A wildland fire is a dynamic event, and it is imperative to stay alert to changes in fireline conditions. As a supervisor, stay calm, or the crew will pick up on the excitability and react to it. Think clearly and take into account fatigue factors and rest when necessary. Also limit the amount of people supervised and the number of radio channels monitored. If overload occurs, then a firefighter will not function appropriately.

Once you make a decision that is based on firm facts, act decisively on it.

Eighteen Situations That Shout "Watch Out!"

The 18 situations that shout "watch out" are warning signs that alert to a potentially dangerous situation that is developing. They are as follows:

1. The Fire Has Not Been Scouted and Sized Up. In this case, tactics are developed based on no information and failure is bound to result. You have *no* idea about the fire's size or intensity levels. Should you do direct or indirect? What resources are required? Where should the safe zones be? How can any of these questions be addressed without scouting the fire?

2. You Are in Country That You Have Not Seen in Daylight. At night you will not be able to see all the factors, such as topography or fuel types, necessary to develop a sound plan. Dangerous terrain features where people or equipment can become trapped, such as drainage ditches, may not be seen.

3. Safety Zones and Escape Routes Are Not Identified. Sometimes the fire behavior factors change and make it necessary to use an escape route and safe zone. This is not the time to make these decisions, and most often injury or death results if the plan did not include escape routes or safe zones.

4. You Are Unfamiliar with Local Weather and Other Factors That May Influence Fire Behavior. Weather is one of the most significant influences on a fire. If a local area has unique weather patterns, then be aware of them. When working a fire in an unfamiliar area, ask the locals about their weather patterns. Also ask fire crews that are coming off line about the fireline conditions and weather that they experienced.

5. You Are Uninformed on Strategy, Tactics, and Hazards. Do not freelance on a fire. Understand the overall objectives and plan of action before starting work on the fireline. The action taken can affect crews working in close proximity. A good example of this would be a poorly coordinated firing operation that jeopardizes adjacent crews.

6. The Instruction and Assignments You Were Given Are Not Clear. How can one carry out a plan one does not understand? If incident objectives are not understood, find them out, ask for clarification, then implement the plan.

7. You Have No Communications Link with Crew or Supervisor. Successful communications are the key to having a successful outcome on a wildland fire. Important tactical changes cannot be made if communication with the crew or supervisor does not take place. The fire may have changed direction or a fixed-wing may be preparing to drop on the location. Dangerous situations may be developing, and the firefighter is unaware of them.

8. You Are Constructing Fireline without a Safe Anchor Point. Usually fireline is built from a safe anchor point. If it is not, the fire may cross the open spot in the fireline and catch a firefighter off guard.

9. You Are Building Fireline Downhill and There Is Fire below You. The key here is that there is active fire below you. This situation is dangerous and another tactic should be chosen. The USDA Forest Service has established rules for downhill line construction, however, they do not apply with fire below you.

10. You Are Attempting a Frontal Assault on the Fire. The most intense part of a fire is the head, which is also where the flame lengths are the greatest. Usually, it is better to start from an anchor point and work up the flank and then suppress the head. Exceptions do exist, and, in these cases, remember to be constantly aware of what the fire is doing and have several experienced lookouts. In the case of structure protection, do not ever place yourself between a high intensity fire and the structure being protected.

11. There Is Unburned Fuel Between You and the Fire. One of the safest places to be, generally speaking, is on the fireline, one foot in the burned area and one foot in the unburned area. In low intensity fuel types, you can step into the burned area and be safe, should the fireline condition change.

If you are in the green unburned area with the fire advancing toward the location, then you run the risk of the fire spotting around you. These new spot fires may cause a fire that entraps you and your crew.

12. You Cannot See the Main Fire and Are Not in Contact with Anyone Who Can. How can you make any tactical decision or plan any escape routes or safe zones? You have no idea of the size or intensity of the fire. It is better to move to a safe location and wait until this information is available.

13. You Are on a Hillside Where Rolling Material Can Ignite the Fuel Below You. If this potential exists, post lookouts and constantly monitor the fire's activity. Cup trenches should be constructed to keep rolling material from crossing the fireline.

14. The Weather Is Getting Hotter and Dryer. Fine fuel moistures will be getting lower. Fine fuels are the primary carriers of the fire. Expect to see a change in your fireline conditions as it gets hotter and drier.

15. The Wind Is Increasing or Changing Direction. Generally speaking, wind gives the fire direction and brings in a fresh supply of oxygen. Increase in wind speed increases the rate of fire spread. Evaluate the fire and make the necessary tactical changes.

16. You Are Getting Frequent Spot Fires Across the Line. This situation is another indication of a change in fire behavior. Evaluate the fireline factors present and make appropriate changes in your tactics. It may be necessary to abandon the original plan and move to the secondary control line.

17. Terrain and Fuels Make Escape to Your Safety Zone Difficult. This is a real safety problem. Either construct new escape routes and safe zones or move out of the area.

18. You Feel Like Taking a Nap Near the Fireline. This is a potential problem for you or your crew. If you are in the unburned fuels, then the potential exists to get burned should the fire spot, rekindle, or jump the control line. Additionally, heavy equipment such as bulldozers or tractor plows may be working and the potential exists for an injury from either of these pieces of equipment. When you feel this way, consider fatigue factors. It may be time to rest yourself and your crew.

LCES

lookouts
persons designated to detect and report fires from a vantage point

communications
an exchange of thoughts, messages, or information that is the key to a successful outcome on a wildland fire

LCES is a mnemonic device developed by Paul Gleason, a former zig-zag hot shot superintendent. Paul took the most important aspects of FIRE ORDERS and the 18 Situations That Shout "Watch Out" and condensed them into an acronym that would be easy for firefighters to remember.

L stands for **lookouts.** The use of lookouts cannot be overstated. They should be experienced and competent crew members, as they need to be able to recognize that a problem is developing and communicate it. They need to update the fire officer on changes in fire behavior and, if a problem develops, sound the alarm early so the fire officer can change tactics.

C stands for **communications.** On any large wildland fire, communications can be a problem. Establish backup radio frequencies early because the primary radio frequencies quickly become overloaded during a fire. Also make sure to have an alternate way to communicate with each other should the radio system fail. This method may be word of mouth from firefighter to firefighter or the use of hand signals. The most important thing is to have a plan.

escape routes
preplanned and understood routes firefighters take to move to a safety zone or other low-risk area

safe zone
an area cleared of flammable materials used for escape in the event the line is outflanked or in case a spot fire causes fuels outside the control line to render the line unsafe

E stands for **escape routes.** An escape route is a preplanned route that firefighters take to get away from the fire to the safe zone. More than one escape route should be planned, and every crew member should know the location of both. Escape routes should be easily accessible and close to the safe zone. If the escape routes deviate from a well-defined path, then they should be marked with flagging tape.

S stands for **safe zone.** A safe zone is an area in which one can survive without a fire shelter. It needs to be large enough for the entire crew. Safe zones can be clean burned areas, rocky areas, water features, or other natural features. Manmade features such as roadways, well-built houses, or other constructed sites may act as safe zones.

Upon arrival to the assigned area, evaluate fireline intensity levels as well as other fireline factors when considering where the safe zone should be. Remember, downwind and upslope areas will sustain more heat impact.

LCES Checklist

- Remember to preplan LCES with the crew.
- Everyone should have common communications and be on the same radio frequencies.
- Everyone must know the escape routes and safe zones.
- All tactical plans should be discussed and adhered to.
- All firefighters should be informed and understand the tactical plan, escape routes, and how to communicate with their supervisors.

Summary

Safety and an understanding of wildland fire behavior are the most important foundational issues to learn as a firefighter.

Personal protection is the first line of defense against a wildland fire. It must be worn at all times. Nomex fabric reflects 43% of the radiant heat of the fire. The fiber absorbs an additional 39% and the remaining 18% will be transmitted to the skin. Always wear a long-sleeved cotton T-shirt under personal protective equipment, as it will afford an additional layer of protection.

A fire shelter should be worn on the gear at all times while in a fire area. Never enter a fire area without one. The fire shelter will reflect 95% of the radiant heat of the fire. Practice several times, during and before the fire season, the proper way to deploy a fire shelter. A shelter can save your life. Treat it wisely and with care.

At some point it may be necessary to use a vehicle or building as an area of safe refuge. The decision to use either one may not be simple. You may be driving through an area and get caught suddenly in the vehicle by the fire. Do not panic. Place your fire shelter against the window to reflect some of the radiant heat. You should already be in your full personal protective equipment. If the vehicle has a breathing apparatus in the jump seat, use it. This action will provide a clean source of air and prevent inhalation of toxic vapors. Many Type III engines have enclosed cabs and breathing apparatus mounted in the seats. If time permits, it may be more beneficial to leave the cab and deploy a fire shelter. The choice may not be simple. One thing not to do is leave the cab when the flaming front is upon you and the intensity levels are great. You will be injured before getting a chance to get in the shelter.

A well-built house can be used as an area of safe refuge. Always make sure all the doors and windows are closed. While inside, remain in full personal protective gear.

Certain terrain features—chimneys, saddles, narrow canyons, box canyons and midslope roads—are cause for concern. Always look for them in the area being worked. Weather features that cause concern on firelines are thunderstorms, eddy effects, and front passages.

Review the rules for safety when working around a bulldozer. These large pieces of equipment should be respected. Aircraft are usually deployed on any major wildland fire so review the safety rules for working around aircraft. Know what to do if you are about to be dropped on.

Memorize the terms LCES and know your FIRE ORDERS and the 18 Situations that Shout "Watch Out." Knowing these rules may save your life.

Much has been said about safety in this chapter, and a summary does not do the chapter justice. This chapter is so important that you should reread it several times.

Review Questions

1. How often should personal protective equipment be inspected?
2. How does the dynamic fire shelter technique differ from the static deployment technique?
3. Why is it necessary to discard all flammable items prior to deploying the fire shelter?
4. What is a chimney?
5. Why could working on a midslope road be a problem?
6. Why is a thunderhead dangerous?
7. What is a snag?
8. What does L stand for in the term *LCES*?

References

National Wildfire Coordinating Group, *I401 Safety Officer* (Boise, ID: National Interagency Fire Center, 1986).

National Wildfire Coordinating Group, *Fireline Handbook 3* (Boise, ID: National Interagency Fire Center, 1998).

USDA Forest Service, Technology and Development Program, *Wildland Fire Fatalities in the United States 1990 to 1998*. (Missoula, MT: March 1999).

USDA Forest Service, Technology and Development Program, *Your Fire Shelter—Beyond the Basics* (Missoula, MT: 1996).

Water Supplies

Learning Objectives

Upon completion of this chapter, you should be able to:

- Identify water sources for use at a wildland fire.
- Describe the different types of water sources.
- Explain the different equipment used to supply water.

WATER SOURCES

Water is a valuable commodity at a wildland fire, therefore, use it wisely. Make every drop count.

Upon arrival at the scene of a wildland fire, it is necessary to find a water source. Water is necessary to supply engine companies and possibly any water-dropping helicopters in use. Once water sources are identified, make them known so incident resources can be supplied.

The first part of this chapter deals with common water sources and the last part addresses the tools used to gain water from these water sources.

Streams, Lakes, Ponds, Irrigation Canals

■ Note A natural stream, irrigation canal, lake, or pond can be used to supply the engine or water tender with water.

A natural stream, irrigation canal, lake, or pond can be used to supply the engine or water tender with water. The area must be evaluated for ease of access. If the engine can get close enough, these natural water sources can be used. The engine's intake hose (hard suction) can be placed directly into the stream or canal to draft water, provided the water is deep enough and flowing sufficiently.

If the engine cannot get close enough to the water source, then another tool—a portable pump, floating pump, gravity sock, or ejector—must be used to access the water supply.

Swimming Pools

■ Note Swimming pools provide large supplies of emergency water.

Swimming pools provide large supplies of emergency water. Access is not always easy for fire apparatus because pools are usually found in residential backyards. Consider using a floating pump or a portable pump in cases where there are access restrictions.

Cisterns

■ Note A cistern is an underground storage reservoir made of reinforced concrete.

A cistern (Figure 4-1) is an underground storage reservoir made of reinforced concrete. The engine can get water from the cistern by using the hose connection found on the tank. At that point, the engine must either draft from the cistern or, if the tank outlet connection is above the engine, then gravity flow can be used to fill the engine. In some cases it is necessary to remove a cover and drop the intake hose into the cistern and draft from it. Cisterns are found in some rural areas of the country and provide water for firefighting purposes.

Fire Hydrants

Few fire hydrants are found in wildland areas. Fire hydrants are more prevalent the closer you get to the interface area. Fire hydrants (Figure 4-2) come in various sizes and shapes. The threads on the hydrant outlet must conform to those used

Figure 4-1 *A large underground reservoir or cistern.*

Figure 4-2 *Fire hydrant.*

by the local fire agency. It is important to have an assortment of adapters on the engine, especially if responding out of the area.

On a large wildland/urban interface fire, do not become heavily committed to hydrant use. Have an alternative water source. Studies have shown that on most large wildland interface fires where numerous structures are being lost, the water system failed. Fire flow requirements placed too great a demand on the domestic water system.

dry hydrant
permanent device with fire engine threads attached to expedite drafting operations in locations where there are water sources suitable for use in fire suppression

In rural areas, dry hydrants may be available. A **dry hydrant** is simply a pipe that is threaded on one end from which water can be pumped by the fire engine. The pipe is usually placed in lakes or streams that run year round. The pipe may also be used in conjunction with underground storage tanks or cisterns. A dry hydrant is not part of the domestic water supply, but can be a reliable source of water from which to fill water tenders.

Water Tenders

water tender
any ground vehicle capable of transporting specified quantities of water

Water tenders (Figure 4-3), tank trucks that hold a large quantity of water, are widely used on wildland fires to transport water supplies to areas where there are no water systems or where water supplies are inadequate.

Some large fire agencies, with wildland responsibilities, maintain a fleet of water tenders. In the case of smaller departments, most water tenders come from the private sector. They are hired on an on-call basis when needed. When using private contractors to shuttle water, always check to see if the vehicle is mechanically sound. The operator should also have personal protective gear while working on the fireline.

The terrain in which the water tender will be working is an important consideration. In areas where there are narrow winding roads, bridge load limits, or load restrictions on the roadway, a smaller size and tank capacity tanker may be necessary.

Portable Water Tanks

portable water tank
a container, either with rigid frame or self-supporting, that can be filled with water or a fire chemical mixture from which fire suppression resources can be filled

Portable water tanks (portatanks) can be used as a reservoir to store water from which engines can fill. Large capacity open tanks can be used to fill helicopters that have snorkels. Water tenders, or in remote areas, helicopters can be used to

Figure 4-3 *Water tender. Photo courtesy Duncan Todd, North Tree Fire.*

Figure 4-4 *Type I helicopter refilling out of a pillow-shaped portable tank. Photo courtesy Duncan Todd, North Tree Fire.*

Note **Portable water tanks can be used as reservoirs to store water from which engines can fill.**

shuttle water to these portable tanks. If a stream or pond is close to the area where the portable tank is set up, then it can be supplied by a floating pump or a portable pump.

Portable water tanks come in open and closed varieties. They can be self-supporting (Figure 4-4), such as the pyramidal design, or they may be the pillow-shaped variety. Collapsible tanks (Figure 4-5) consist of a tubular metal frame with

Figure 4-5 *Collapsible metal frame portable tanks. Photo courtesy National Interagency Fire Center.*

a synthetic or canvas duck line. Each type is foldable, which allows for ease of storage. Once folded, they are easily transported. Tank sizes vary from those normally used for staged type use to large 600 to 3,000 gallon varieties.

EQUIPMENT TO SUPPLY WATER

■ Note
A gravity sock is used to take advantage of flowing water that is above the fire.

Various appliances and equipment are used to obtain water from a pooled source, stream, lake, or other natural water point.

Gravity Sock

gravity sock
a sock-like device placed in water sources above the fire to which a hose is attached, allowing gravity to feed water to the fire

Used to take advantage of flowing water that is above the fire, a **gravity sock** (Figure 4-6) is placed in a stream and anchored securely. The large opening is placed upstream and the tail end has a hose connection on it. Gravity feed supplies water to the fire.

Ejectors

ejector
a siphon device used to fill an engine's tank when the water source is below or beyond the engine's drafting capability

■ Note
Ejectors are used when the lift required from the water source is greater than the pump's ability to lift the water.

Fire pumps, in good working order, lift or draft water approximately 20 feet at sea level. **Ejectors** (Figure 4-7) are used when the lift required from the water source is greater than the pump's ability to lift the water. In this situation, an ejector can be used to supply water to the engine or nozzle.

The principle by which ejectors operate is the venturi effect. Water is pumped through a small opening at high velocity into a larger chamber, which in turn creates a vacuum. The vacuum created drafts the water through the intake hose. Cotton-jacketed hoses should be used rather than hard lines due to the greater friction loss in hard lines. Connections need be only hand tight as small leaks do not adversely affect the operation of the ejector to any great extent.

Figure 4-6 *A gravity sock used to supply water from above the fire.*

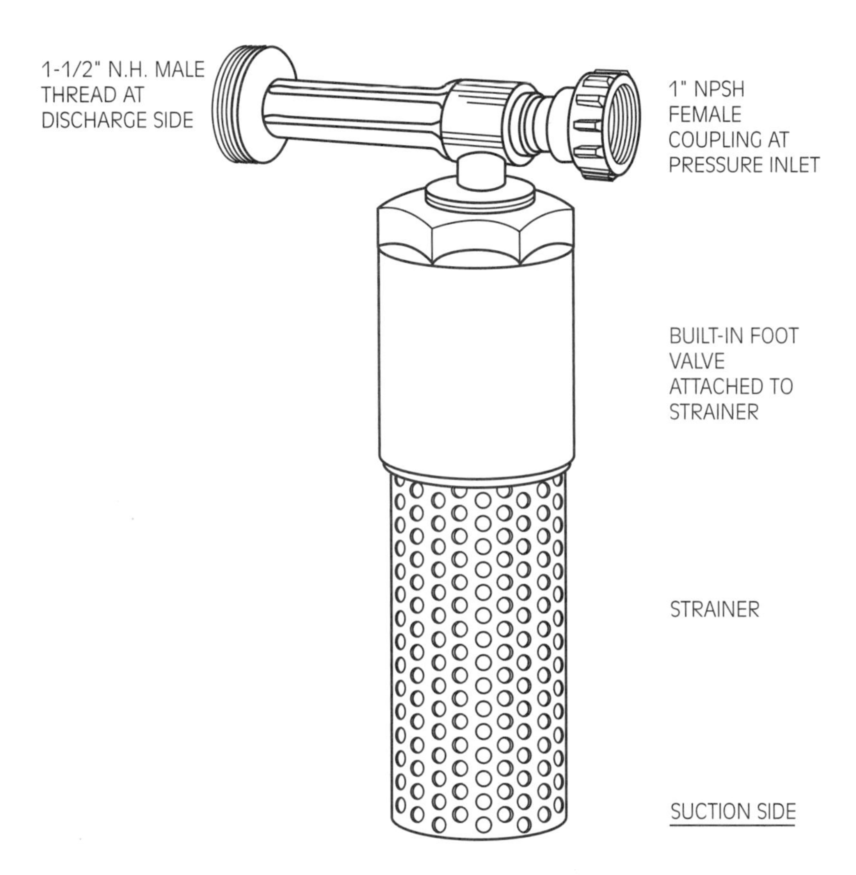

Figure 4-7 *Water ejector.*

Here is a list of situations in which an ejector might be used:

- Inaccessible areas to fire engines caused by terrain obstacles or unstable soil
- Elevation too great for drafting
- Distance from the engine to the water source too far for drafting
- To supply water to nozzles or the engine water tank from a natural water source in a remote area such as a pond

■ **Note**
Portable pumps are used to pump water from pools, lakes, ponds, streams, and other water sources.

Portable Pumps

portable pump
small gasoline-driven pump that can be carried to a water source by one or two firefighters or other conveyance over difficult terrain

Portable pumps (Figure 4-8) are used to pump water from pools, lakes, ponds, streams, and other water sources. They can be used to supply firefighting hose lines, fill portable tanks, or fill engines or water tenders.

For wildland fire use, there are two basic types of portable pumps: positive displacement pumps and centrifugal pumps. They are powered by either a two-cycle or four-cycle pump. Always use a strainer or foot valve on the end of the intake hose to prevent debris from flowing through the pump and causing damage.

Figure 4-8 *Portable pump being used to supply a hose lay.*

Floating Pumps

floating pump
small portable pump that floats in the water source

Floating pumps (Figure 4-9) are portable pumps that float on the water's surface. Some engines carry floating pumps. They work extremely well on interface fires where there are many residential swimming pools or in wildland areas with available lakes or streams.

Figure 4-9 *Floating pump.*

Summary

Water is a valuable commodity on any wildland fire. Soon after arrival at a wildland fire, plans should begin to secure a water source.

This chapter gave ideas on where water can be secured and the equipment that may be needed to supply that water.

Streams, ponds, lakes, and irrigation canals can be used to supply the engine with water. At times, the engine may be able to get close enough to place the hard suction hose directly into the water source. If this is not the case, then a portable pump or floating pump must be used. Should a flowing water source be above the engine, consider using a gravity sock for water supply needs. Ejectors are used to draft water when the lift required from the water source is greater than the pump's ability to lift the water.

Primarily found in interface areas swimming pools and hydrants can be used to supply water. Hydrants are not readily found in wildland areas.

Water tenders are used to transport water supplies in areas where there is no water. Plan to order them early due to extended travel distances.

Pay close attention to water needs early at an incident. This type of planning will pay off in big dividends. Do some prefire planning of the first-in response areas and learn where water sources are located.

Review Questions

1. What is a cistern?
2. Why is it necessary to have an assortment of hydrant adapters on the engine?
3. Are there terrain considerations when using a water tender?
4. Can helicopters fill out of portable tanks?
5. Name the two different varieties of portable tanks.
6. When would a gravity sock be used?
7. What is an ejector used for?
8. Name the two types of portable pumps.

References

International Fire Service Training Association, *Water Supplies for Fire Protection* (Stillwater, OK: Oklahoma State University, 1981).

National Wildfire Coordinating Group, *S-211 Portable Pumps and Water Use* (Boise, ID: National Interagency Fire Center, 1993).

Tactics and Strategy

Learning Objectives

Upon completion of this chapter, you should be able to:

- Explain the difference between tactics and strategy.
- Describe the size-up process.
- Identify the factors that should be evaluated upon arrival at the fire.
- Describe how to form an action plan.
- Explain how the fire officer establishes initial fireground command.

INTRODUCTION

strategy
general plan or direction selected to accomplish incident objectives

Strategy can be defined as, the art of devising or employing plans or stratagems toward a goal. It is the general method or methods that you will employ to achieve the incident objectives. Tactics describe what procedures must be accomplished to achieve the incident objectives. They are framed by the strategies that are set; they describe how you are going to do it. Tactics are the operational aspects of fire suppression. The **Incident Commander** (Figure 5-1) or **Operation Section Chief** is responsible for establishing the tactical direction on the incident.

■ Note
Tactics describe what must be accomplished to achieve the incident objectives.

incident commander
individual responsible for the management of all incident operations at the incident site

operation section chief
on a larger wildland fire, an individual who reports directly to the incident commander. That individual is delegated the responsibility of managing all the tactical operations spelled out in the Incident Action Plan

SIZE-UP

size-up
evaluation of the fire to determine a course of action for suppression

Size-up is an orderly process in which one develops a picture in one's mind en route to the fire of what is happening, as well as the factors present on arrival at the fire scene. It also deals with understanding the probabilities of what the fire will do and where it is going. Size-up is an ongoing process, as fire is dynamic and needs to be constantly reevaluated. The first component of a good size-up is knowing and understanding the facts.

■ Note
Size-up is an ongoing process.

Figure 5-1 *A Battalion Chief takes command of an initial fire attack.*

Dispatch to Arrival-on-Scene Factors

■ **Note**
On the way to work, observe the weather factors.

Size-up should start on the way into the fire station before the shift assignment begins. On the way to work, observe the weather factors. Are they different from the day before?

■ **Note**
When the alarm goes off, look at the fire area on the map and compare it with a mental picture of that area.

When the alarm goes off, look at the area where the fire is burning on a map, and compare it with a mental picture of that area formed from previous knowledge of that location. The fire officer will try to recall the terrain features of the area, the fuels that are present, and the daily wind patterns.

■ **Note**
En route to the fire the fire officer looks again at the weather factors.

En route to the fire, the fire officer looks again at the weather factors present. Which way is the wind blowing? What is the wind speed? Are the winds gusty or steady? Are there thunderheads or other cloud types forming that indicate unstable conditions? Does the humidity seem to match that forecast for the day?

■ **Note**
The fire officer checks the smoke column for direction, shape, size, height, and color.

While coming closer to the fire, the fire officer looks at and evaluates the smoke column. The fire officer checks the smoke column for direction, shape, size, height, and color. If the smoke column is leaning, it is a good indication that the fire is wind driven. The wind gives the fire direction and increases its rate of spread. There is also a potential for spotting activity.

plume-dominated fire
wildland fire whose activity is determined by the convection column

If the column has an anvil top to it, high winds aloft are present. A large, well-developed column can indicate a **plume-dominated fire,** where the thermal energy being released upward overpowers the local winds. The fire's rate of spread and direction will be unpredictable. There will be strong indrafts and strong downbursts. This fire will be dangerous. A well-defined smoke column that rises to great heights indicates that the air mass is unstable. An unstable air mass contributes to intense fireline conditions.

■ **Note**
The smoke color gives a good indication of the type of fuel in which the fire is burning.

The smoke color gives a good indication of the type of fuel in which the fire is burning. Light-colored smoke indicates that the fire is burning in light flashy fuels, such as grass. Dark-colored smoke indicates that the fire is burning in heavier fuels, such as brush or timber.

■ **Note**
"The fire is the truth—observe it." Look at the flame lengths.

Once the company officer is in a position to observe the fire, it should be evaluated. As a friend, Doug Campbell, once said, "The fire is the truth—observe it." Look at the flame lengths. Observing flame lengths gives an indication as to whether the fire is a direct attack fire or whether an indirect attack method should be used. Flame length observations help define the intensity of the fire.

Look at the rate of spread of the fire and the direction it is burning. Will it be moving into a different type of fuel soon? Are there any natural or man-made fire barriers? Will there be a change in topography? How large a fire is it? Is it spotting?

Arrival at the Fire

Upon arrival at the fire scene, the fire officer is still gathering facts and also starts doing some forecasting; forecasting deals with probabilities. Once the initial size-up has been completed, the company officer makes a report on conditions. This gives incoming fire units information on the size of the fire, fuel type, rate of spread, direction of fire spread, exposures involved, and the mode of operation (command or fast attack) that the first-arriving unit is in. Additional resource requests are also placed by radio. A typical report on conditions would sound like this, "Orange County Engine 318 on scene of a vegetation fire in light fuels with a rapid rate of spread. The fire is approximately 1/4 acre and moving easterly toward a group of homes on Overhill Drive. Engine 318 will be Peters Canyon I.C. I would like to request one additional Type III strike team, two handcrews, two dozers, and two helicopters."

! Safety

After giving a report on conditions and ordering any additional resources, the first consideration is to evaluate the area for a safe zone.

After giving a report on conditions and ordering any additional resources, the first consideration is to evaluate the area for a safe zone. Once the safe zone is identified and communicated to the crew, it is time for the fire officer to evaluate factors pertinent to the initial attack fire operations. At this time, the fire officer either takes command of the fire or works with the fire crew to extinguish the fire. If the fire officer is working with the crew, command is usually taken by the next arriving officer. The decision to take command of the fire is based on size-up factors. If the first-in fire officer feels that his involvement would allow for extinguishment of the fire, then he or she should remain with the crew and suppress the fire. Should the fire require multiple alarms (extended attack fire), then the fire officer usually becomes the incident commander.

■ Note

The critical key facts that must be looked at to complete the fire size-up are based on the observed fire behavior, the fuels present, the weather, topography of the area, and the time of day.

The critical key facts that must be looked at to complete the fire size-up are based on the observed fire behavior, the fuels present, the weather, topography of the area, and the time of the day. By observing the flame lengths, you can judge the kind of fireline intensity levels to expect.

Fuel Factors

Fuels provide the best energy for the fire. The fuels must be looked at for the following factors:

- What type of fuel is the fire burning in and what type will it be moving into?
- What is the continuity of the fuel bed?
- Is it an old age class fuel bed with a large dead component (live to dead ratio)?
- What ladder fuels are present?
- Can the fire step up into the crowns of tall brush or trees?
- What is the horizontal continuity of those crown fuels?
- Are there snags or widowmakers in the area?
- What is the fuel moisture of the 1-hour fine dead fuels?
- Is there a reburn potential, especially if in a previous year there was an understory burn only?
- Is it a fuel that contains flammable oils, such as chaparral, palmetto, gallberry, pine, or eucalyptus? These fuels produce greater fireline intensities.

Topography Factors

Topography factors must also be evaluated, as they can influence the rate of spread and direction of the fire.

These topography factors need to be evaluated:

- Steepness of the slope.
- Origin of the fire on the slope (bottom of the hill, midslope, or ridgetop).
- Aspect. Is it a hot slope or cold slope? What time of the day is it?
- Terrain features that could intensify or increase the rate of spread (chimneys, saddles, box canyon).
- Location. Is the fire in a narrow canyon? Can it spot to the other side easily?
- Elevation of the fire.

Weather Factors

Weather factors really have an effect on a fire. Because they are the most changeable, they must be evaluated constantly. Evaluate these weather factors:

- What is the wind speed and direction? Are the winds gusty or steady?
- Can any thunderstorm activity be observed?
- Are there any indicators of a frontal passage?
- Are there indications of instability of the air mass? Is the visibility clear? Is the smoke column well defined and rising to great heights?
- Can you see firewhirls or dust devils?
- What is the temperature?
- What are the humidity levels?
- Are there indications of an inversion layer?
- What are the normal wind patterns for the area?
- Did the winds suddenly become calm or are they battling?

EVALUATING THE TOOLS FOR THE JOB

Next, the Incident Commander must evaluate the equipment available to do the job. Is it the right kind or type? Are there enough resources present? How long will it take to get additional ones?

The Incident Commander also needs to evaluate the crew and determine if it is ready to do the job. Has it just finished another fireline assignment without a period of rest? Has the crew traveled a great distance to the fire? When was the last time the crew was fed?

Next, the fire officer needs to evaluate the water supplies. How far are they from the fire? What kind of source and how much water is there?

Finally, the Incident Commander looks at the actions that were already taken. Can the operation that has started be supported or does it need to be scrapped and a new plan developed?

FORECASTING

■ **Note**
The fire officer also needs to do some forecasting and understand where the fire is going.

The fire officer also needs to do some forecasting and understand, based on the previous facts, where the fire is going. What will its intensity levels be? Where will it be in 30 minutes or one hour (depending on the size before arrival)?

One of the most important decisions is evaluating the life hazards involved. Will the fire be moving into homes soon? Have they been evacuated? If not, should evacuation take place? The decision to evacuate residents or shelter them in place is based on a size-up of the current observed fire behavior and the expected future fire behavior. The fire officer also must determine if the evacuation route is safe from the fire while the evacuation is in progress.

■ **Note**
One of the most important decisions is evaluating the life hazards involved.

Part of the forecasting that is done involves the availability of critical support items, such as aircraft, bulldozers, handcrews, or additional engine companies. The incident commander also needs to know if there are restrictions on suppression tactics, such as the restriction on use of bulldozers in wilderness areas.

Once the evaluating and forecasting are done, the incident commander can start developing a plan.

FORMING AN ACTION PLAN

Developing an action plan is the next step. It is based on:

- Knowing the facts
- Looking at the available tools
- The forecasting done

The Incident Commander must decide where to attack the fire. The first consideration and the highest priority in developing an action plan is always life safety. Stabilization of the incident should be the second priority. The third priority would be conservation of the homes in the fire's path.

■ **Note**
An objective is a statement that defines what is to be accomplished.

From these incident priorities, the Incident Commander sets an incident objective or objectives. An objective is a statement that defines what is to be accomplished. Objectives need to be specific, measurable, reasonable, attainable, and timely.

Next, the Incident Commander decides on a strategy. How can the incident objectives be attained? Time can be used as a control tool and tactics time tagged. After a given time if the strategy is not working, then it may be time to switch to another tactic.

■ **Note**
Tactical priorities are tasks that must be performed to meet incident objectives.

Now the Incident Commander must develop tactical priorities. They are listed in the order of priority. Tactical priorities are tasks that must be performed to meet incident objectives. The situation and the progress being made must constantly be evaluated, accounting for personnel and supporting their needs.

OPERATIONAL MODES

After the tactical objectives are developed, the Incident Commander deploys resources in one of three operational modes: offensive, defensive, or combination mode. The operational mode is determined by the types and amounts of resources available and the fire size and behavior. If ample resources are readily available, then offensive mode can be used to accomplish the strategic goals of the incident. Fire behavior, intensity levels, and the kind of resources available determine whether the fire is attacked directly or an incident method is used. If resources are limited, then a defensive mode of operation is often used. This mode minimizes losses while accomplishing some of the incident strategic goals until ample resources arrive. Combination mode refers to operations that combine both offensive and defensive modes during fire attack.

head
most rapidly spreading portion of a fire's perimeter

flanks
parts of a fire's perimeter that are parallel to the main direction of the spread

base of fire
rear portion of the fire

PARTS OF THE FIRE

Before an explanation is given on where to attack the fire, the parts (Figure 5-2) of the fire must be identified. Knowing the parts of the fire is necessary to understanding the remainder of the chapter.

The **head** of the fire is the most rapidly spreading, most intense part of the fire. The **flanks** are the sides of the fire, and the intensity level is approximately half that of the head. The **base** of the fire is the rear portion of the fire.

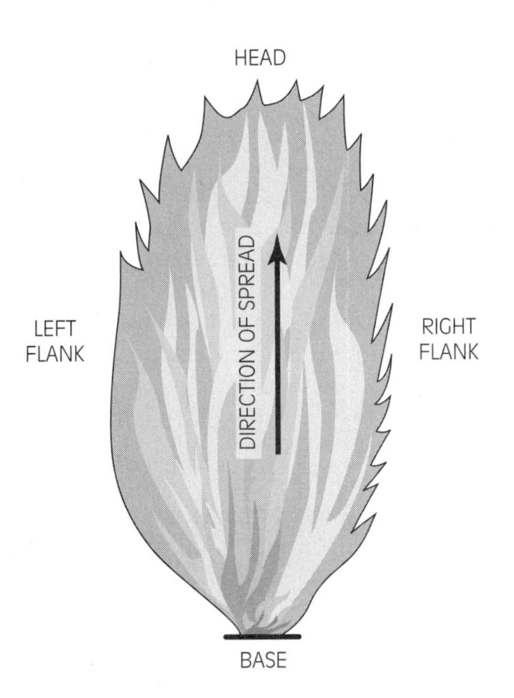

Figure 5-2 *The parts of a fire.*

WHERE AND HOW TO ATTACK THE FIRE

anchor point
advantageous location, usually a barrier to fire spread, from which to start constructing a fireline

After size-up, the plan must be communicated to the crew and an **anchor point** selected from which to start the fire attack from. Anchor points are physical barriers to fire like a roadway surface.

The Incident Commander must then decide where to attack the fire. Fires are generally attacked where they are most likely to escape. In the case of an interface fire, the Incident Commander needs to evaluate whether to try to contain the fire first or protect a house or group of houses. The fire attack may require resources to attack the fire at the flanks, base, or head of the fire or a combination of all three. This decision is determined by the fireline intensity level and rate of spread of the fire. A good way to judge a fire's intensity is to look at the flame lengths. **Flame length** (Figure 5-3) is a measurement from the midbase of the advancing flame to the tip. An evaluation of the flame length helps the incident commander determine which part of the fire to attack in both initial attack fires and larger fires.

■ **Note**
Fires are generally attacked where they are most likely to escape.

■ **Note**
A good way to judge a fire's intensity is to look at the flame lengths.

A good tool to help the Incident Commander pick an attack method is the USDA Forest Service's Table of Fire Suppression Limitations Based on Flame Length, given here as Table 5-1.

The Incident Commander always has an alternate plan for fire control when the fire cannot be contained promptly. He or she has a secondary control line in mind.

flame length
distance between the flame tip and the midpoint of the flame depth at the base of the flame

Based on the fire's intensity levels, its rate of spread, and the crew's ability to access the fire, the Incident Commander must decide how to attack the fire. There are basically three methods of fire attack: direct, parallel, and indirect. These methods are discussed in the following chapters. Chapter 6 discusses direct, parallel, and indirect methods of attack for engine companies. Handcrew tactics are discussed in Chapter 7. Tactics for bulldozers and tractor plows are discussed in Chapter 9. How aircraft are tactically deployed is discussed in Chapter 10.

■ **Note**
The incident commander always has an alternate plan for fire control when the fire cannot be contained promptly.

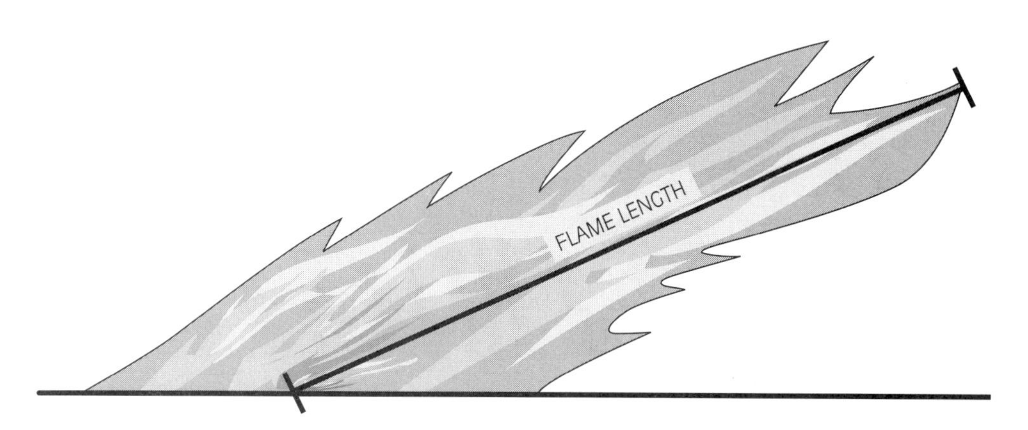

Figure 5-3 *Flame length from the midbase of the advancing flame to the tip.*

Table 5-1 *Fire suppression limitations based on flame length.*

Flame Length	Intensity	Method of Attack
FL 0–4 ft	100 BTUs/ft/sec	Direct attack handtools
FL 4–8 ft	500 BTUs/ft/sec	Hose lays, retardant, bulldozers
FL 8–11 ft	1,000 BTUs/ft/sec	Direct control difficult, control efforts at head may be ineffective
FL more than 11 ft	1,000 BTUs/ft/sec	Indirect attack, control efforts at head ineffective, spotting, major runs

Summary

This chapter discussed how a fire is sized up, how a plan is developed, and finally, how it is implemented. You need to read Chapters 6, 7, 9, and 10 before you will fully understand tactics on a wildland fire. Each chapter contains additional pertinent information of a tactical nature to enable you to best use these resources. It is also suggested that you find an experienced mentor in your agency to monitor on several fires and see how he or she develops an attack plan. Remember, training is not a substitute for experience.

Review Questions

1. What does the term *strategy* mean?
2. Define the term *tactics*.
3. What is the first component of a good size-up?
4. Why should the smoke column be evaluated?
5. Name three fuel factors that should be evaluated.
6. Name three topography factors that should be evaluated.
7. Name three weather factors that should be evaluated.

References

National Wildfire Coordinating Group, *Planning of Initial Attack!* (Boise, ID: National Interagency Fire Center, 1985).

National Wildfire Coordinating Group, *Wildland Fire Suppression Tactics Reference Guide* (Boise, ID: National Interagency Fire Center, 1996).

Engine Company Operations

Learning Objectives

Upon completion of this chapter, you should be able to:

- Describe how to use a wildland engine effectively.
- Explain the different types of fire attack methods.
- Explain the different types of hose lays.
- Describe how to mobile pump.
- Explain engine company tactics on both wildland and urban interface fires.

INTRODUCTION

This chapter discusses engine company operations and how to use this resource effectively. Early in the chapter the focus is on the use of the engine as a wildland resource. The later part of the chapter is devoted primarily to operations in the urban interface.

WILDLAND ENGINES

■ Note A fire engine is one of the most versatile pieces of fire apparatus in use on a wildland fire.

A fire engine is one of the most versatile pieces of fire apparatus in use on a wildland fire. It carries personnel and equipment—hoses, water, handtools, and sometimes class A foam—to a wildland incident. These tools and equipment allow fire crews to attack fires by direct or indirect attack methods. They can mobile pump fires where terrain allows this type of operation.

Wildland engine companies also do mop-up operations and patrol firelines. On a wildland interface fire, wildland engines are used to provide structure protection and to apply Class A foam or fire-blocking gel to structures.

■ Note A wildland engine usually has high ground clearance and a short wheelbase.

A wildland engine usually has high ground clearance and a short wheelbase. The engine should have the ability to mobile pump. In order for the engine to be able to mobile pump, it must have a separate pump engine or a power takeoff unit.

■ Note Type I engine companies (urban fire companies) do not generally operate off road.

Type I engine companies (urban fire companies) do not generally operate off road. They are not designed for that purpose; they are generally used to provide structure protection. At times, fire crews will attack fires from the roadway with hose lays.

Limitations

Wildland engines have limitations to their use. They must be properly managed on a wildland fire. Here is a list of six situations that may limit the use of a wildland engine:

■ Note Engines should not be used to control the fronts of fast-spreading fires where the fire intensities are too great.

1. Steep or rugged terrain
2. Water supply too far from the fire
3. Incapability of water tenders or other engines to support the engines doing the pumping
4. Fast-spreading fires where the fire intensities are too great
5. Poor access situations where an engine can become trapped, for example, an entry road with large stands of heavy fuel next to the roadway
6. Areas with access problems, such as narrow substandard bridges that could collapse

HOW TO USE A WILDLAND ENGINE

■ Note
One of the primary purposes of a wildland engine is to pump hose lays.

A wildland engine company is a very versatile tool. It can be used for hose lays, structure protection, handcrew work, or to perform a mobile attack.

One of the primary purposes of a wildland engine is to pump hose lays. Two different types of hose lays are presented in this chapter: the simple hose lay and the progressive hose lay.

Hose Lays

■ Note
The simple hose lay has no junctions between the nozzle and the pump.

Safety
A hose lay should be started from an anchor point.

Safety
Do not leave active fire behind you as you extend the hose lay.

The simple hose lay (Figure 6-1) has no junctions between the nozzle and the pump. With no lateral junctions in the hose lay, there are no safety provisions to pick up flare-ups behind the nozzle operator. A progressive hose lay (Figure 6-2) is usually done with 1½-inch single jacket hose in 100-foot sections. As sections of hose are added, water is discharged on the fire. Laterals (Figure 6-3) are put in about every 200 feet. They have a 1-inch hose fitting on the top and are commonly called "Ts." The 1-inch hose is used to pick up hot spots on the fireline and mop-up.

A hose lay should be started from an anchor point and progress up the flanks of the fire toward the head of the fire. The fire should be completely extinguished as the hose line is advanced up the flank of the fire. Do not leave active fire behind as you extend the hose lay. You should have one foot in the burned area and one foot in the green unburned area during the operation. Use a broken stream pattern on the nozzle to knock down the fire. Also, remember to use the nozzle in such a

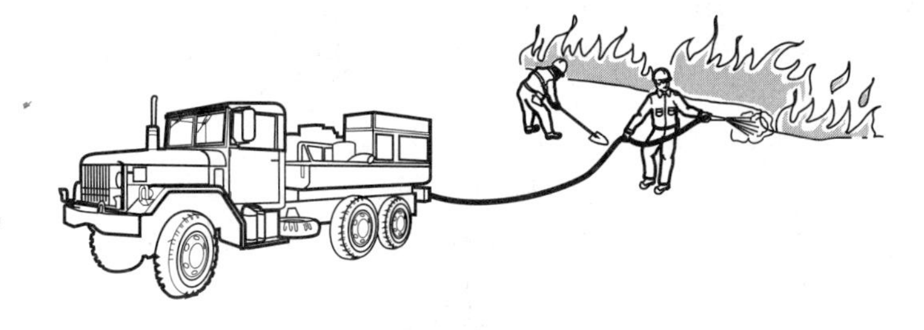

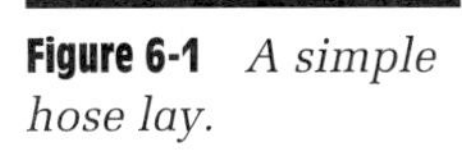

Figure 6-1 *A simple hose lay.*

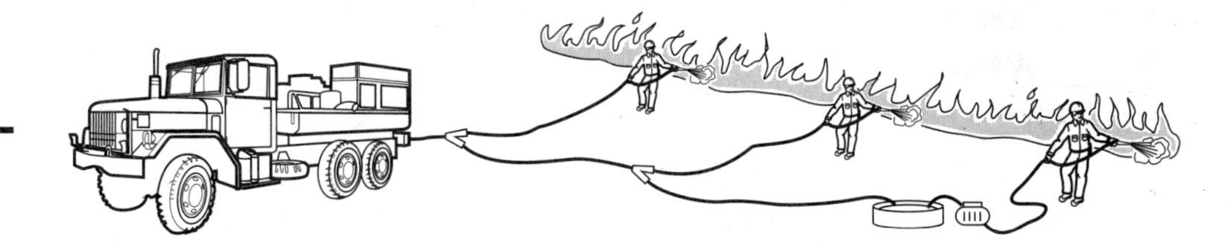

Figure 6-2 *A progressive hose lay.*

Figure 6-3 *A forestry* T *is an appliance used in a hose lay so a lateral hoseline can be run off the main hose lay.*

way that you push the fire into the burned area and not across the line. Nozzle flow is dependent on the fuel type and intensity of the fire. If the fire is burning hot, then the nozzle needs to flow at least 30 gallons of water per minute. Generally speaking, fire attack should be done with 1½-inch single jacket hose. During mop-up operations, 1-inch single jacket hose is used.

■ Note

Once the hose lay is started, it needs to be supported or you will soon run out of water.

Once the hose lay is started, it needs to be supported or you will soon run out of water. Support can come from engines shuttling water to the pumping engine or using water tenders.

ENGINE COMPANY TACTICS

Engine companies attack wildland fires in basically three ways: direct attack, parallel attack, or indirect attack.

Class A foam
foam intended for use on Class A or woody fuels

fire-blocking gel
a super absorbent polymer used as a firefighting agent

Direct Attack

Engines are effective with direct attack (applying water directly on the fire) in light fuel types. They are mobile and offer firefighters the ability to suppress the fire quickly with either water, **Class A foam,** or **fire-blocking gels.**

When using engines for direct attack on the fireline, consider the terrain, access, point of attack (head or flank), escape routes, safe zone, fire behavior factors (rate of spread and intensity), and water supplies.

Direct attack tactics are: *tandem, pincer, flanking, envelopment, mobile pumping,* and an *inside out* application.

■ Note

The purpose of mobile pumping is to knock down the fire quickly and keep it from spreading.

Mobile Pumping Mobile pumping (Figure 6-4) is done in light fuel types on terrain that allows the engine to move along the fire's edge while pumping. The purpose of mobile pumping is to knock down the fire quickly and keep it from spreading.

Figure 6-4 *Fire engine mobile pumping.*

Figure 6-5 *Tandem action.*

Water is applied quickly and efficiently by an engine that can pump and roll at the same time. A short section of hose is used by the nozzleman while the engine is driven on the fire's edge. Water is applied to the fire at the base of the flames. The nozzleman walks with the hose line and remains in full view of the engine operator during the operation.

In heavier fuels it is beneficial to use two nozzlemen. The first nozzleman knocks down the intense areas of the fire (hot spots) and the second nozzleman completes the extinguishing process.

Note
A tandem action involves two engines that are mobile pumping on the fire's edge.

Tandem Action A tandem action (Figure 6-5) involves two engines that are mobile pumping on the fire's edge. It is effectively used on the flanks of fast-moving fires. The first engines usually knock down the heavier volumes of fire, while the second engine follows and totally extinguishes the fire.

Tandem actions are safer and more effective than mobile pumped single engine operations. Two engines working together are able to extinguish more fireline than single resource units. From a safety standpoint, if one engine breaks down, the second can then take over.

Note
A pincer action is a direct attack method in which two engines start from an anchor point on opposite sides of the fire and proceed up the flanks.

Pincer Actions A pincer action (Figure 6-6) is a direct attack method in which two engines start from an anchor point on opposite sides of the fire, proceed up the flanks, and eventually extinguish the head of the fire. It is generally used on smaller fires, however, it can be used on any size fire.

Note
Flanking actions are done on the flank of a fire to keep it from moving into heavier fuels.

Flanking Actions Flanking actions (Figure 6-7) are done on the flank of a fire to keep it from moving into heavier fuels. This type of fire attack is used on moderately hot fires that are moving at a moderate rate of spread. As a general rule, concentrate the majority of resources on the most active flank of the fire first. After good progress is being made on the active flank, additional resources can be placed on the inactive flank to secure it.

Note
Envelopment actions are very effective on small fires as the fire is aggressively attacked at many points along its edge.

Envelopment Actions Envelopment actions (Figure 6-8) are very effective on small fires as the fire is aggressively attacked at many points along its edge. This type of action requires real coordination of resources, good communications, and a tight command structure. Engines must work from good anchor points during envelopment actions or they may become entrapped by the fire. Deployed resources will move along the fire's edge in many different directions. Envelopment action suppresses the fire quickly, if it is done correctly.

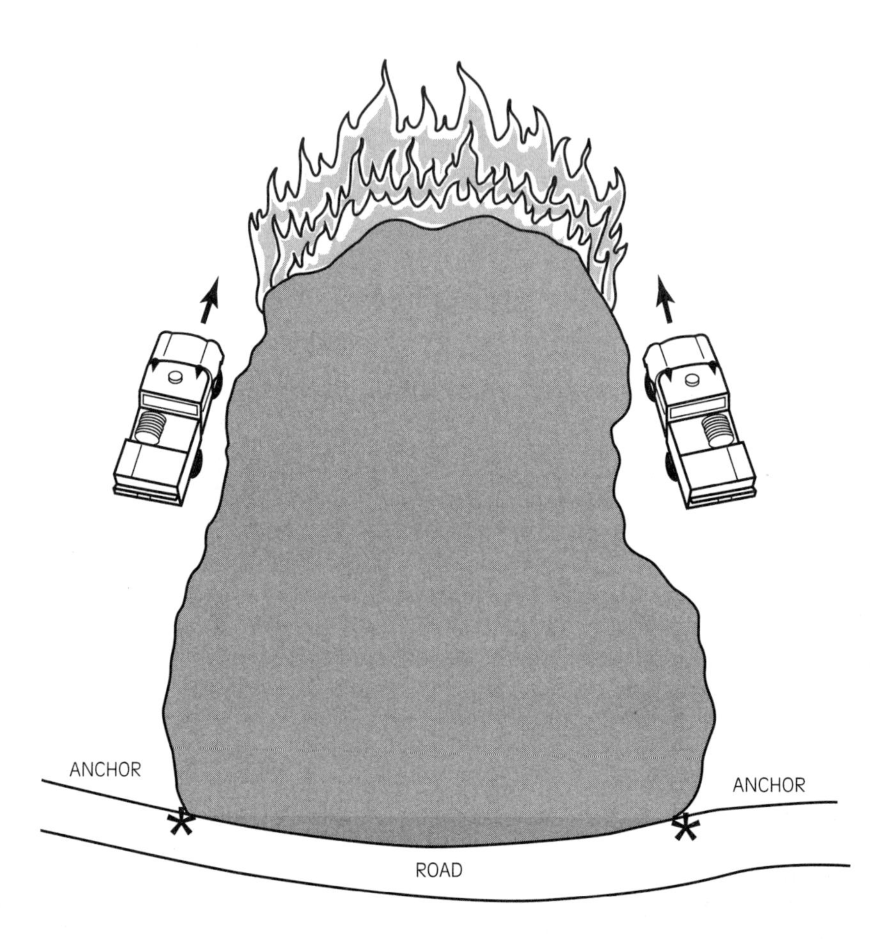

Figure 6-6 *Pincer action.*

Inside Out Inside out tactics (Figure 6-9) are done where the terrain allows mobile pumping of the fire. Inside out tactics are used on smaller fires and afford great safety for the engine suppression crews working on the fire. When working in the burn area, it is also easier to see any obstacles as the fuels are usually gone.

■ Note
Engines are driven into the burn from the origin. Then the back side of the head of the fire is attacked from the burn.

Engines are driven into the burn from the origin. Then the back side of the head of the fire is attacked from the burn. When the engines reach the head, suppression action is started, and the engines move in opposite direction(s), working their way from the head, along the flanks, to the point of origin. This attack works well on smaller fires, as it allows you to attack the head of the fire and stop its forward spread. Then the slower spreading flanks can be worked on.

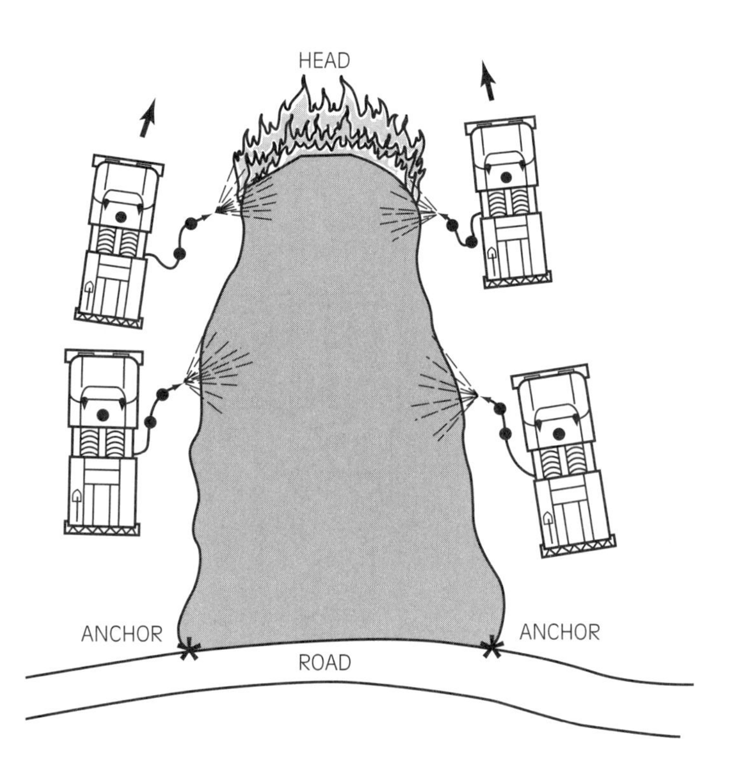

Figure 6-7 *Flanking action.*

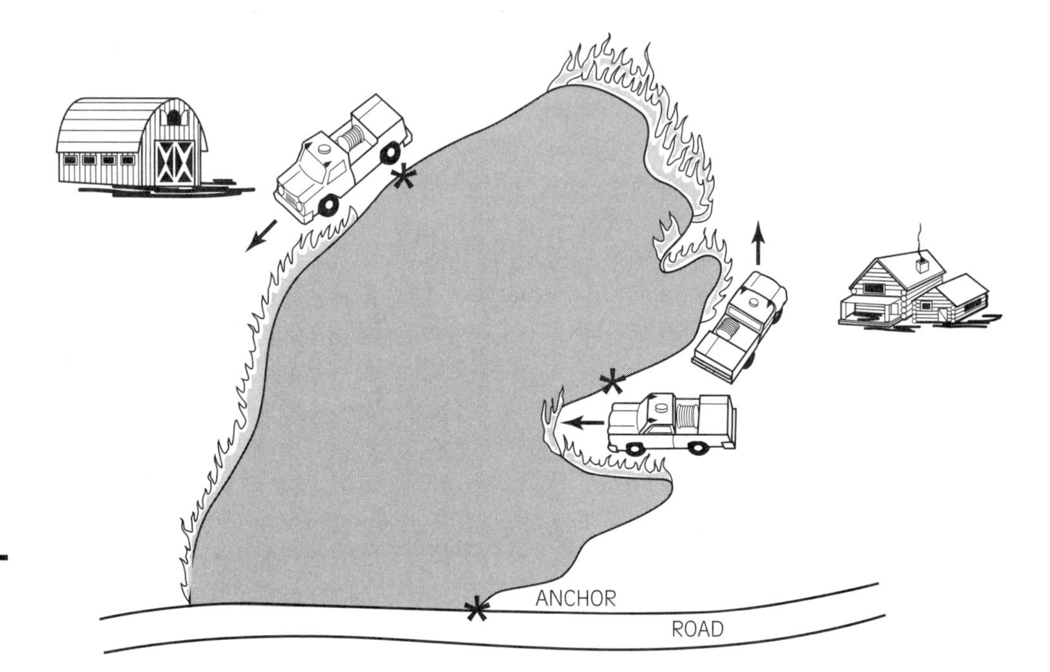

Figure 6-8 *Envelopment action.*

Figure 6-9 *Inside out attack.*

Parallel Attack

Parallel attack (Figure 6-10) allows control of the fire at a predetermined line. The advantage to this method is that firefighters are able to work close to the fire and not be exposed to the heat and products of combustion. The disadvantage to this method is that it puts firefighters in a situation where they are working in unburned fuels near the fire.

burn out
to set fire inside a control line to consume fuel between the edge of the fire and the control line

Here is an example of a parallel attack method that can be used on a fire burning in light fuels. An engine or engines working in tandem lay a wet line down and **burn out** from it. This wet line is a short distance from the fire. The wet line can be made from water, retardant, or class A foam. Figure 6-11 shows a good example of a wet line. As the wet line is being laid, firefighters follow and immediately burn out from the wet line. It is usually best to start the operation from an anchor point and work along the flank of the fire. Fire intensity levels are not as great on the flanks.

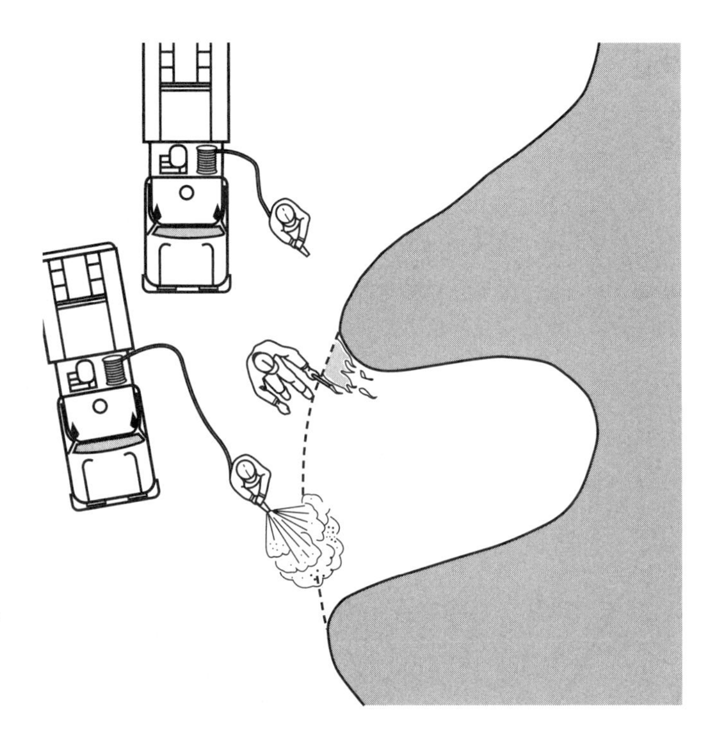

Figure 6-10 *Parallel attack with a wet line.*

Figure 6-11 *Water tender putting down a foam line and firefighters burning out from it. Photo courtesy Duncan Todd, North Tree Fire.*

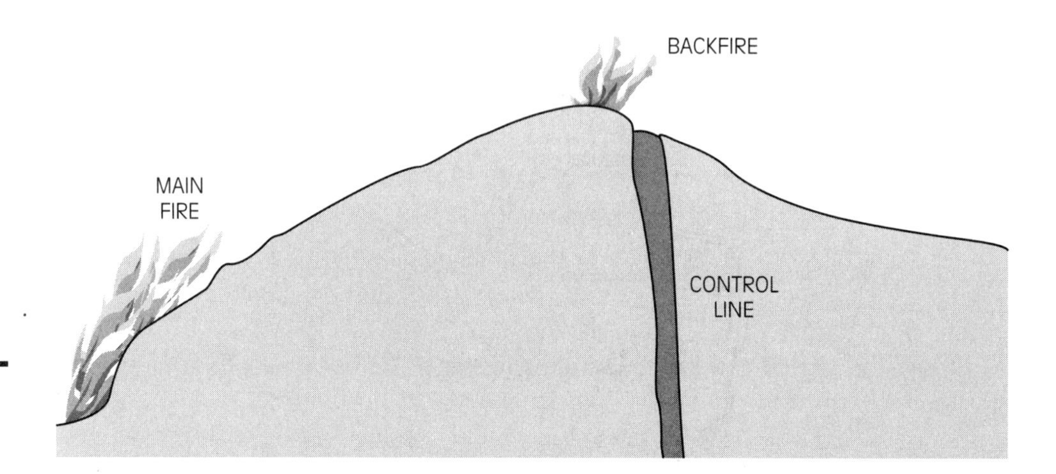

Figure 6-12 *An indirect attack on a high-intensity fire.*

This attack method works best where two engines work in tandem. The first engine mobile pumps and puts the wet line down and burns out. The second engine acts as the holding team and reinforces the set line. It also extinguishes any slopovers that may occur.

Indirect Attack

Indirect attack is a suppression method in which the control line is located at a considerable distance from the fire's edge. It is generally done on a fast-spreading or high-intensity fire. The control line is constructed using favorable breaks in topography. Natural or constructed fire breaks can also be used for this purpose. To reinforce the control line, it is usually backfired (Figure 6-12). Occasionally, the main fire is allowed to burn to the control line, depending on how big the control line is and the escape threat that exists.

WILDLAND/URBAN INTERFACE TACTICS

In recent years more homes have been built along the wildland/urban interface. Also in recent years, more firefighters have made poor tactical decisions and have placed themselves between a high-intensity fire and the structure they were trying to protect.

■ Note
While responding to an interface fire, note the fire's behavior on the way to your assignment area.

Size-up

While responding to an interface fire, note the fire's behavior on the way to your assignment area. If the factors that drive the fire (fuel, topography, and weather) are the same at your assigned location, then the observed fire flame lengths and intensities should be the same when the fire arrives. Use this information to help develop strategic goals and tactical objectives. These fire factors were discussed in Chapter 5. On the way to the structures you are going to protect evaluate the following:

- Evaluate access to the home.
  - Is the road one way or narrow?
  - Will the width of a bridge to the home allow passage of fire equipment?
  - Will the bridge support fire apparatus weights?
  - Is there high fuel loading or tree canopies along the side of the road that could cause engines to become entrapped by fire?

Upon arrival on scene, report on conditions and make any requests for additional resources that may be needed initially. This communication should be done from a safe location where there is full view of the situation. Before anything else is done, an evaluation must be made to locate all the additional safe zones.

Post a lookout at a vantage point where the entire area being worked can be seen.

Then start working on tactical objectives or structure preparation concerns. Concurrently, post a lookout at a vantage point where the entire area being worked can be seen. The next thing to do is pick what we call a trigger point, a location in the fire's future path, which is used to alert the fire crew of the impending fire. The lookout is made aware of this point and communicates to the fire crews when the fire has reached this point.

safe refuge area
area that can be used to shelter firefighters or residents in place until the flaming front passes

Once you have established that the area of structures you have decided to protect has safe zones or **safe refuge** areas, you have posted a lookout, and located the trigger point, it is time to go to work.

Now, evaluate what the fire is doing at your location. Observe the fuels present, the weather, topography of the area, and the time of day. These factors were discussed in Chapter 5 in the section called Arrival at the Fire.

After the fire behavior factors have been evaluated, the items on the following structure protection list must be evaluated:

- Decide whether to evacuate residents or shelter them in place.
- Locate water supplies and evaluate the dependability of the water source.
- Determine the amount of clearance around the structure.
- Evaluate the construction features of the house.
  - Does it have wood siding and a wood roof that has not been treated to make it fire resistant? Untreated roof coverings are probably the greatest single hazard.
  - Does it have deck and eaves projections that can trap heat?
  - Does it have exposed attic vents?
  - Are rain gutters full of debris?
  - Does it have large single-pane glass windows on the fire's side?
  - Are there hazardous materials located on the property?

    - Where is the propane tank located and are the fuels cleared around it?
    - Will overhead powerlines fall close to the structure?
    - Is there a place to park the engine away from the power line?
    - Does the house have a wood pile next to the building that needs to be scattered?
    - Where is the septic tank? Fire apparatus should not be parked on top of a septic tank.
    - Are there any outbuildings that may house hazardous materials, such as gasoline or pesticides?
    - What is the location of the house? Is it in a chimney or saddle? At the top of the hill, midslope, or where?
- Are there enough resources to do the job safely? If not, can I get more before the fire arrives?

This evaluation is a part of structure triage that involves identifying houses that are defendable and those that are not. Homes that are identified as not being defendable will not have resources deployed around them.

Developing a Tactical Plan

Developing an action plan is the next step in the process, after size-up. Development of an action plan was discussed at length in Chapter 5. As stated in that chapter, the priorities of a good action plan are:

1. life safety considerations
2. stabilization of the incident
3. conservation of the properties in the path of the fire.

Then set strategic goals and define tactical objectives. Once this is done, put incoming resources to work.

Structure Protection Tactics

If the flaming front is still a distance off, you may have time to better prepare the structure and make it more fire resistive. Here are some areas to look at:

- Clear the roof and gutters of debris.
- Remove flammable window awnings.
- Remove flammable native and ornamental vegetation from around the structure's propane tank and outbuildings.
- Remove woodpiles close to the structure.
- Place all flammable patio furniture in the garage or house.
- Close all doors and windows.

- Cover all vents, exterior ducts, and under floor screen vents.
- Turn off the LPG tank.
- Cover windows on the fire side with plywood (if any is on the property) or sheetmetal.
- Park vehicles that are left on the property in the garage with the keys in the ignition. Make sure the doors and windows are closed on the vehicle. If the property has no garage, then place the vehicle in a cleared area on the side of the house away from the advancing fire.
- Move the furniture away from large picture windows on the fire side of the house.
- Close heavy drapes or blinds that are nonflammable.
- Remove light curtains.
- Turn off any fans in the house.
- Place homeowner's ladders against the roof overhang for roof access.
- Leave the water pump on and turn off the blower if the home has an evaporative cooler.
- Connect garden hoses and test the water pressure.
- Place a garden hose in the top tank fill or the engine if you have water pressure.

If there is no time to prepare the structure because the fire is close, then using the fire behavior factors you observed, start developing a tactical plan. Evaluate the fire's flame lengths and intensity levels. Where is the fire and what is it doing? Generally speaking, one of the following situations will be occurring.

1. *The fire is spotting at the location.* In this case, the fire is spotting around the house being protected, however, the main fire is a distance away. The main flaming front may or may not pass through your location. If it does, then tactics will be different.

If embers in the convection column are reaching the ground and starting spot fires close to the house being protected, then remain mobile and do not have long hose lines out. The objective of the operation is to quickly extinguish all spot fires and post lookouts to advise of any new spot fires. If you cannot keep up with the spot fires as the flaming front arrives, then you must decide on new tactics.

2. *The fire has arrived at the trigger point.* When the fire has reached the trigger point, it is a time to heighten awareness to the fact that the fire is close. It is now time to make some decisions. Again, base the decisions on the flame lengths and intensity levels.

If the fire is burning in light fuels, is not driven by wind or slope, and the house has some clearance around it, then put it out as it reaches the outer edge of the yard where the wildland fuels start. The fire's intensity is the key here. If it is a low-intensity fire, then this tactic will work. If there are limited water supplies and resources and conditions are appropriate, then consider a firing operation. This is discussed later in this chapter and in depth in Chapter 8.

If the flame lengths and intensity levels are great, as the fire reaches the trigger point, then retreat to the safe zone and take the fire on when the flaming front passes and the intensity levels have diminished. This can be done in two ways and requires the use of some fire prediction skills.

If the fire is too intense, then stay in the safe zone and wait for the flaming front to pass. Once it has passed, then start extinguishing any residual fire.

In the second case, the fire intensity level will be high for only a short duration. After the passage of the flaming front, levels will dissipate quickly. In this case, before the flaming front arrives, pull the hose back behind the house (side away from the flaming front) and stay low. As the intensity levels start to diminish, advance the hose lines out and start putting out the fire where it is moving toward the structure. If it starts to get too hot, then retreat behind the structure again. If necessary, use the house as an area of safe refuge to ride out the fire. Before using this tactic, make sure that the house is well built, has proper clearance, and can safely be used for this tactic.

DO NOT try to put yourself between a high-intensity fire and the house being protected.

The important thing to remember is *do not* try to put yourself between a high-intensity fire and the house being protected. If in doubt, stay in the safe zone until the main fire passes.

3. *Burning out is being done around a structure.* Burning out, using fire to remove the intervening fuels between the house and main fire, can be used effectively in structure protection operations. Firing operations, though, should only be done by experienced and trained firefighters who understand fire behavior.

Burning out removes fuels in the fire's path and gives firefighters more defensible space around the structure. This action is done in situations where the flaming front will impact the house being protected. For information on how this is done refer to Chapter 7 and review the section on firing operations in structural areas. Also note in Chapter 7 an ignition technique called *ring firing*.

STRUCTURE PROTECTION SAFETY

When working around structures, base your tactics on the weather, fuels, and topography present in the area. Evaluate the fire behavior you are observing and have a plan that puts safety first. Always have a safe zone and know its location.

In addition, remember to follow these guidelines:

- Be aware of possible toxic fumes, and stay upwind and out of the smoke.
- Always wear full personal protective equipment.
- Do not wet down ahead of the fire. Conserve the water supply for when it is needed. Keep at least 100 gallons of water in reserve in the engine tank to act as a water supply for protection of the engine crew.
- Always back the engine in; it may be necessary to exit quickly.

- Use 1½-inch hoselines whenever possible.
- Do not lay long hose lays as it cuts down mobility, and the potential exists to burn up a lot of hose.
- Use foam or fire gel to coat the structure, if time permits.
- Do not park under power lines.
- Do not park next to propane tanks.
- Do not park in terrain features such as chimneys or saddles.
- Do not enter a burning structure unless you are properly equipped and trained.

Structures exposed to a wildland fire can or should be considered another significant fuel source. They can burn with great intensity. In addition, the presence of homeowners, the media, traffic, livestock, and pets present other hazards that must be dealt with. Follow the Standard Fire Orders and the 18 Situations that Shout "Watch Out," use good judgment, and plan carefully during wildland/urban interface operations.

Summary

A fire engine is a very versatile piece of fire apparatus because of all the things it can do. It carries the crew to the incident location and has the water, hose, and handtools necessary to do the job. Engines do, however, have use limitations. They must be properly managed. This chapter listed six limitations of their use. Know and understand these items.

Two types of hose lays are done by engine companies. The simple hose lay is a hose lay that has no junctions between the nozzle and the pump. A progressive hose lay is usually done with a 1½-inch single jacket hose in 100-foot sections. As sections of hose are added, water is discharged on the fire. Forestry Ts are added about every 200 feet to allow for the connection of lateral hose lines, should they become necessary.

Engine companies attack wildland fires in basically three ways: direct attack, parallel attack the fire with set lines, or use an indirect attack method when the fireline intensities are too great.

Wildland interface fires have been on the increase as the population increases and encroaches on nature. There has been an increase in firefighter deaths and injuries on interface fires. Firefighters have made errors in judgment and have been sent to burn centers throughout the country because they failed to recognize changes in fire behavior or tried to put themselves between a structure and a high-intensity fire. This was the case in California at an entrapment in the Malibu hills. Reread this chapter and know the presented concepts. They will help you operate safely on your next interface assignment.

Review Questions

1. What makes a fire engine a versatile piece of equipment?
2. Name three use limitations of a wildland engine.
3. What is a simple hose lay?
4. What is a progressive hose lay?
5. Describe mobile pumping.
6. Describe a pincer action.
7. What is one of the first things you should do when arriving at an interface fire?
8. What does the term *trigger point* mean?

References

California Department of Forestry and Fire Protection, *CDF Fire Protection Training Handbook 4300 Series* (Sacramento, CA, 1993).

California Fire Service Training and Education System, *Wildland Firefighting Essentials Fire Control 6* (Sacramento, CA, 1992).

National Wildfire Coordinating Group, *Introduction to Wildland Fire Suppression for Rural Fire Departments* (Boise, ID: National Interagency Fire Center, 1985).

National Wildfire Coordinating Group, *S-205 Fire Operations in the Urban Interface* (Boise, ID: National Interagency Fire Center, 1991).

National Wildfire Coordinating Group, *S330 Strike Team Leader—Engine* (Boise, ID: National Interagency Fire Center).

National Wildfire Coordinating Group, *S336 Fire Suppression Tactics* (Boise, ID: National Interagency Fire Center, 1990).

Handcrew Operations

Learning Objectives

Upon completion of this chapter, you should be able to:

- Identify the different types of handcrews.
- Explain the different methods of attack a handcrew uses.
- Describe the rules for working around inmate crews.
- Identify the different handtools a handcrew uses.
- Describe how to tool up.
- Explain how to use various handtools.

INTRODUCTION

■ **Note** Handcrews often operate on remote parts of the fireline that are inaccessible to motorized equipment.

Handcrews play an important part in the suppression of a wildland fire. They become a key resource when it is necessary to construct firelines in steep or rugged terrain. Handcrews often operate on remote parts of the fireline that are inaccessible to motorized equipment. Handcrews are also used in areas that are protected by environmental regulations, which prohibit the use of mechanized equipment.

The removal of fuels by hand is a preferred method when working close to structures. Bulldozers may have problems working close to structures and may leave unsightly scars that are the result of the deep soil disturbance they create.

HANDCREW STANDARDS

■ **Note** Handcrews are made up of between 12 and 20 firefighters who use handtools to actively suppress low flame production fires.

direct attack
any treatment applied directly to burning fuel, such as wetting, smothering, or separating the burning from unburned fuel

indirect attack
a method of suppression in which the control line is located a considerable distance away from the fire's active edge; utilizes natural or constructed firebreaks or fuelbreaks and topography

Handcrews are made up of between 12 and 20 firefighters (Figure 7-1) who use handtools to actively suppress low flame production fires. Handcrews are used to construct handlines directly along the fire's edge, where intensity levels allow their use. This is a **direct attack** method. Handcrews also perform **indirect** fireline operations ahead of the fire, such as line constructed some distance from the fire's edge. They then **backfire** or burn out the unburned fuels between the control line and the fire's edge. Handcrews use this tactic where fireline intensity levels are too

Figure 7-1 *A handcrew tooling up.*

backfire
a fire set along the inner edge of a fireline to consume the fuel in the path of a wildfire and/or change the direction of force of the fire's convection column

parallel attack
method of suppression in which fireline is constructed approximately parallel to, and just far enough from the fire edge to enable workers and equipment to work effectively

■ **Note**
Type One crews are the most experienced with the highest levels of training.

great for a direct attack on the fireline. **Parallel attack** (Figure 7-2) is another method of attack employed by handcrews. Fireline is constructed parallel to, but further from, the fire's edge. The fireline is usually started on a flank from an anchor point. The advantage to this tactic is that it may shorten the fireline by constructing line across the unburned fingers and, thus, burn out the intervening fuels.

Handcrews are typed depending on their training and experience levels. Standards for the typing of crews are found in the *Fireline Handbook*. Type One crews are the most experienced with the highest levels of training. They receive more than 80 hours of training annually, and at least 80% of the handcrew must have been on the crew for at least one season. They have no fireline use restrictions. Type Two handcrews are not trained to the same levels and have limitations on their use. They do not generally perform direct attack operations on the fire's edge. It is best to know what type of crew you are working with.

INMATE CREWS

Many states use inmate labor for fire operations. Use these general guidelines if working with inmate fire crews:

- Inmate crews are not to be split up.
- In fire camp, separate sleeping areas are provided.

Figure 7-2 *Parallel attack by a handcrew.*

- Do not offer food, drinks, or cigarettes to inmates.
- Do not talk to or befriend inmates; maintain work related contact only.
- Do not offer to take messages back to their friends or family.
- If an inmate is injured, make sure a correctional officer or sheriff accompanies that individual to the hospital.
- Keep male and female inmate crews separate.
- Check the rest and feeding requirements of inmate crews.
- Inmate crews are accompanied by trained correctional officers accustomed to supervising inmates.
- The use of inmate crews is usually limited to the state where they are based.
- Contact with inmates shall be made through the corrections officer in charge of the crew.
- Inmates are the responsibility of the correctional officer in charge.

Be sure to consider these requirements when planning inmate use on the fireline.

HELITACK CREWS AND SMOKE JUMPERS

Some paid handcrews have additional training that allows them to be deployed from helicopters or fixed-wing aircraft. United States Department of Agriculture (USDA) Forest Service and Bureau of Land Management smoke jumpers are deployed from fixed-wing aircraft by parachute to suppress wildfire in remote areas. Agencies that deploy handcrews from helicopters either land their crews on the ground before deployment or they rappel from the helicopter. They are called helitack crews. Helitack crews are usually deployed on hot spots or ahead of the fire to construct handlines and are supported by water drops after their helicopter lifts off. The water drops are provided by the fixed tank on the crew's helicopter.

HANDCREW TOOLS

■ Note Handcrew tools consist of hand and power tools necessary to construct handlines.

Handcrew tools consist of hand and power tools necessary to construct handlines. Handcrews use a variety of tools: some are cutting tools, others are used to scrape.

A frequently used power tool in heavier fuel is the **chain saw** (Figure 7-3), which can be used to cut a variety of fuels, from large trees to heavier brush types, rapidly. **Double-bitted axes** (Figure 7-4) were used before the advent of the chain saw. They still remain part of handcrews' inventory. The **brush hook** (Figure 7-5) is a great tool to use where the brush is heavier. It cuts brush 1 to 3 inches in diameter.

The **Pulaski** (Figure 7-6) is an excellent tool for cutting or grubbing. One end of this tool looks like an ax, while the other end has a sharp, small, hoelike end that is used to cut or remove flammable items at ground level or below.

chain saw
a portable power saw with a continuous chain that carries the cutting teeth

Figure 7-3 *Chain saws.*

double-bitted axe
a cutting tool with a long handle and double-bladed head used on the fireline to chop with

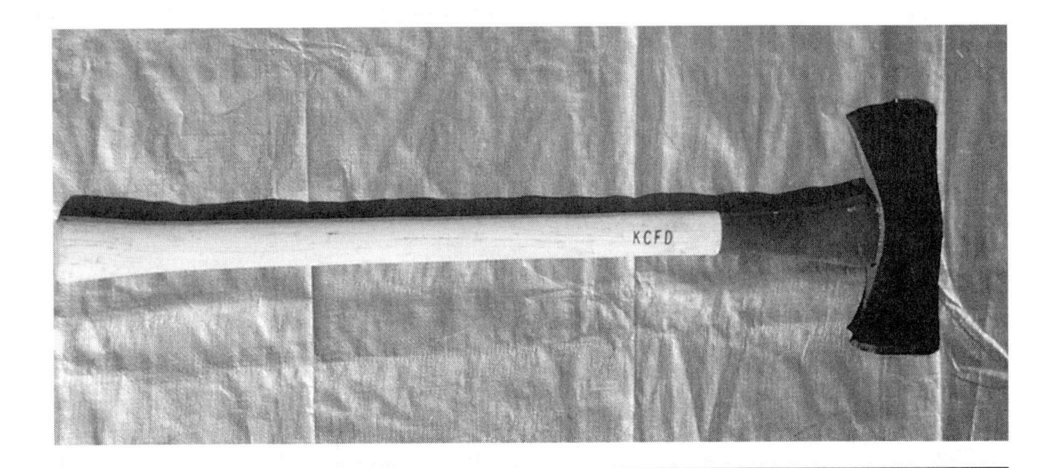

Figure 7-4 *Double-bitted axe.*

brush hook
a heavy cutting tool designed primarily to cut brush at the base of the stem

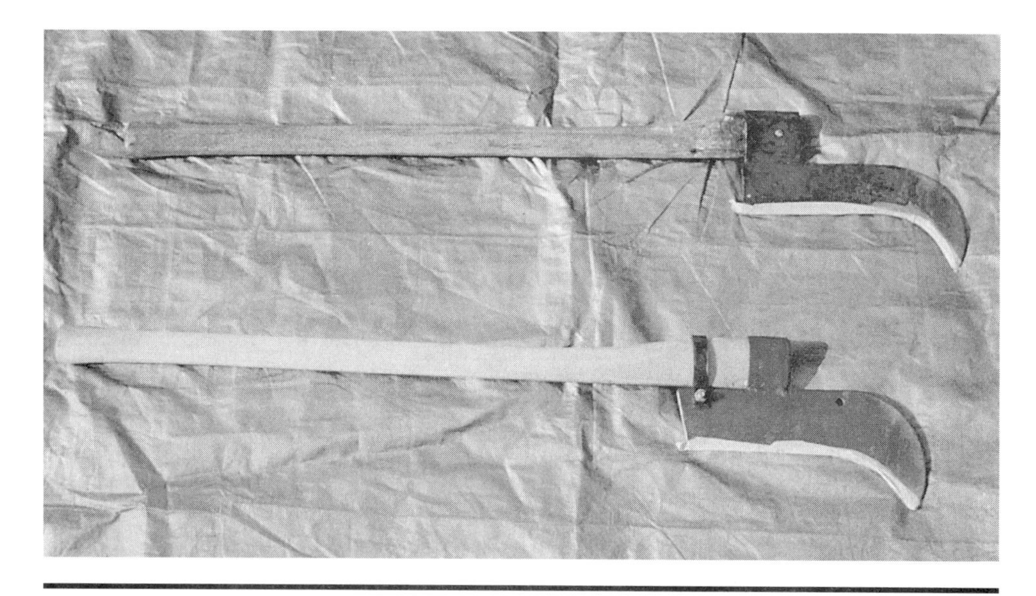

Figure 7-5 *Two types of brush hooks.*

Pulaski
combination chopping and grubbing tool

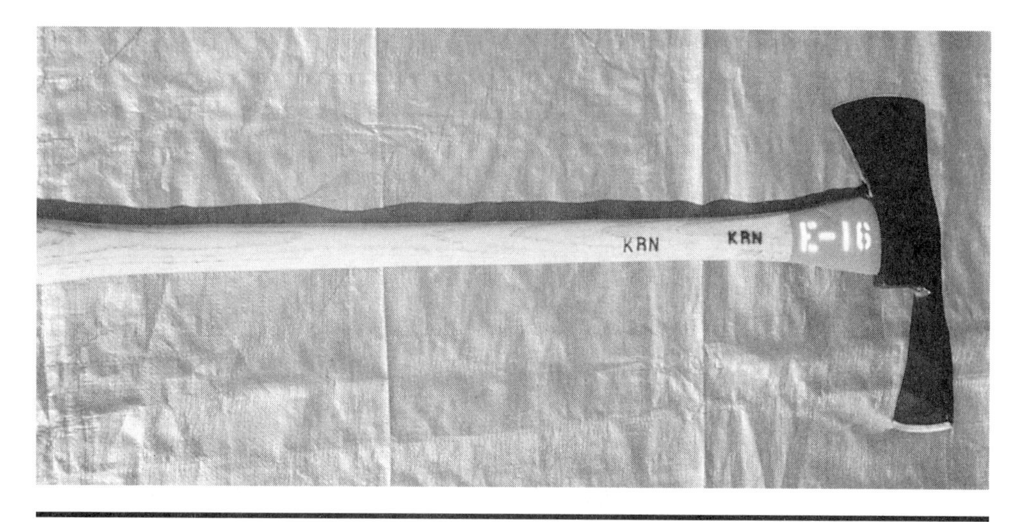

Figure 7-6 *A Pulaski.*

The **McLeod** (Figure 7-7) is basically a scraping tool with a rake on the other end. It is used to scrape the line down to mineral soil. The rake side is used to rake up the lighter fuels that have been cut. The rake end is also used to break up piles of light fuels close to the constructed control line.

One of the most versatile tools to have on a fireline, the **shovel** (Figure 7-8) is an all-around tool that can be used to cut smaller diameter fuels or to scrape with. It can also be used to throw dirt in order to knock down the fire.

A **fire broom** (Figure 7-9) is used in rocky areas where short grasses protrude through the rocks. It is used in a sweeping motion to brush burning fuels. The bristles of the broom pick up the burning fuel. Care must be taken not to spread the fire with this tool.

McLeod
combination hoe or cutting tool and rake, used to scrape away fire fuels and rake away litter and other loose materials from the fireline

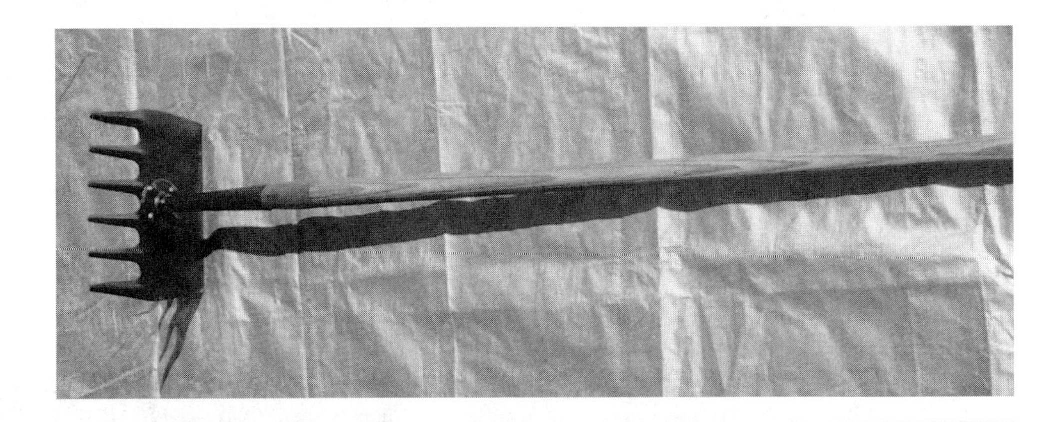

Figure 7-7 *A McLeod.*

shovel
a tool with a broad scoop and long handle used to scrape with, to cut smaller diameter fuels, and to throw dirt to knock down a fire

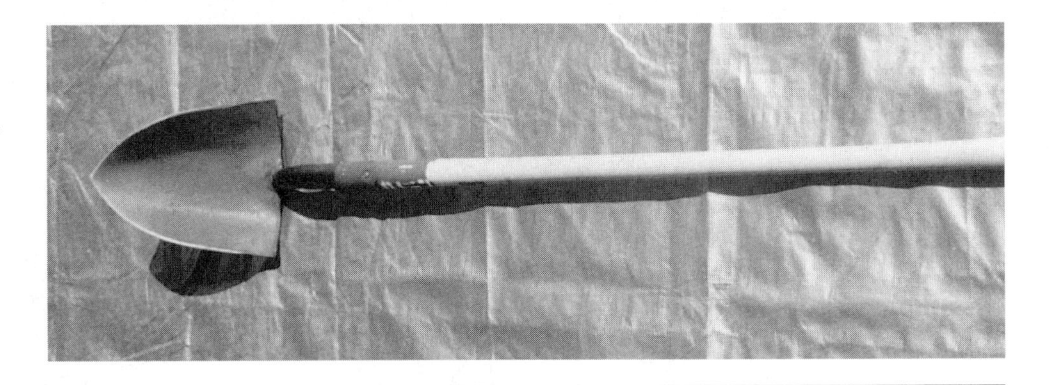

Figure 7-8 *A shovel.*

fire broom
a wire bristled, long handled broom used to brush out short burning grasses from rocky areas

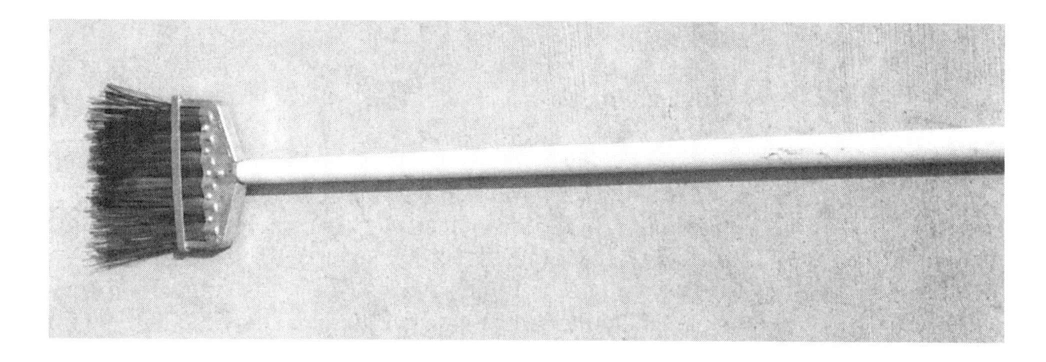

Figure 7-9 *A fire broom.*

HANDCREW ARRIVAL PROCEDURES AND USE OF HANDTOOLS

crew-carrying vehicle
vehicle used to transport handcrews

A fire crew usually arrives on the fireline in a **crew-carrying vehicle** (Figure 7-10), also called a CCV. The following is an *example* of a tool-up procedure. Crew members have assigned seats in the vehicle according to the handtool they will be using. When the crew arrives on the fireline, the swamper, who sits in the front passenger seat next to the crew leader, unloads from the vehicle first. His or her

Figure 7-10 *Crew carrying vehicle.*

cutting section
the section of the handcrew responsible for all sawing, chopping, or grubbing of woody fuels and removing it from the fireline

scraping section
the section of the handcrew responsible for completing the handline by removing smaller fine fuels from the fireline

■ Note
The cutting section responsibilities on fireline include all sawing, chopping, or grubbing of woody fuels.

■ Note
The scraping section completes the handline.

■ Note
The crew leader makes the decision on the attack method to use and the tool order.

first duty is to place wheel chocks under the vehicle. After that is done, the crew leader unloads and instructs the swamper on which tools to issue. The safety person, usually the last crew member with a shovel in the handcrew lineup, unloads from the CCV. He or she prepares to issue canteens.

The crew leader then gives the orders to unload the rest of the crew from the CCV. The crew leader then gives the tool order. This establishes the order in which handtools are placed during a line-cutting operation. The tool order is dependent on the fuel type in which the handline will be cut.

After the crew members receive the tool order, they line up, in full safety gear, at the rear left corner, on the driver's side, of the vehicle. The swamper then issues the handtools to the crew. After crew members receive the tools, they proceed to the right side of the CCV and receive a canteen. When all handtools and canteens have been issued, the crew proceeds to the hookline position. Crew members keep a 10-foot distance between each other.

Placement of handtools falls basically in two different sections: the **cutting section** and the **scraping section.** The cutting section responsibilities on fireline include all sawing, chopping, or grubbing of woody fuels. Once the fuel has been cut, it is then removed from the fireline. The scraping section completes the handline. It removes duff, grass, and smaller brush from the fireline. McLeods, shovels, and fire brooms are used in the scraping section. The crew leader makes the decision on the attack method to use and the tool order. Again, this decision depends on the type of fuels in which the crew will be working.

HANDLINE CONSTRUCTION PRINCIPLES

Handline is constructed progressively with each member of the handcrew placed 10 feet apart. This is called "your dime." Each member of the handcrew takes a stroke with his handtool and then moves one step forward. With each member of the crew removing some fuel, the line progresses forward more rapidly. Toward the end of the line, clean line that is down to mineral soil can be seen. The last person in the line (the drag shovel) ensures that the line is clean and complete.

Fireline should be constructed no wider than necessary to stop the fire. Time and energy are best used to construct more handline. The general guideline for fireline width is to make it 1½ times the dominant fuel height. In timber, the object is to construct line that will stop burning surface fuels as well as the lower aerial fuels. The rule is to construct line 20 to 30 feet wide and then provide a scraped line between 3 and 4 feet. Most firelines are unsuccessful at stopping an active crown fire.

Tool Order

Tool order is established by the crew leader and is dictated by the fuel type the crew is operating in. In woody fuels, such as shrub, the tool order is as follows:

Note
The crew leader is near the front of the operation.

Note
Handcrew members that follow the lead hook or lead saw are spaced at least 10 feet apart.

Note
All previously burned or burning materials need to be scattered into the burned area.

Brush Hook and Chain Saw The crew leader is near the front of the operation. He or she is followed by the lead hook or first saw. This crew member is the first person that operates a chain saw or brush hook. He or she is the lead person on the crew and locates the line as instructed by the crew leader. The first hook and first saw only cut the brush wide enough to walk through. He or she usually disposes of the brush that has been cut.

Handcrew members that follow the lead hook or lead saw are spaced at least 10 feet apart. This is done for safety reasons as persons will be using handtools. The brush hook operators that follow the lead hook widen the line to the specifications set by the crew leader. Each brush hook operator cuts some brush and leaves some to be cut by the next brush hook operator.

Disposal of Cut Brush After the brush has been cut, it must be disposed of properly. All previously burned or burning materials need to be scattered into the burned area. Unburned cut fuels are scattered in the unburned areas.

The three methods used to disperse or dispose of brush are as follows:

1. Toss method. The brush is thrown and scattered from where it is cut. This only works if the surrounding brush is sparse or low in height.
2. Window method. If the surrounding brush is too tall and dense for the toss method, then you need to pull brush to an opening, called a *window,* and dispose of it. If a natural opening is not available, then you may need to create one.
3. Bank method. This method is used when the surrounding brush is too tall or dense for the toss or the window method. In this case, a double width line is cut and all the cut brush is compacted against the standing brush on the side away from the burned area.

Pulaski Use Pulaskis are normally placed behind the last brush hook or chain saw operator. These grubbing tools are used to remove stumps and stobs (previously cut and pointed branches and stumps close to the soil) and to cut roots that cross the control line. They are also used to trench (Figure 7-11) where it is necessary. The mattock edge of the tool is used for this purpose.

Trenching is done on steep hills during handline construction. The purpose of trenching is to prevent any rolling firebrand materials from crossing the control line.

Pulaskis have an ax on the end opposite the mattock edge. Therefore, this tool is sometimes moved up in front of the lead brush hook to cut down heavier fuels. In harder soils, Pulaskis are used to loosen soil, especially if dirt is needed to be thrown on the fire.

McLeod Use McLeods are scraping tools and they generally follow the crew members carrying Pulaskis. The tool has two basic parts. One side has a straight cutting edge

Figure 7-11 *A cup trench is used to catch rolling material on a slope.*

used to scrape away fine fuels. The other side has tines and resembles a large rake. This side of the tool is used to rake away litter, duff, and other loose materials.

The tool is used to break up rat nests, piles of woody debris this animal uses to build its nest, and windows of piled brush.

Shovel Use Shovels are a universal tool used to complete the fireline construction. The edge can be used to cut small brush, to scrape fine fuels, to dig with, and to throw dirt along the fire's edge to attack low-intensity fires. Shovel operators usually follow the McLeods. Shovels are used to finish the control line and remove fuels that were missed by the McLeod operators.

Additional shovel operations are as follows:

- Hot materials can be carried or thrown back in the burn.
- Trenches can be cleared out.
- Windows can be scattered with the use of a shovel.

Again, as with any of the tools mentioned, keep a 10 foot spacing between workers.

Other Fuel Types In lighter fuels, the handtool order changes to fit the fuel type. In this case, there may be two Pulaskis as the first handtools, then a mix of scraping tools with a Pulaski in the middle of the crew.

Additional Fireline Construction Rules

In order to have a complete understanding of fireline construction, it is necessary to study the following additional rules.

- Clean all fireline to mineral soil for all or part of the width, except in fuels such as bog, peat, or tundra.
- Cover stumps and logs just outside the fireline with dirt to protect them from radiant heat when the fire reaches the fireline.
- Cut low hanging limbs from trees on either side of the fireline that could cause fire spread across the line.
- The hotter and faster the fire burns, the wider the control line must be. Seven factors used to determine fireline width are:
  1. Slope.
  2. Fuel.
  3. Weather and wind direction: Is the wind steady? Is it blowing toward or away from the control line?
  4. Part of the fire where the control line will be constructed: Head versus flank.
  5. Possibility of cooling.
  6. Size of the fire.
  7. Fireline conditions and intensity levels.
- Make fireline no wider than necessary. Time can be better spent constructing more fireline and encircling the fire.
- Always watch for rolling hot materials starting spot fires below the crew.

CHAIN SAW SAFETY

Since the advent of chain saws, line construction is much easier. The following is a list of safety rules to be followed:

- Always wear your personal protective equipment, including ear and eye protection, and chaps when operating a wildland power saw.
- A chain saw should be started on the ground (Figure 7-12) or between the knees (Figure 7-13).
- When operating a chain saw, check your stance. Your feet should be spread apart in a wide balanced stance, with feet and legs away from the saw (Figure 7-14).
- Always grip the saw with both hands, thumbs and fingers encircling the handle.
- The operator should move his or her hands from one side of the saw to the other on the handlebar. Do not cross your hands over each other.

Figure 7-12 *Starting a chainsaw while it is on the ground.*

- Before cutting, analyze overhead obstructions, widow makers, and other loose debris.
- Alert others that you will be using a chain saw.
- Have escape routes planned and know them before starting a cutting operation with your saw.
- Maintain a safe working area clear of debris.
- The saw team should be safely spaced away from adjoining personnel.
- Good communications must be maintained between the saw team and adjoining personnel.
- Visually account for adjoining personnel before starting the operation.
- Ensure control of the cutting area.
- Post lookouts.
- Stop the motor when carrying, making adjustments, repairing, or cleaning a chain saw.
- Use a blade guard when carrying a saw in rough country.
- Always cool the motor before refueling. Fill the saw on bare ground and move a safe distance from the fueling area before restarting the saw.

Figure 7-13 *Starting a chainsaw between the knees.*

Figure 7-14 *Correct stance while operating a chainsaw.*

HANDCREW PRODUCTION

Handcrews are an important fireline tool. They should train constantly and keep in good physical condition. They should know how to use their tools, especially chain saws. They are a team and teamwork gets the job done.

The basic mission of a handcrew is to suppress the fire and construct handlines by using a combination of hand and power tools. Handcrew operations should be supported tactically and logistically on the fireline if they are going to be used to their maximum potential. Also, remember to rest these crews, as the physical demands placed on them are great. Handcrew production rates will decrease as the operational shift goes on.

Table 7-1 *Handcrew production, initial attack (chains per hour per person).*

Fire Behavior Fuel Model	Type	Conditions Used In	Construction Rate (chains per hour)
1	Short grass	Grass	4.0
		Tundra	1.0
2	Open timber Grass understory	All	3.0
3	Tall grass	All	0.7
4	Chaparral	Chaparral	0.4
		High pocosin	0.7
5	Brush (2 feet)	All	0.7
6	Dormant brush/	Alaska black spruce	0.7
	hardwood slash	All others	1.0
7	Southern rough	All	0.7
8	Closed timer litter	Conifers	2.0
		Hardwoods	10.0
9	Hardwood litter	Conifers	2.0
		Hardwoods	8.0
10	Timer (litter and understory)	All	1.0
11	Light logging slash	All	1.0
12	Medium logging slash	All	1.0
13	Heavy logging slash	All	0.4

Handcrew Production Rates

An estimate of handcrew production rates is important for a fire manager to understand. This rate is used during planning to estimate how long it will take to complete a section of line, with the available handcrew resources, before the fire arrives. It can also be used as a basis for ordering handcrews for the fire.

Table 7-1, Handcrew Production, is furnished as a guide only. It gives the reader an idea of the capabilities of a Type One crew. The table defines production rates in chains per hour; a chain is 66 feet.

Summary

Handcrews are made up of between 12 and 20 firefighters whose primary purpose is to construct firelines at a wildland fire. Inmate crews are handcrews made up of prisoners wearing orange personal protective equipment.

Smoke jumpers parachute from fixed-wing aircraft, whereas helitack crews are deployed from helicopters, and their operations are supported by the aircraft that carried them.

Handcrews use a variety of different handtools, depending on the fuel types they are going to work in. Arrival and tool-up procedures may vary from agency to agency. Consult your local agency to find out what its policy is.

Handline construction rates vary depending on the type of fuel in which the line is being constructed. Heavy fuels slow the progress.

Handcrews are an important resource on a wildland fire. They are especially beneficial at constructing handlines on steep slopes, where bulldozers cannot work.

Review Questions

1. How many people are in a handcrew?
2. Explain the direct attack method of line construction.
3. When working with handtools, how far apart should crew members be spaced?
4. What is a smoke jumper?
5. Explain the cutting section responsibilities when constructing handlines.
6. When cutting lines, where is previously burned material disposed of?
7. Explain the toss method of brush dispersal.
8. Name the seven factors used to determine fireline width.

References

California Department of Forestry and Fire Protection, *Fire Crew Firefighting Handbook 4200* (Ione, CA: 1993).

National Wildfire Coordinating Group, *Fireline Handbook* (Boise, ID: National Interagency Fire Center, 1998).

National Wildfire Coordinating Group, *Crew Boss (Single Resource) S230* (Boise, ID: National Interagency Fire Center, 1996).

National Wildfire Coordinating Group, Video—*Wildfire Handtools* (Boise, ID: National Interagency Fire Center, 1986).

Backfire/Burnout Basics

Learning Objectives

Upon completion of this chapter, you should be able to:

- Define the terms *backfire* and *burnout.*
- Explain the difference between backfire and burnout.
- Explain the steps that must be taken before undertaking a firing operation.
- Identify the tools used in firing operations.
- Explain three different ignition techniques used in firing operations.

INTRODUCTION

In wildland firefighting we sometimes fight fire with a fire (Figure 8-1) that we light. In certain situations—when the flame lengths and British thermal unit (BTU) output are too intense for direct attack with hose lines, it is appropriate to either backfire or burn out an area.

Strategic values must never take precedence over safety concerns.

Firing operations always present some risk, especially if done incorrectly. Strategic values must never take precedence over safety concerns. As a wildland firefighter, it is essential that you understand the concepts presented in this chapter and wildland fire behavior before getting involved in any firing operation. Backfire/burnout skills are another tool to put in the wildland toolbox for use when appropriate.

FIRING OPERATION APPROVAL

Some states have provisions in their government codes that address backfire or burnout operations and who should do them. For example, California addresses

Figure 8-1
A California Department of Forestry firefighter burning out with a drip torch.

division supervisor an operations supervisor responsible for all suppression activities on a specific division of a fire

■ Note
Every firing operation should be run by one person who focuses his or her total attention on supervising the firing team.

backfire operations in its Public Resource Code (Section 4426) and identifies who should do this operation. Check with your state to see if there is a government code that addresses firing operations.

In many firefighting organizations in this country, it is necessary to acquire approval, through the chain of command, from the **Division Supervisor** before proceeding with a backfire or burnout operation. The only exception would be an emergency situation to save a life or property. Some agencies permit the crew boss to initiate burnout operations. Before going out on the fireline, determine the local agency policy regarding firing operations.

Every firing operation should be run by one person who focuses his or her total attention on supervising the firing team. That person should be experienced and trained in such operations. He or she should be constantly aware of weather changes, problem areas developing on the line, and fireline intensities being created by the firing operation. He or she must always be thinking safety and fire behavior.

BACKFIRE

Backfire is defined as a fire intentionally set along the inner edge of a control line with the expectation that it will be influenced (indraft effect) by the main fire and, thus, burn out the intervening flammable vegetation and/or change the direction of the fire's convection column (Figure 8-2). It is a tactic usually used only when other fire control methods are judged impractical. It is usually a preplanned event with sufficient resources available to hold the fireline during the firing op-

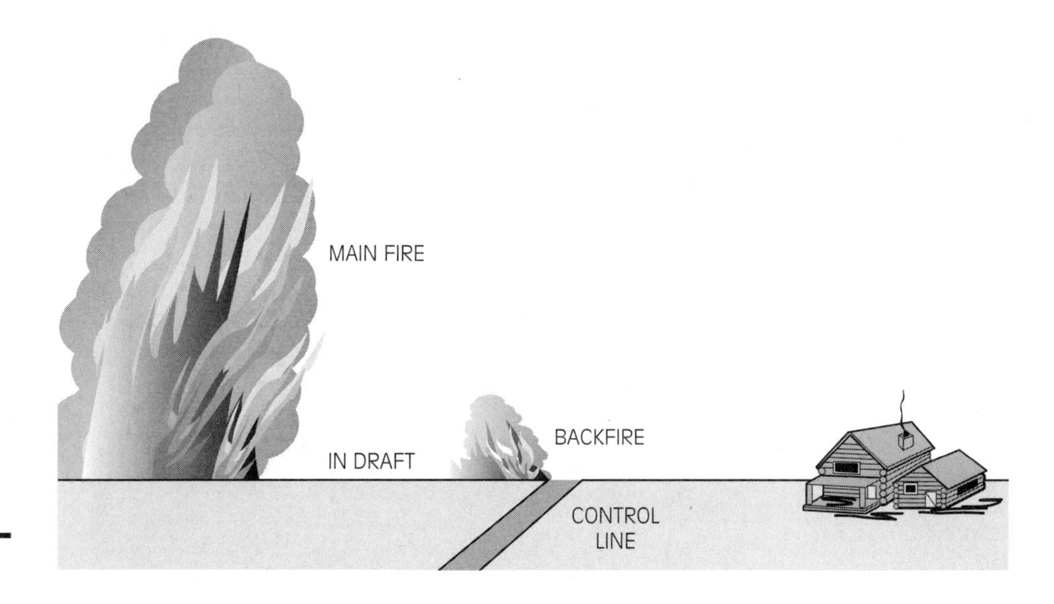

Figure 8-2 *Backfire.*

indraft
the process whereby air is drawn into the larger main fire

eration and sufficient resources held in reserve to rapidly attack any slop over and spot fires.

When we talk about the backfire being influenced by the main fire, we mean that an **indraft** effect takes place as the main fire approaches. The main fire has a strong need for oxygen, as it generally has a strong convective column. Thus, the fire being lit will be drawn toward the base of the approaching larger main fire (see Figure 8-2). This is called the *indraft* effect.

Note
When you backfire, allow enough time (based on fuel factors, topography, weather, and fire behavior) to have the backfire develop and move away from the control line.

When you backfire, you must allow enough time (based on fuel factors, topography, weather, and fire behavior) to have the backfire develop and move away from the control line. If the distance between the backfire and main fire are too great, then consider the use of a counterfire. Give yourself enough time; the worst thing that can happen is that both fires meet next to the control line, jeopardizing that which you are trying to protect. Always remember to proceed at a rate at which the fire you are lighting can be held. Never create more fire than you can control.

Never create more fire than you can control.

There is no firm set of rules as to when to light the backfire, as this is based on the fuels present, the topography, existing fire conditions, and the weather. These factors must be evaluated and wildland fire prediction skills used to determine when to light the backfire, so it is important to use the most experienced fire crew member for this operation.

Note
The purpose of a counterfire is to hasten the spread of the backfire.

The purpose of a counterfire or auxiliary fire is to hasten the spread of the backfire. It is a fire set between the main fire and the backfire. A counterfire is used when the distance is too great between the main fire and the backfire (Figure 8-3). A counterfire requires the aid of firing devices such as a flare launcher, a helitorch, or a Premo Mark III.

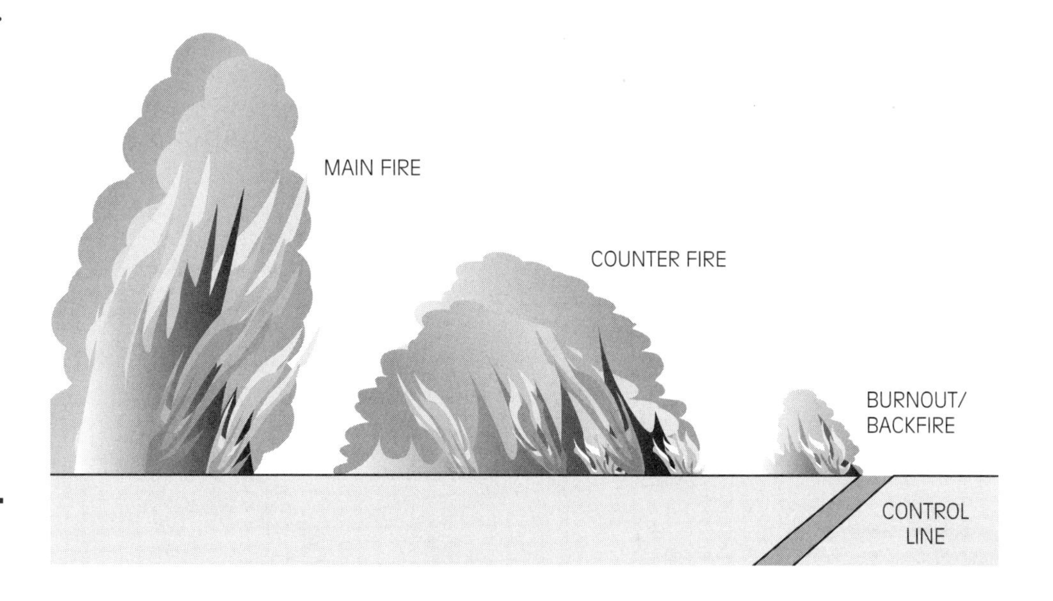

Figure 8-3
Counterfire.

BURNOUT

Burnout is defined in the *Fireline Handbook* as "setting fire inside a control line to consume the fuel between the edge of the fire and the control line" (Figure 8-4). The control line is not complete until all fuels have been removed between the fire's edge and the control line. Large islands of unburned fuel within the fire area that are close to the control line need to be removed before the control line can be considered secure.

■ Note
Burnout takes advantage of line construction techniques by cutting across fingers and around concentrations of spot fires.

Handcrews also burn out as they cut line from fingertip to fingertip as this reduces the total amount of line that needs to be constructed. They cut line and burn out at the same time. Fire is carried from finger to finger while line is being constructed.

Burnout takes advantage of line construction techniques by cutting across fingers and around concentrations of spot fires. It also takes advantage of natural barriers.

BEFORE STARTING

■ Note
The operation should be organized and planned.

■ Note
A complete briefing should be given to those crews involved in the firing operation.

The operation should be organized and planned unless it is done as a result of an emergency. A complete briefing should be given to those crews involved in the firing operation and appropriate assignments made. Crews need to be assigned to work as firing forces, holding forces, mop-up forces, and reserve forces.

Before the operation is started, objectives need to be defined. Objectives spell out what is to be accomplished with the operation. At that time, time frames also need to be determined for completion of the operation. Consideration needs to be given to the ignition sequence, when the firing operation is to commence, and what firing method will be used. A topographic map should be shown to the crew doing the firing operation, designating where the main fire is.

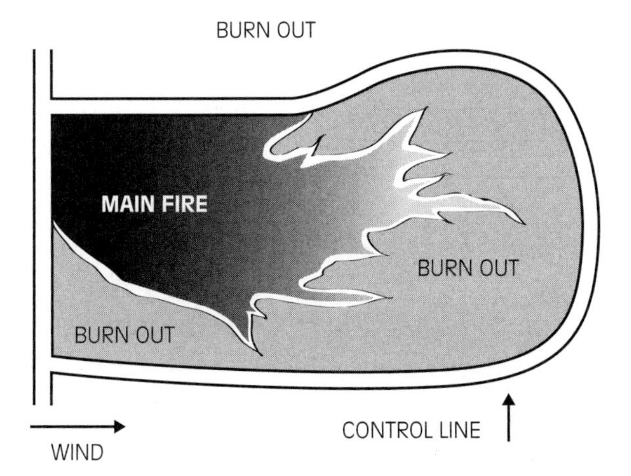

Figure 8-4 *Burnout.*

drop point
an identified area on fireline where supplies and personnel are dropped

A discussion of safe zones and escape routes should be included in any briefing.

Incident Action Plan (IAP)
written or oral plan containing objectives reflecting the overall incident strategy and specific tactical actions and supporting information for the next operational period

Consideration should be given to the amount and type of supplies needed to complete the operation. Planning also needs to be done to anticipate future needs of those doing the firing operation, and **drop points** need to be established.

A discussion of safe zones and escape routes should be included in any briefing. Every crew member needs to think LCES at all times.

The topography map is used to show anticipated problem areas to the fire crews who will be doing the burning. A problem area can be defined as an area that could delay the operation, has the potential for slop over, or just cause confusion in the operational area. Examples may be topographic features, such as saddles; sharp turns along a roadway; or narrow canyons. The area should be scouted and problem areas identified.

Before the firing operation begins, a decision should be made about the chain of command and what crews are in the division. Any fire weather forecasts and fire behavior predictions should be reviewed. These can be found in an **Incident Action Plan.** Communications on any wildland incident are a problem. The communications plan should be reviewed and an alternate plan should be in place. A plan also needs to be in place for medivac procedures in case someone is hurt.

If these points are covered, the operation will go smoother, and the objectives will be met and accomplished safely.

FUEL, TOPOGRAPHY, AND WEATHER CONSIDERATIONS

Fuel, topography, and weather need to be evaluated before the firing operation begins. These factors are the key to the intensity levels and flame lengths that will be created. *Never* create more fire than can be controlled. Regulate the heat.

Light Fuels

Light fuels are easily ignited by the tools at hand and generally produce low heat intensities on the fire line. They generally burn clean and spread rapidly. They can be held with a smaller control line than other fuel types. Of all the fuel types, light fuels (1-hour time lag fuels) are the most susceptible to a change in relative humidity. Therefore, their moisture content can change rapidly. This fuel type has a greater tendency to spot than other fuel types.

Medium Fuels

Medium fuels produce greater heat intensity and flame lengths than light fuels. They can be ignited by the tools at hand; however, they normally require a drip torch. This type of fuel is generally less susceptible to spotting than light fuels. Medium fuels are more readily ignited at night than light fuels due to the slower effect of humidity on them. These fuels require increased control line widths and an increased resource deployment to hold the line.

Heavy Fuels

■ Note There is probably no real advantage to firing heavy fuels.

There is probably no real advantage to firing heavy fuels. These fuel types produce great fireline intensities and are a problem for fire crews. They need large control lines and are very susceptible to blowing embers across the control line due to the fuel height. Heavy fuels often require ignition sources that are not readily available, such as a helitorch or a terra torch.

Topography

■ Note Topographic features present some of the greatest obstacles to successful firing.

Topographic features present some of the greatest obstacles to successful firing. Saddles present special problems, because they are areas of intense air movement. They have to be fired simultaneously downhill from the high points to the middle. Narrow canyons present special problems, because the canyon walls are so close together and there is great potential for spotting. Hooks and sharp bends are a cause for concern, as they have the potential to expose the line to slopover. Midslope roads call for extreme care as unburned fuel is either above or below the operation and there is a higher likelihood of spotting. Some topographic features can work for you, such as natural barriers, while others are cause for concern. Evaluate these features before proceeding with the firing operation.

Weather

Winds need to be evaluated when planning the firing operation. Firing with the wind, where the wind blows away from the control line, produces an increased fire intensity and a more rapid rate of spread. More heat builds causing a strong convective column and perhaps more spotting. The flammable vegetation in the fuel's path can be removed by fire more quickly. It will take less time to complete the backfire/burnout operation. A smaller control line is needed to hold the firing operation.

When firing against the wind, it will take longer to complete the operation because the fire will spread at a slower rate. It will also be harder on crew members, because the smoke and heat will be blown back toward them.

When the firing operation is done from the fire's flank, fire against the wind from a well-anchored point. The wind should be steady, but watch for wind shifts.

Other Weather Concerns to Watch

- Be aware of any thunderstorm activity developing in the area.
- Evaluate the smoke column. It will provide information on atmospheric stability and burning conditions.
- Monitor high level winds, as they may become surface winds later during the day.
- Watch for eddies created by topographic changes.

■ Note Evaluate the smoke column.

FIRING OPERATIONS IN STRUCTURE AREAS

Special precautions must be taken with firing operations in or around structures. The firing operation may require additional resources assigned strictly to protect the structures. The citizens living there should have been evacuated prior to starting the operation, because the firing operation may cause traffic congestion and reduced visibility.

Always evaluate, or take into consideration, the amount of defensible space around the structure, the construction type, where the houses are located in relation to the topography, the fuel loading, and proximity to the houses that you are trying to protect with the firing operation.

GENERAL RULES ABOUT FIRING OPERATIONS

Review the following general rules to ensure an effective and safe firing operation.

termination point
a planned point on the fireline where a firing operation is terminated

- Every firing operation needs an anchor point, a safe place to start the operation from, and a predetermined **termination point.**
- Generally firing operations should start at higher elevations and work down so that the intensity level of the fire being lighted can be controlled. Firing operations that are started at the bottom of the hill can produce uncontrollable fire intensities because there is more available fuel to burn and the slope affects the fire spread. There are two exceptions to this rule as follows:
  1. When the upper control lines are secure enough that there is not a possibility of an escape or slopover.
  2. When there are strong downslope winds driving the fire and they are overcoming the effect of the slope.

Other Potential Firing Problems

- Areas with lots of dead fuel mixed with the live fuels.
- Areas with heavy fuel buildup.
- Power lines.
- Aerial retardant drops. Coordinate these so they do not interfere with the firing operation.

FIRING DEVICES

Three commonly used firing devices are discussed in this section. These devices may be used in a firing operation, therefore, become familiar with the characteristics of these devices.

Fusee/Backfire Torch

Safety

Once lit, keep fusees away from your clothes and body.

These types of fusees have extension alternatives in the form of a cardboard sleeve on the rear portion of each fusee (Figure 8-5) that allows them to be stacked and allows you to perform an extended operation. To light them, have full protective clothing on. Pull off the cap protecting the striker, and strike away from the body. Once lit, keep them away from your clothes and body. Remember also that fumes produced by a burning fusee should not be inhaled.

Fusees must be kept dry. Whenever possible, transport them in their original containers. They are light, portable, and produce a hot flame. The phosphorus material burns at about 1400°F. Fusees can be broken and thrown and can also be relit. A good thing about fusees is that they are usually readily available on all fires. To extinguish them, place the tip into the dirt and snuff them out.

Drip Torch

A drip torch is a canister that drips ignited fuel from a spout (Figure 8-6). It works on all fuel types. The contents of the canister are a mixture of one part gas to four

Figure 8-5
Fusee/backfire torch.

Figure 8-6
Firefighter using a drip torch.

parts of diesel fuel and last about one hour. This firing device is to be used with care, as it can create a lot of fire very quickly.

The use of full protective clothing is required before operating a drip torch.

Instructions for Operating a Drip Torch

- The use of full protective clothing is required before operating this firing device.
- Secure the sealing plug (Figure 8-7), and remove the spout with the fuel trap from the canister (Figure 8-8).
- Position the spout in the upright position with the fuel trap and wick in an opposite direction from the handle. When using this device, the fuel mix should pass over the wick. Secure the spout in place (Figure 8-9).

Figure 8-7 *Properly secured sealing plug.*

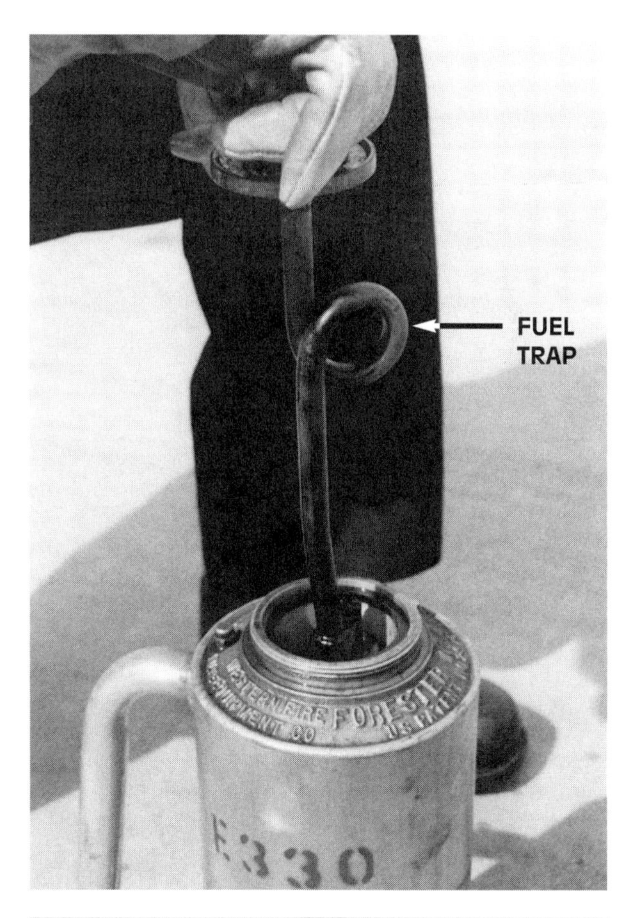

Figure 8-8 *Removing the spout from the fuel trap.*

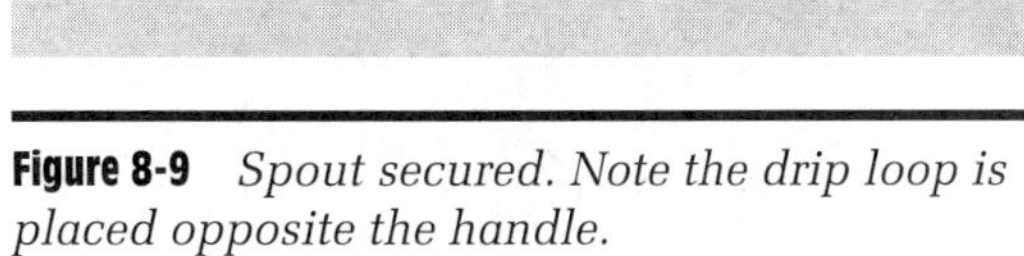

Figure 8-9 *Spout secured. Note the drip loop is placed opposite the handle.*

Figure 8-10 *Opening the breather valve.*

- Open the breather valve (Figure 8-10).
- Pour a small amount of fuel mix on the ground and light it. Then bring the canister wick in contact with the small amount of fuel that is burning on the ground (Figure 8-11). The drip torch is now ready for use (Figure 8-12).
- Hold the drip torch by the handle in an upright position until you are ready to use it.
- To extinguish the drip torch, place the canister in an upright position and let the wick burn dry (Figure 8-13).

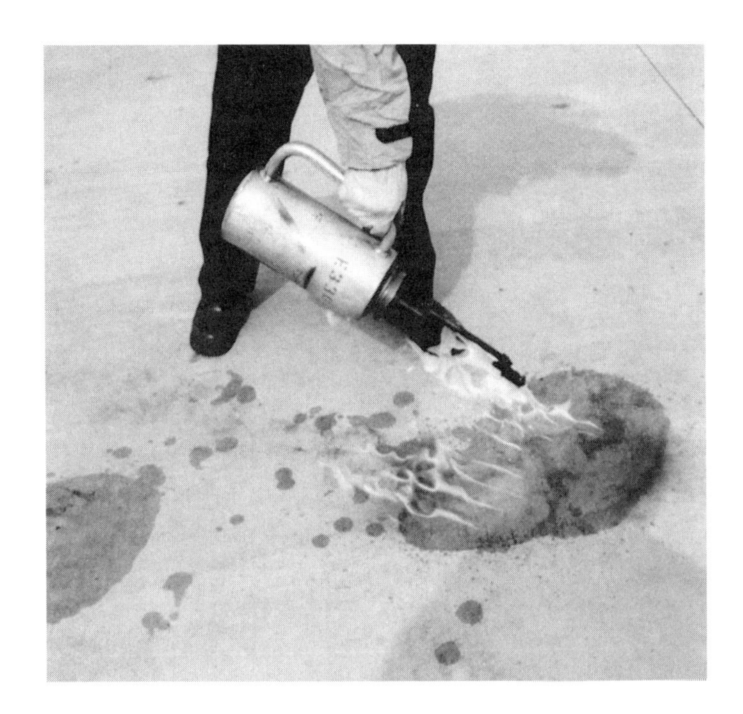

Figure 8-11 *Bringing unlit drip torch to small amount of burning fuel mix on the ground.*

Figure 8-12 *Firefighter using the drip torch after it is lit.*

Figure 8-13 *Letting the wick burn out at the finish of the operation.*

Flare Launcher

A flare launcher (Figure 8-14) requires some special field training for proper use. It is used in situations where a fire needs to be started at a distance from the control line.

The 1-inch flare system consists of a flare, a launcher, and blank cartridges (Figure 8-15). The launching system was designed in conjunction with the USDA Forest Service. It is the only fire-starting flare system designed specifically for that agency. The system is certified by the Bureau of Alcohol, Tobacco, and Firearms (BATF) as an industrial tool and not as a firearm. It is found in most Forest Service caches.

Preparing the Launcher for Use Wipe down the launcher with an absorbent cloth, removing any excess oil and dirt. Holding the launcher on its right side in the palm of your right hand, press on the cylinder release button and remove the cylinder

Figure 8-14 *Flare launcher.*

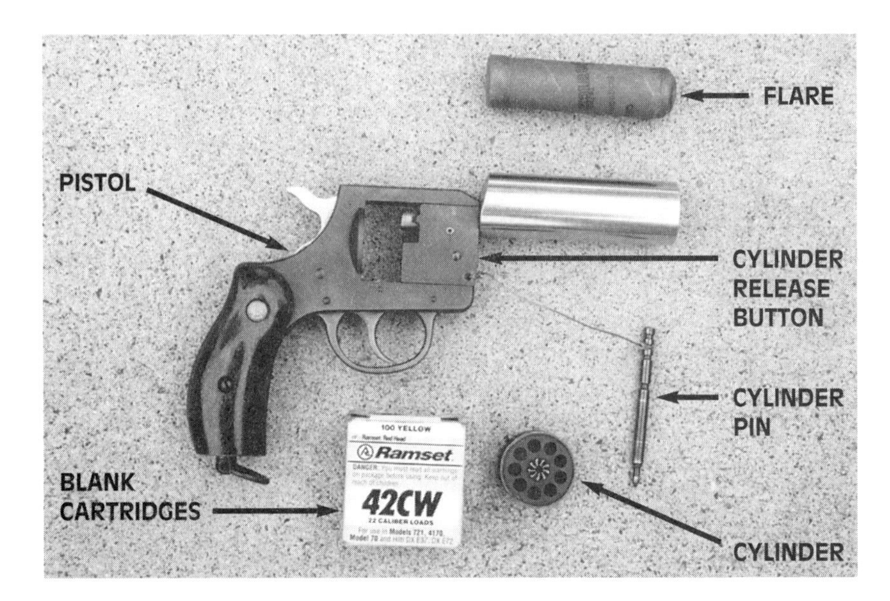

Figure 8-15 *Components of a flare launcher include the launcher, cylinder, blank cartridges, and flare.*

pin allowing the cylinder to fall into your right hand. With the cylinder removed, load the cylinder with blank cartridges.

With the launcher in the palm of your right hand with the barrel pointing away from you, replace the cylinder in the launcher and insert the cylinder pin while pressing the pin release button. Release the button and ensure the pin is locked in place by attempting to remove the pin without pressing the release button.

Loading the Flare Do not cock the hammer or place your finger inside the trigger guard while loading the flare. Grasping the launcher by the grip with one hand, insert the capped end of the flare into the muzzle of the launcher. The flare will extend out the muzzle about ½ inch when fully inserted into the barrel. The flare will fit tightly and may need extra pressure to be fully inserted into the barrel.

Holding the launcher in one or both hands, point the launcher in the proper direction, elevate the muzzle about 15° or angle for maximum range and ignition after the flare hits the ground. Cock the hammer and pull the trigger to fire the launcher. The launcher can also be fired by just pulling the trigger in double-action mode; however, without the hammer being cocked there may not be enough room inside the trigger guard for a gloved finger. Practice firing the flares to gain confidence in your ability to hit the target area.

Safety Precautions Follow these safety precautions when using the flare launcher:

- Always wear ear and eye protection when using the launcher.
- The launcher is *not* classified as a firearm by the BATF; however, the launcher should be treated as a firearm and general firearm safety rules are to be followed.
- Always treat the launcher as if it were loaded.
- Always check the cylinder for cartridges when you first handle the launcher.
- Never transport the launcher while it is loaded with cartridges.
- Never point the launcher at anything you do not want to destroy.

Unloading the Launcher Place the right side of the launcher in the palm of your hand. While pressing on the cylinder release button, remove the cylinder pin, allowing the cylinder to fall into your right hand. Using the cylinder pin as a punch, remove the spent cartridges from the cylinder and replace the cylinder in the launcher.

Maintenance After each firing session, spray the launcher down with gun oil or light machine oil and return the launcher to its storage compartment. Inspect the launcher periodically for paint buildup in the barrel. Use a lacquer thinner to clean the barrel of paint buildup.

Trouble Shooting The most common malfunction for the launcher is the sticking of the cylinder when firing a flare is attempted. This indicates that the cylinder pin is not locked in place and is working loose. Keeping the launcher pointed in a safe direction, center the cylinder and replace the pin, ensuring it is locked in place by trying to remove it without pressing on the release button. Should the launcher become inoperable for any reason, contact the manufacturer for recommended actions. Replacement cylinders are available, if required.

Air Transport The launcher can be taken aboard a passenger aircraft in checked baggage if it is unloaded, in a hard case, and the ticket agent is notified.

System Facts The following list indicates the flare launcher system facts:

Description:	1-inch Fire Quick Flare,[1] 1 inch × 3¼ inch / 1.5 ounces
Effective Range:	Approximately 100 yards
Fuel Types:	Light to medium
NFES Numbers:	0371 for flares, 0372 for cartridges
GSA Contact Numbers:	GS-07F-9454Guide

Other Firing Equipment

There are many other firing devices, such as hand-thrown flares (Figure 8-16), Terra Torch, Helitorch, Premo Mark III, and the Aerial Flare Igniter Dispensing System. These devices require special field training and are not discussed in this book.

Figure 8-16 *Two-inch hand-thrown flare sprays 4,000° material out and starts a fire.*

[1] This flare launcher is manufactured by Quoin, Inc., located at 139 North Balsam in Ridgecrest, California. Contact the technical support division at 760-384-5035 for questions of a more technical nature.

CREW ASSIGNMENTS

Firing teams do the actual lighting and must focus their attention on the actual firing operation. They must also be aware of what is happening around them and be thinking safety at all times. *Holding forces* follow the firing personnel and protect them from injury from the fire they are lighting. They also prevent the fire from the firing operation from sloping over the control line and are used to pick up spot fires. Holding forces also strengthen critical points on the control line. *Mop-up forces* must remain behind to secure the line and do mop-up. These forces may start out as part of the holding forces. *Reserve forces* are held in reserve to be used when needed. They can act as a separate tactical unit to pick up slopover, rapidly attack a spot fire, or provide structure protection when needed.

FIRING TECHNIQUES

Now some basic ignition techniques are discussed.

line firing
setting fire to only the border fuel immediately adjacent to the control line

Line Firing

The first and most commonly used ignition technique is **line firing.** Line firing is simply burning along the inner edge of the control line to a termination point (Figure 8-17). The person doing the lighting works on the control line or slightly in the

Figure 8-17 *Line firing.*

green and his or her escape route is to step onto the control line. The safe zone is the control line or a planned area that has been identified. This is the safest of all the firing methods discussed here.

Strip Firing

strip firing
setting fire to more than one strip of fuel and providing for the strips to burn together

The second method used quite often is **strip firing** (Figure 8-18). The firing team generally consists of three crew members. They are placed in the green at a distance that varies depending on several factors: fuels, topography, wind, and the need to return to the control line. They move parallel to the control line in a staggered fashion, igniting strips and letting them burn together to a termination point. It is very important that all the firing team members stay in visual contact and maintain safe spacing.

When using the strip firing technique, always evaluate what will be driving the fire you are lighting. It will either be the wind or the topography and this will dictate the firing order of the team members. Study Figure 8-19 through Figure 8-22 for the correct placement of the firing team members in adverse winds (Figure 8-19), favorable winds (Figure 8-20), adverse slopes (Figure 8-21), and favorable slopes (Figure 8-22).

Figure 8-18 *Strip firing operations. Note the other strip of fire left by the firefighter not in the picture.*

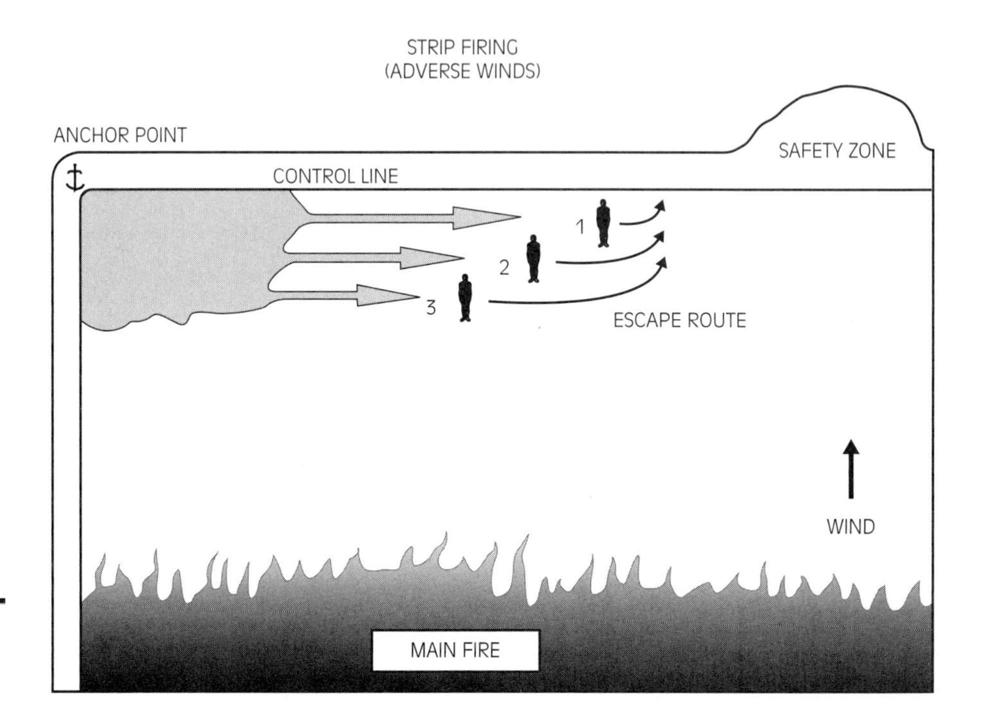

Figure 8-19 *Strip firing in adverse winds.*

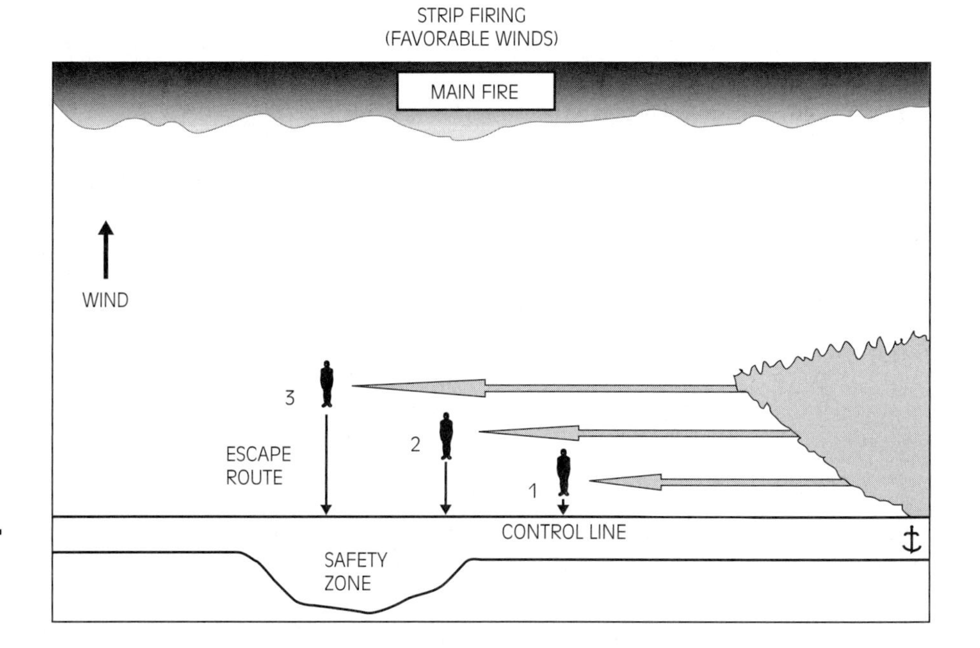

Figure 8-20 *Strip firing in favorable winds.*

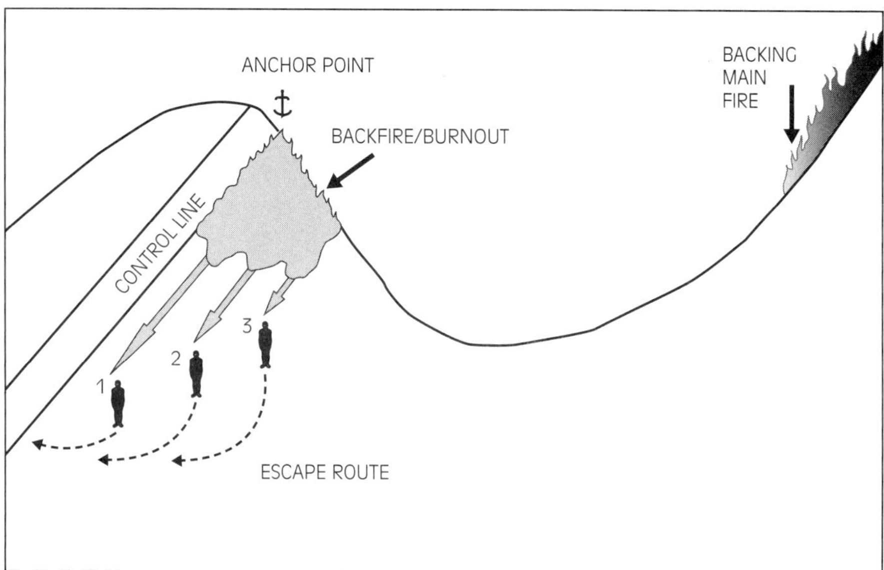

Figure 8-21 *Strip firing on adverse slopes.*

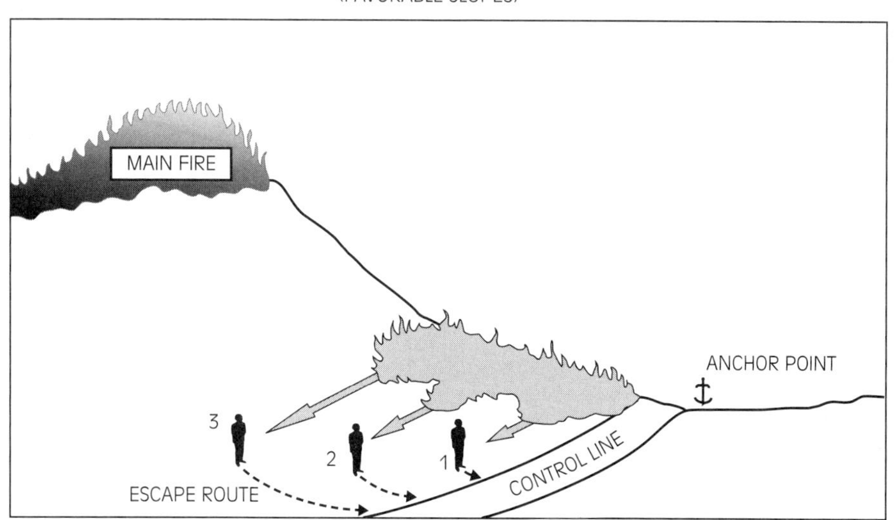

Figure 8-22 *Strip firing on favorable slopes.*

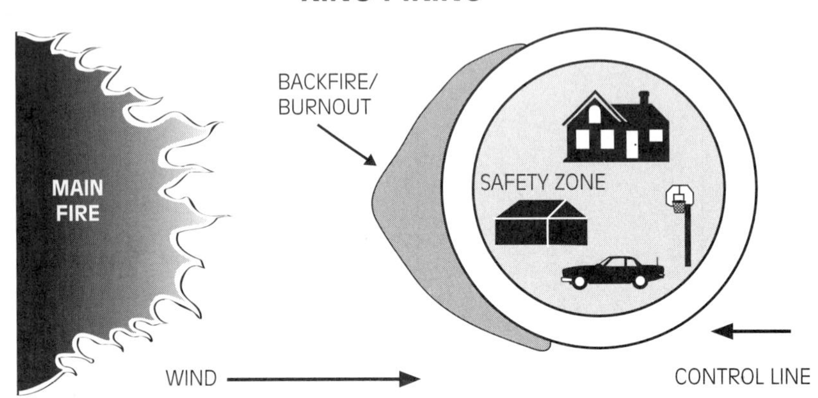

Figure 8-23 *Ring firing.*

Strip firing works best in light continuous fuels. In strip firing, as with any firing operation, it is imperative that everyone know the termination point, escape routes, safety zones, and location of all the firing forces.

Ring Firing

ring firing
a technique generally used as an indirect attack and backfire operation involving circling the perimeter of an area with a control line and then firing the entire perimeter

A technique that works well in structure protection situations where there are pockets of houses that have large expanses of vegetation between them and an approaching large main fire is **ring firing.** This firing technique splits the head of the fire, thus moving it around the structures being protected (Figure 8-23).

There are several other ignition techniques not covered in this text, such as chevron burning, spot firing, and burn strip. To further develop your skills in firing techniques it is suggested that you take an S-234 Ignition operations class.

To get some actual field exposure in controlled situations, find a prescribed fire school, like the Interagency School in Tallahassee, Florida, and attend the hands-on training. Another option would be to find out when a controlled burn is going on in your area. Contact the local agency representative who is in charge of the burn and ask to observe the operation. Pick a mentor at the burn site and talk with him or her about the operation and technique to be used during the burn.

Each year the USDA Forest Service also conducts fire school. This is an excellent way to learn firing operations in addition to other wildland skills. In the southern California area, it is held annually at the Camp Pendleton Marine Base near Oceanside, California.

Summary

Firing operations always present some risk. These operations should be done by experienced fire personnel who understand fire behavior and firing operations.

Firing operations are another tool for use in your wildland toolbox. You need to learn this skill if you are going to be involved in wildland fire suppression. Firing operations work well in situations where resources are limited or where fire-line intensities are too great for direct attack.

This information has been included in the text to help you understand firing operations and the tools used to start them. Reread this section after you have an active fire season behind you. You will have had a chance to observe actual fire behavior and be better prepared to learn firing operations.

Review Questions

1. Explain the term *backfire.*
2. Explain the term *burnout.*
3. What is a counterfire?
4. Generally speaking, in mountainous terrain, where should the firing operation start?
5. If you fire with the wind, what type of rate of spread would you expect?
6. Define the term *line firing*?
7. What is strip firing?
8. What is ring firing?

References

California Department of Forestry and Fire Protection, *Wildland Fire Fighting* (North Highland, CA: State of California, 1987).

National Wildfire Coordinating Group, *Fireline Handbook* (Boise, ID: National Interagency Fire Center, 1998).

National Wildfire Coordinating Group, *S234 Firing Methods and Procedures* (Boise, ID: National Interagency Fire Center, 1991).

National Wildfire Coordinating Group, *Wildland Fire Suppression Tactics Reference Guide* (Boise, ID: National Interagency Fire Center, 1996).

Bulldozers and Tractor Plows

Learning Objectives

Upon completion of this chapter, you should be able to:

- Identify the different types of bulldozers.
- Explain the different types of control systems and blades used on a bulldozer.
- Describe how to use a bulldozer.
- Identify a tractor plow.
- Explain how to use a tractor plow.

INTRODUCTION

Bulldozers and tractor plows are effective initial attack tools on wildland fires. These powerful heavy pieces of mechanical equipment are used to construct firelines within a variety of fuel and soil types. They must be closely supervised and integrated into your plan.

This chapter discusses how to best manage and use these resources. If they are used correctly, much fireline can be constructed quickly.

BULLDOZER TYPES

As noted in Chapter 1, bulldozers (Figure 9-1) are typed by size. Light bulldozers (Type III) are very maneuverable in tight situations and are best used in soil types that have few rocks and light fuels. Light bulldozers work well in wet soil if they are equipped with wide tracks and are effectively used on level ground and moderated slopes. Bulldozers must be supervised closely, and control lines must be well planned. If not, they can cause serious long-term damage to fragile soils. Some jurisdictions require special permission before bulldozers can be utilized for fireline construction. Generally speaking, state parks, national parks, and national forest wildernesses require special permission before using bulldozers in fire suppression.

Medium bulldozers (Type II) perform well on moderately steep slopes, are maneuverable, and are usually considered the best size for all-around use on

Figure 9-1 *Bulldozer on a transport ready for dispatch to a fire.*

wildland fires. They are well suited for the average fuel and terrain features found in mountainous areas. They can be fitted with wider tracks that enable them to work well in wet or boggy areas.

Heavy bulldozers (Type I) do not maneuver well in tight situations, especially in steep terrain. Generally speaking, they are too big to be used to construct firelines in average situations. They are best used as lead bulldozers in heavy fuels on moderate slopes.

CONTROL SYSTEMS AND BLADE TYPES

■ Note
The hydraulic system of blade control enables the bulldozer to dig in hard ground and can be used to brake with while going downslope.

■ Note
Straight blades are best used to establish original firelines as well as finished lines.

water bar
a shallow channel or raised barrier that leads off water, particularly storm water

Newer bulldozers, built within the past 30 years, have hydraulic systems that control blade operations. Older bulldozers had cable control systems. The advantage to a hydraulic blade control system is that pressure can be exerted either up or down on the blade; cable systems cannot exert downward blade pressure. Bulldozers with hydraulic systems are best used on firelines and are more versatile.

The hydraulic system of blade control enables the bulldozer to dig in hard ground and can be used to brake with while going downslope. If the bulldozer gets stuck or high-centered, downward pressure on the blade can be exerted and used to raise the bulldozer and materials can be placed underneath it to free the bulldozer.

Understanding the different blade types is also a consideration when picking a bulldozer. The following three types of blades are in use:

1. U-shaped blades are used to move earth in road construction. Bulldozers with these types of blades are best suited to pioneer firelines, dig stumps, or construct roads.
2. Brush blades are best suited to first build a fireline in brush areas, clean and pile logging slash, and use on mop-up operations.
3. Straight blades can be set at an angle to push vegetation and soil to either side of the dozer. They are best used to establish original firelines as well as finished lines. They can also be used to put in **water bars,** lines that are used for erosion control that run from one side of the finished line to the other, and for road construction.

USE OF BULLDOZERS

■ Note
The maximum percentage of slope for downhill bulldozer operations is 75%.

Most of the larger wildland fire agencies use bulldozers as an initial attack resource, primarily to construct lines on the flank of the fire while working toward the head. Bulldozers are best used in tandem (Figure 9-2) with the lead bulldozer pioneering the line and the second bulldozer widening, strengthening, and completing the fireline. Bulldozers can also be used to construct safe zones or open up roads so that engines or CCVs can access the area. As the fire starts to come to a completion, bulldozers do mop-up and rehabilitation work.

Figure 9-2 *Two California Department of Forestry and Fire Protection bulldozers working in tandem to construct a fireline.*

Table 9-1, fireline production rates for bulldozers, shows single pass line construction rates in chains per hour for several variables. The table was developed from a series of field tests with different type of bulldozers. Some generalities that can be concluded from this table are: (1) production rates drop as the fuel loading increases and (2) slope has an effect on production rates, particularly traveling upgrade, and (3) some bulldozer sizes are better suited to select jobs than others.

■ Note
Bulldozers can work uphill to a maximum of 55% slope.

Also consider the maximum percentage of slope that a bulldozer can operate on. It is not only important to consider uphill and downhill operations, but also sidehill operations. The maximum percentage of slope for downhill bulldozer operations is 75%. Bulldozers can work uphill to a maximum of 55% slope. Sidehill operations should not exceed 45%. These percentages are dependent on soils, fuels, and other factors. If these percentages are exceeded, then consider the use of handcrews.

■ Note
Sidehill operations should not exceed 45%.

Management of bulldozers is vital. Remember, they can produce a lot of finished fireline, but if not properly managed, they can cause considerable damage to the environment. The bulldozer operator must understand the plan as he or she is the vital link between the bulldozer and what you are trying to accomplish. Include him or her in your planning.

■ Note
Management of bulldozers is vital. Remember, they can produce a lot of finished fireline, but if not managed properly, they can cause considerable damage to the environment.

Here are some general rules for operation of bulldozers:

- Be aware of the steepness of the slope—up-, down-, and sidehill.
- Consider soil type—sandy, rocky, hard, boggy, or loose.
- Consider the fire intensity level. Bulldozers can work directly on firelines if the intensity level is low. As fuel gets larger and the intensity level of the fire increases, then bulldozers must be worked at a distance from the fire.

Table 9-1 *Fireline production rates[a] (single pass) for bulldozers manufactured since 1975.*

Fire Behavior Fuel Model	Slope Class 1 (0–25 %)		Slope class 2 (26–40 %)		Slope class 3 (41–55 %)	
	Up	Down	Up	Down	Up	Down
Small Bulldozers (Type III)						
1,2,3	63	88	36	88	14	16
4	22	29	12	30	3	22
5	63	88	36	88	14	61
6	39	59	22	62	8	42
7	39	52	22	56	8	35
8	63	88	36	88	14	16
9,11,12	22	30	12	30	3	11
Medium Bulldozers (Type II)						
1,2,3	88	118	58	112	35	73
4	32	47	18	53	5	31
5	88	118	58	112	35	73
6	51	75	26	78	9	48
7	51	75	27	78	9	48
8	88	118	58	112	35	73
9,11,12	32	47	18	53	5	31
10,13	17	23	10	25	3	11
Large Bulldozers (Type I)						
1,2,3	91	124	62	118	35	83
4	43	60	27	62	12	40
5	91	124	62	118	35	83
6,7	63	91	41	90	22	57
8	91	124	62	118	35	83
9,11,12	43	60	27	62	12	40
10,13	27	38	15	34	4	16

[a] In chains per hour (a chain is 66 feet)

■ Note

Generally, this is 1½ times the height of the fuel in the brush and not less than one-half the height of the fuel in timber.

Safety

Bulldozer operators need to be in personal protective clothing and have a fire shelter.

- Production rates drop off in nighttime bulldozer operations.
- Construct only as much line as necessary. The line needs to be only wide enough to hold the fire. Generally, this is 1½ times the height of the fuel in the brush, and not less than one-half the height of the fuel in timber.
- Sometimes it is necessary to make some sections of the line wider. Have a reason for this. Maybe a safe zone is needed or maybe the fuel type has changed from grass to brush.
- The best use of bulldozers is generally in pairs or tandem. The larger machine takes the lead and the smaller one follows and cleans up the line. If one gets stuck, the other can assist. They can also help each other if one needs help.
- Use larger bulldozers for large fuels and smaller bulldozers for lighter fuel types, such as grass and brush.
- On a large fire, bulldozer organization (Figure 9-3) will vary with the size of the fire, amount and type of fuels, resource availability, and local practices.
- Always use anchor points when starting line construction.
- Think safety and follow the LCES guidelines.
- Operators need to be in personal protective clothing and have a fire shelter.
- As with any line construction operation, make sure unburned fuels are pushed away from the fireline and disbursed as practical.
- Construct the fireline on ridges, where possible. If the fireline is constructed on the midslope or canyon bottom, there is danger that the fireline may not hold due to spotting and radiant heat.
- In some situations, dirt and soil must be pushed to the inside of the fireline. These bulldozer piles must be broken up and scattered well into the burn.
- In fuel models with downed timber, such as slash fuel models, it may be necessary to put a saw team out in front of the bulldozer operation. When the lead bulldozer is large enough, this will probably not be necessary.
- Where snags are present, evaluate whether the bulldozer can bring them down safely or whether a saw team is necessary.
- Line should be located well in advance of the lead bulldozer. The **swamper** must evaluate the terrain, fuel type, and fire control strategy when locating the control line for the dozer to follow.
- Dozer operations need to be followed up with handtools and/or engine hose lines to finish the line.
- Assign a swamper to each bulldozer. His or her job is to work as a spotter for the bulldozer operator. He or she is also responsible for cutting

swamper
worker on a bulldozer crew who pulls winch line, helps maintain equipment, and generally speeds suppression work on a fire

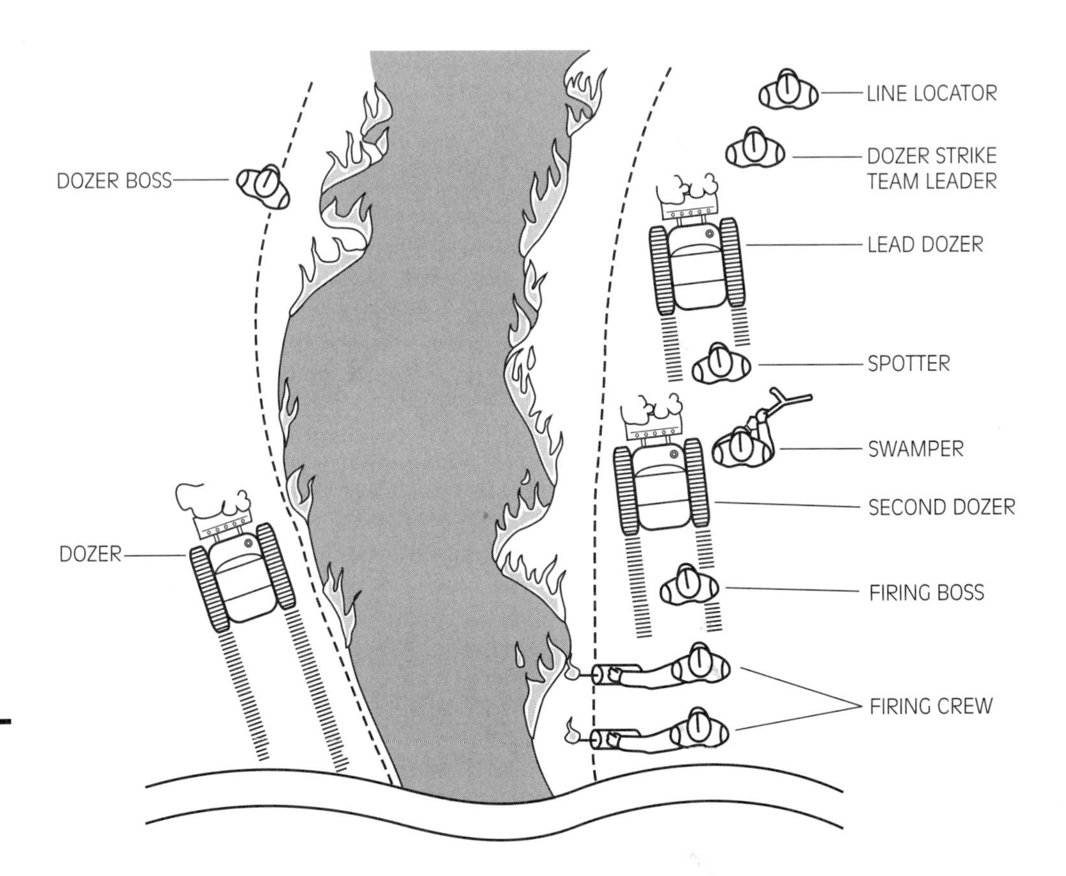

Figure 9-3 *Typical bulldozer organization on a large fire.*

away projecting parts of trees or heavy brush that may injure the bulldozer operator or his machine. The swamper assists with maintenance of the bulldozer and is required to position the blade or handle the winch line and choker. He or she communicates with the operator with hand signals (Figure 9-4) and also acts as an alternate operator.

- A cleanup crew of normally three to six persons should follow behind the last bulldozer in heavier fuels. The crew's job is to reduce the amount and kind of fuel along the control line, thus reducing the chances for slopover. This is also done to ready the control line for burnout operations.

BULLDOZER USE IN THE UNITED STATES

Bulldozer use is widespread in the United States.

In southern and central California, bulldozers are plentiful and many fire agencies respond with them as an initial attack tool. They are extensively used on

Figure 9-4 *Bulldozer hand signals.*

these fires. However, they may be precluded from use in federal responsibility areas because of land management policies. Topography may also limit their use in steep mountain areas. In northern California and the rest of the northwest, bulldozers are especially effective working in tandem in heavy fuels. The first bulldozer pushes over the standing material and the second unit pushes it aside.

In the northern Rocky Mountains, bulldozers are found in constant use building main firelines because of the heavy fuels found in the area. Bulldozer use in the Great Basin and southern Rocky Mountains has decreased in recent years due to sensitive environmental concerns.

In the southwestern United States we find a limited supply of bulldozers in use. At lower elevations use is sometimes restricted by rock, steep topography, and environmental concerns.

A mix of bulldozers and tractor plows is used in the northeast region of the United States. In this 20-state region, which includes Maine—the most forested area in the United States—Pennsylvania, and New Jersey, we find bulldozers in use.

The southeast section of the United States is represented by three geographic regions, the Appalachian Mountains, the Piedmont Plateau, and the coastal plains. In the coastal plains and Piedmont Plateau, bulldozers are used as a primary tool.

TRACTOR PLOWS

Tractor plows (Figure 9-5) are used extensively in the South, Southeast, and Florida. They are found in use in the coastal plains and flat woods in the region. Plow units are also in general use in the Midwest and Great Lakes states, such as Michigan, Minnesota, and Wisconsin. Tractor plows are also used in parts of the northwest United States.

Figure 9-5 *Tractor plow. Photo courtesy Florida Division of Forestry.*

Figure 9-6 *A USDA Forest Service tractor plow unit in Florida's Ocala National Forest. Note the straight blade on the front.*

A tractor plow is best described as a bulldozer pulling a plow unit. The blade on the bulldozer is used to clean downed material, slash, and debris out of the way while the plow unit, being pulled, constructs the fireline (Figure 9-6). A unit with a straight blade is also useful for clearing a pathway for a small brush engine. The brush engine then supports the tractor plow's operation. Some tractor plows, however, have C blades (Figure 9-7), which do not disturb the soil and are

Figure 9-7 *A tractor-trailer plow unit with a* C *blade. This unit is used in the sand pine forested area of the Ocala National Forest in Florida.*

Figure 9-8 V *blade on a tractor plow.*

used to push small trees out of the way. In some parts of the country tractor plows have V blades (Figure 9-8), which cut and push small trees out of the way.

■ Note
Tractor plows are best used in areas that are relatively flat or have rolling hills and in soils that crumble easily or are sandy.

Tractor plows are best used in areas that are relatively flat or have rolling hills and in soils that crumble easily or are sandy. Plows do not work well in rocky areas. In areas of standing timber, tractor plows should not be used unless the spacing of the trees is such that they do not interfere too much with the plow unit.

Generally speaking, a tractor plow can construct lines at about 3 miles per hour providing that the fuels, soils, and terrain are suitable for their use. The speed drops off as the slope increases or the plow unit encounters boggy sections, stumps, trees, and poor soil conditions.

Shallow lines should be constructed with the plow as it puts less drag on the tractor and enables it to put in line more rapidly. The line should be deep enough to obtain clean firelines to mineral soil. The fireline needs to be clean and continuous.

Line construction should be followed up with burnout operations. Two tractor plow units working in tandem are best, however, the operating principles for single units remain the same.

Figure 9-9 is a drawing of a typical tractor plow organization.

In areas of trees and heavy brush, it is best to have a bulldozer take the lead and follow it with a tractor plow. The bulldozer takes out the heavy brush and trees and pushes them to the outside of the fireline.

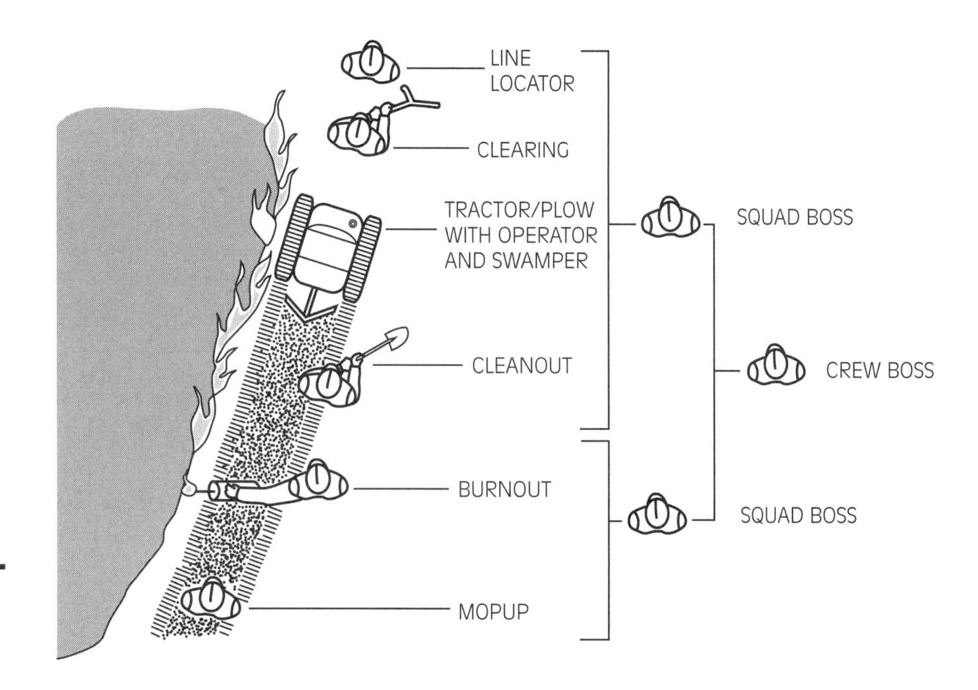

Figure 9-9 *Typical tractor plow organization.*

When using tractor plows, be aware of the rate of fire spread, the plow rate of line construction, the distance the unit is to the fire, the fire's intensity levels, and the time it is taking to complete the burnout operations.

Safety concerns about tractor plows are similar to those expressed about bulldozers.

Summary

Bulldozers and tractor plows are effective tools used to construct firelines rapidly. Their use must be constantly monitored, as these powerful pieces of mechanized equipment are capable of causing much damage to the environment. Bulldozers can also be used to create safe zones, open up roads so engines or crew buses can get into the fire, and do rehabilitation work after the fire is over.

A tractor plow is best described as a bulldozer pulling a plow. Tractor plows are used extensively in the South, Southeast, and Florida. Plow units are also in general use in the Midwest and the Great Lakes States.

They are best used in areas that are relatively flat or have rolling hills. Plows do not work well in rocky areas.

You need to understand how to best use these resources as they can construct a lot of fireline quickly. If used properly, they are a great initial attack resource.

Review Questions

1. What type of bulldozer is best suited for all-around use on a wildland fire?
2. New bulldozers, built within the past 30 years, have what kind of control system?
3. What type of bulldozer blade is best suited for constructing original firelines?
4. What is the maximum percentage of slope for downhill bulldozer operations?
5. Sidehill operations for a bulldozer should not exceed what percent?
6. For what type of terrain is a tractor plow best suited?
7. In what type of soil conditions do tractor plows work best?

References

National Wildfire Coordinating Group, *Tractor Use/Tractor Boss S-213* (Boise, ID: National Interagency Fire Center, 1979).

National Wildfire Coordinating Group *Wildland Fire Suppression Tactics Reference Guide* (Boise, ID: National Interagency Fire Center, 1996).

Firefighting Aircraft

Learning Objectives

Upon completion of this chapter, you should be able to:

- Explain the different types of aircraft used on a wildland fire.
- Describe the differences between fixed-wing aircraft types.
- Explain the capabilities of air tankers.
- Explain the different coverage levels used by fixed-wing aircraft.
- Explain the capabilities of helicopters.
- Identify the tactics used by helicopters.

INTRODUCTION

retardant
a substance or chemical agent that reduces the flammability of combustion

suppressant
an agent that extinguishes the flaming and glowing phases of combustion by direct application to the burning fuel

This chapter focuses on the types and uses of different kinds of firefighting aircraft and describes the use of **retardants, suppressants,** foams, and water in aerial operations.

AIRCRAFT OPERATIONS

■ Note
Aircraft resources complement ground attack forces.

■ Note
Costs must always be analyzed versus the benefits gained.

Aircraft resources complement ground attack forces. They are of greatest value during initial attack on wildland fires while the fires are small. They are quickly able to make an initial attack on fires in remote areas where ground forces would take too long to access the area. The use of aircraft during initial attack helps keep fires small. In California, where there are some of the busiest air attack bases, an air tanker can be on a wildland fire within 20 minutes or less (Figure 10-1). Aircraft support ground attack forces, however, ground forces ultimately have to go in and extinguish the fire.

Costs must always be analyzed versus the benefits gained. Aircraft are a very costly resource. There is an hourly fee as well as a per gallon cost for the retardant. Incident commanders must therefore know how to use this tool effectively and to its full potential.

Figure 10-1 *California Department of Forestry and Fire Protection air tanker ready for dispatch.*

FIXED-WING AIR TANKERS

air tactical group supervisor
person responsible for directing and coordinating airborne aircraft operations and managing air space for an incident

Fixed-wing air tankers are classified by the retardant payload they carry. Retardant and its use is discussed in the final section of this chapter. The fixed-wing air tankers are typed as shown in Table 10-1.

Decisions on tactical use of aircraft are best made as a result of a team effort among the **air tactical group supervisor**, operations section chief, and the incident

Note
Decisions on tactical use of aircraft are best made as a result of team effort among the air tactical group supervisor, operations section chief, and the incident commander.

Table 10-1 *Types of fixed-wing air tankers.*

Tanker	Max. Gallons	Cruise Speed (kts)	No. of Doors
	Type I Tankers, 3,000+ gallons retardant		
Martin Mars	7,200	165	26
KC-97	4,500	210	16
C-130	3,000	250	8 or constant flow
P3	3,000	275	8
DC-7	3,000	235	6–8
	Type II Tankers, 1,800–2,999 gallons retardant		
DC-6	2,450	215	6–8
P2V	2,700	187	6
SP2H	2,000	195	Constant flow
PB4Y2	2,200	184	8
DC-4	2,000	178	8
DC-4 Super	2,200	200	8
	Type III Tankers, 600–1,799 gallons retardant		
B-26	1,200	200	2–6
Super PBY	1,400	105	2
CL-215	1,400	160	2
CL-415	1,600	179	4
AT802	800	145	Constant flow
S2	800	180	4
S2T[a]	1,100	230	Constant flow
Twin Otter	600		
	Type IV Tankers, 100–599 gallons retardant		
Turbine Thrush	350	140	1–2
Ag-Cat	300	100	1–2
Dromadear	400	110	1–2
Beaver	108	100	1

[a]New version of S2T is being built for CDF. These are estimates.

commander. The air tactical group supervisor is orbiting the fire and from his vantage point, can furnish the incident commander or operations section chief with information that will be helpful in their decision making process.

Capabilities of Air Tankers

■ Note
Air tanker tactics are similar to the tactics employed by ground forces.

Air tanker tactics are similar to the tactics employed by ground forces. If there is a fast-moving fire with a broad front, then the best use of the resource is probably to attack the active flanks (Figure 10-2). The tanker pilot starts dropping his long-term retardant from an anchor point and proceeds along the active flanks. Long-term retardant has a red colorant in it so the next tanker pilot can see where the previous drop ended and can build on that previous drop. Remember that continuous retardant coverage needs to be provided or your efforts will be ineffective. Have enough air tankers to provide one drop every 5 minutes on the fire. Air drops also need to be supported by ground holding actions or again your efforts may be wasted.

If the fire has a narrow head because it is on a steep slope or is wind driven, then the tanker pilot may opt to attack the head. Again the head must be narrow enough to contain the fire with the retardant payload on board or the result may be a splitting of the head of the fire.

Figure 10-2 *An air tractor with computer-controlled drop doors in operation. Photo courtesy Air Tractor Inc.*

Figure 10-3 *Tanker 115—a consolidated PBY5A with a 1,000 gallon water capacity.*

■ **Note**
Modern air tankers have a computer-aided drop system that regulates the retardant coverage level and the drop pattern.

salvo
dropping by an air tanker of its entire load of fire retardant at one time

trail
dropping an air tanker load in sequence, causing a long unbroken line of retardant

split
dropping made from one compartment at a time from a multi-compartment air tanker

Author's Note: When I was part of the flight crew on a PBY tanker I saw this operation work (Figure 10-3). We were dispatched from a tanker base loaded with 1,000 gallons of long-term retardant. Upon arrival at the fire, we made one pass across the head with one door open. Then we immediately turned back toward the fire and opened the other drop door, thus delivering the rest of our retardant payload. The retardant pattern looked like an X. *This would hold the forward spread of the fire. Immediately we flew to a lake close by and scooped 1,000 gallons of water (a PBY is a water-scooping seaplane) and flew back to the fire and dropped water directly on the head of the fire. The retardant kept the head of the fire in check, while water directly applied to the head, with successive drops, put it out.*

On higher intensity fires with higher rates of spread, air tankers pretreat a ridge ahead of the fire. This tactic combines the effect of the retardant material with a terrain feature that causes a natural decrease in fire activity.

Modern air tankers have a computer-aided drop system that regulates the retardant coverage level as well as the drop pattern. The basic drop patterns are as follows:

- **Salvo**—Dropping the entire load of retardant at once or dropping a combination of tanks simultaneously.
- **Trail**—Dropping tank loads in sequence, causing a long unbroken line of retardant. This is the normal drop sequence for an air tanker.
- **Split**—Dropping a partial payload on the fire and then delivering the second part of the payload to another area of the fire. Or coming back for a second pass on the same fire area.

Table 10-2 *Retardant coverage table.*

Description of Fuel Type	Coverage Level
Annual and perennial grasses, tundra	1
Conifers with grass understory, short needle closed conifer, summer hardwood, long needle conifer, fall hardwood	2
Sagebrush with grass, sawgrass, intermediate brush (green), light logging slash	3
Short needle conifer with heavy dead litter in the understory	4
Southern rough, intermediate brush (cured), Alaska black spruce	6
California mixed chaparral, high pocosin, medium logging slash, heavy logging slash	Greater than 6

■ Note
The amount of retardant coverage, on the ground, is governed basically by the fuel type.

The amount of retardant coverage, on the ground, is governed basically by the fuel type. Heavier fuels require greater coverage levels. Table 10-2 is a guide to coverage levels. If a coverage level 4 is desired, 4 gallons of retardant needs to be delivered to each 100 square feet of ground area. Remembering this and knowing the amount of retardant carried on different aircraft will help you order the correct number of aircraft needed to do the job.

Checklist for Fixed-Wing Aircraft Use

- Determine your tactics based on fire behavior size-up and available resources.
- Determine direct or indirect application of retardant.
- Make sure that aircraft use the proper drop height.
- Always establish an anchor point and work from it.
- Apply the proper coverage levels.
- When conditions allow, drop downhill and away from the sun.
- For best accuracy, drop into the wind.
- Maintain good communications between the ground and air so you can receive an honest evaluation of tactical air operations.
- Use direct attack only when ground support is available.
- Plan drops to extend the retardant line.
- Constantly evaluate the effectiveness of the retardant. If it is not working, then make the necessary adjustment. Do not misuse this resource, because it is very costly.

Conditions That Limit Air Tanker Usage

- Dense smoke in the fire areas make air tanker use hazardous and ineffective.
- Air tanker use is cut off one-half hour before sunset, limiting air tanker use to daytime only. At night the fire activity normally lessens and fire crews make better progress toward containment.
- Wind speeds over 20 miles per hour reduce air tanker effectiveness.
- High velocity winds over 40 miles per hour, turbulent air, and shifting winds may restrict or eliminate air tanker use.
- Topography features, such as steep slopes, seriously reduce fixed-wing effectiveness.
- Deep canyons are problems for air tankers, because they have difficulty getting in and out safely. This limitation may preclude their use.
- At certain times of the day, such as early morning and late afternoon, deep shadows are produced by the sun being low on the horizon. Certain aspects are then in deep shade, thus making it difficult for the pilot to see obstructions on the ground or the fire target.
- Tall timber with a closed canopy may cause higher retardant drops. These factors limit the amount of retardant that reaches the ground fuels that help carry the fire.

Air tanker effectiveness improves when the wind speed decreases; as slopes become less steep; as fuels on aspect become shaded and colder; as fuel types, such as grass or light brush, become lighter; where there is light tree covering over surface fuels; and as the distance from the air attack base is lessened.

HELICOPTERS

Note **Helicopters are used on the fireline usually for direct attack.**

Note **Helicopters are advantageous because of their short turnaround time.**

Helicopters are used on the fireline usually for direct attack applications. They normally carry water, although they can be loaded with class A foam or retardant. Water is applied directly on the flames to extinguish them (Figure 10-4).

Helicopters are advantageous because of their short turnaround time. After they drop water on the fire, vertical takeoffs and landings allow them to utilize refill sites close to the fire. Lakes, ponds, the ocean, portable tanks, and engines placed next to conventional water sources are used as refill sites.

Helicopters drop water from either fixed tanks or portable buckets. They can carry as little as 75 gallons to just over 2,600 gallons of water.

On large fires where both fixed-wing aircraft and helicopters are working on the same fire, they are separated. They are separated by altitude differences and they may also be separated geographically. Helicopters orbit in a clockwise or right-hand pattern; whereas fixed-wing aircraft orbit in a counter clockwise direction. The air tactical supervisor orbits above both of these resources.

Figure 10-4 *Florida Division of Forestry helicopter drops water on a fire. Photo courtesy Bruce Ackerman, Florida Division of Forestry.*

If the air tanker and helicopter are not separated geographically, then the helicopters should clear the airspace before the tankers come in.

Helicopters are also a valuable resource for transporting supplies to and from remote locations on the fireline, such as helispots. A helispot is a temporary location where supplies, equipment, or personnel are transported or picked up. Helispots are numbered and are close to the fireline. Helicopters also fly other logistical or support needs. For instance, they may be used to establish a cellular phone site on a mountain peak and they can provide aerial reconnaissance on wildland fires.

Table 10-3 *Minimum standards for helicopters.*

Components	Type I	Type II	Type III	Type IV
Seats, including pilot	16	10	5	3
Card weight capacity in pounds	5,000	2,500	1,200	600
Tank/bucket water or retardant capacity (in gallons)	700	300	100	75
Examples of helicopters	Boeing 234 S-64 and 61 AS-332 Bell 214	Bell 204, 205, 212, BK-117, S-58T	Bell 206B-III, Bell 206L-3, Lama, MD 500	Bell 47, Hiller 12E

As mentioned in Chapter 7, helicopters also carry helitack crew to the fireline. These handcrews are dropped off to construct handline near the head of the fire and are supported by water drops from the helicopter that transported them.

On many large fires a helicopter is dedicated for medical evacuation purposes. If a firefighter is injured he is immediately flown to the local hospital.

As shown in Table 10-3, helicopters are typed by the seating capacity, allowable payload at sea level at 59°F, and the water or retardant capacity.

Fire personnel that work around or are transported on helicopters must understand helicopter safety rules. Remember to always follow any directions given by helicopter personnel as they are responsible for the aircraft and its occupants. Always make sure you approach the aircraft in a crouching manner and in full view of the pilot. Make sure you approach or leave the aircraft on the downhill side and avoid the main rotor blade. While waiting for the helicopter to land or take off, remember to stay back and have your full protective gear on. Stay at least 50 feet away from a small helicopter and at least 100 feet away from a larger helicopter.

AERIAL RETARDANTS AND CHEMICALS DROPPED FROM AIRCRAFT

Suppressants

A suppressant is usually applied directly to the fire and usually at the base of the flames. It is used to suppress the flames not just to prevent their spread. A suppressant is a chemical mixture or formulation, but water also falls into this classification.

Retardants

Retardants are chemical mixtures generally dropped ahead of the flaming front to slow the rate of fire spread or reduce the intensity of a wildfire.

Short-term Retardant As the name implies, its effectiveness is of a short duration. Its effectiveness relies on the moisture in the formulation to reduce or inhibit combustion of the burning fuels. Once the moisture has evaporated, the retardant becomes ineffective.

Long-term Retardant Most fixed-wing air bases load long-term retardant into their aircraft. Long-term retardant has the ability to inhibit or reduce combustion even after the water has evaporated. These formulations contain chemical salts. Some of the more commonly used are ammonium sulfate, monammonium phosphate, diammonium phosphate, and ammonium polyphosphate.

Guidelines for Retardant Use

- Air tanker bases usually load only long-term retardant, however, some have the ability to load both short- and long-term retardant. Learn the options.
- Short-term retardant costs about one-third as much as long-term retardant.
- Short-term retardant can be effective in low intensity fires where the flame lengths are less than 4 feet. Use long-term retardant if flame length is greater than 4 feet.
- Ground forces must immediately back up short-term retardant applications. If this is not possible, then use long-term retardant.
- High intensity fire, with flame lengths greater than 8 feet, cause aerial retardant operations to be ineffective. If fire intensity levels are such, then find other fire targets. Look for topographic features, where the fire will back downslope and intensity levels will lessen, to lay down the retardant lines. At times, it may even be necessary to shut down the operation until fire conditions improve for retardant use.
- Retardant is very effective when used on control line slopovers (where a small portion of fire jumps the control line) or isolated spot fires.
- Retardant is usually ineffective on active crown fires. It is difficult to use enough retardant.
- If torching is not widespread, then retardant can be effective.
- Retardant is effective on isolated spot fires. Use of retardant on widespread spotting is ineffective due to the fireline intensities.

Class A Foam

Class A foam has two basic properties. First, it contains surfactants, which are chemical formulations that break down the surface tension of water and allow it to better penetrate ground fuels. Second, it contains liquid concentrates that form tiny air-filled bubbles. When dropped on ground fuels, it adheres to the fuels. The foam blanket then cuts off oxygen, prevents formation of combustible gases, and cools the flammable surface. Also as the bubble structure starts to break down and it returns to a liquid, it still retains its surfactant properties.

Fire-Blocking Gels

Fire-blocking gels are now entering the war against wildfires. They are polymer-based chemical formulations that form a gel coating when dropped or applied. These Fire Gels contain water-filled bubbles instead of the air-filled bubbles that we find in Class A foams. These water-filled bubbles allow the product to resist higher temperatures than Class A foams. The gel also seems to have better adhesion qualities. Ground fuels that are coated with fire gels resist fire for up to 24 hours. At the time of this writing, only helicopter application is available for dispensing gel products by air (Figure 10-5). See Chapter 11 for a discussion of ground applications of fire gels.

Figure 10-5 *Fire gel being applied with Baker Aerial Snow System.*

Summary

Aircraft are a very valuable resource on a wildland fire, however, they are very costly. Understand the capabilities and limitations of aircraft to maximize their use. In addition, understand the different types of payloads they are capable of delivering. With this understanding, better tactical decisions on aircraft use can be made.

Aircraft can be vital resources on a wildland fire, because they complement the ground attack resources. They are of greatest value during initial attack applications on fires in remote areas; ground resources have extended travel times during which the fire can grow larger.

Fixed-wing air tankers are classified by the retardant payload they carry. Most of the time, fixed-wings carry long-term retardant. Long-term retardant is dropped just ahead of the flaming front and slows the spread of the fire. Ground resources then move into the area and support the aircraft drop by starting suppression actions. The amount of retardant coverage, on the ground, is basically governed by the fuel type. Heavier fuels require greater coverage levels.

Helicopters are used on firelines usually for direct attack applications. They are classified by the amount of water or Class A foam they carry and the number of seats available on the aircraft. The advantage of helicopters is their short turnaround time; they do not have to return to a base, if water sources are close by. On larger fires, where both fixed-wing aircraft and helicopters are working on the fire, both types of aircraft are separated geographically and by altitude.

Again, aircraft resources are extremely costly; use them wisely. Learn the capabilities of both types of aircraft for effective management of these resources.

Review Questions

1. Describe the minimum amount of retardant carried on a Type III air tanker.
2. Why is it important to back up a retardant line with ground support?
3. Name the three basic drop patterns used by fixed-wing aircraft.
4. What is meant by the term *coverage level*?
5. What is the cut off time for fixed-wing aircraft?
6. What attack method is generally used by helicopters?
7. Do helicopters always have to return to a base to refill with water?
8. Can helicopters be used to fly logistical or support missions?

References

National Wildfire Coordinating Group, *S270 Basic Air Operations* (Boise, ID: National Interagency Fire Center, 1991).

National Wildfire Coordinating Group, *Aircraft Chief of Party/Flight Manager Workbook* (Boise, ID: National Interagency Fire Center,1998).

National Wildfire Coordinating Group, *S217 Interagency Helicopter Training Guide* (Boise, ID: National Interagency Fire Center, 1993).

Class A Foam and Fire-Blocking Gels

Learning Objectives

Upon completion of this chapter, you should be able to:

- Explain what Class A foam is and how it works.
- Identify the difference between Class A foam solution types.
- Describe how foam is generated.
- Differentiate between compressed air foam systems and low-energy foam systems.
- Describe the different types of nozzles used to apply Class A foam.
- Explain the tactical applications of Class A foam.
- Explain what a fire blocking gel is.
- Explain fire-blocking gel tactical applications.

INTRODUCTION

■ Note
Class A foam is a hydrocarbon-based surfactant.

Class A foam and fire-blocking gels are discussed in this chapter. Class A foam (Figure 11-1) has been around since the early 1980s and has shown itself to be a very effective tool for wildland fire suppression. Fire-blocking gels (Figure 11-2) are just recently being used at wildland fires and show great promise. They were used very effectively during the 1998 Florida fires in the wildland interface areas. Gel technology is relatively new and product enhancements are still taking place. Discussion in this chapter focuses on the application aspects of both of these suppression aids.

surfactant
any wetting agent; a formulation that, when added to water in proper amounts, materially reduces the surface tension of the water and increases penetration and spreading abilities of the water

CLASS A FOAM

Class A foam is a hydrocarbon-based **surfactant** or detergent intended for use on Class A or woody fuels. These surfactants tend to bond with the carbon elements in charred fuels. Class A foam is an agent that causes water to work more efficiently, which is what surfactants do. A surfactant reduces the surface tension of water and allows it to better penetrate the fuel. The reduction of surface tension caused by surfactants allows the water to form bubbles. Spreading capabilities of water are also enhanced by water draining from the foam.

■ Note
A surfactant reduces the surface tension of water and allows it to better penetrate the fuel.

Figure 11-1 *Class A foam. Photo courtesy National Interagency Fire Center.*

Figure 11-2 *Mixing Barricade Fire Gel for aerial delivery to the fire.*

Class A foam extinguishes a fire basically in five ways:

1. Absorbs heat (water held in the foam solution acts to absorb heat)
2. Separates oxygen from the fire and acts as a vapor suppressant (the foam blanket)
3. Isolates the fuel (the foam bubble structure insulates)
4. Acts as a reflective barrier (the white color reflects radiant heat)
5. Interrupts the chemical chain reaction.

■ Note
Class A foams are designed to drain water from the bubble structure in order to wet wildland fuels.

Class A foams are designed to drain water from the bubble structure in order to wet wildland fuels. The rate at which this happens is called *foam drain time*. Fuels are wetted by the free liquid; therefore, a regulated released rate from foam to liquid is necessary.

■ Note
Class A foam is documented as being three to five times more effective than plain water.

The environment is not harmed by the use of Class A foam at the recommended application rates. Class A foam is biodegradable.

Costs are relatively low for Class A foam because of the low mixing ratio. Class A foam should be used at the following mixing ratios:

- 0.2% for mop-up and overhaul operations
- 0.5% for direct attack applications on the fire
- 1% when applied to homes while doing exposure protection

expansion
the ratio of the volume of the foam in its aerated state to the original volume of the nonaerated foam solution

To give you a better idea of what that means, a 5-gallon container of Class A foam is used as an example. Five gallons of foam will treat the following amounts of water:

- at 0.2%—5 gallons will treat 2,500 gallons of water
- at 0.5%—5 gallons will treat 1,000 gallons of water
- at 1%—5 gallons will treat 500 gallons of water

■ Note
Low-expansion foams are wetter and are used for direct fireline applications.

■ Note
High-expansion foams have more air, are drier, and stick better to vertical surfaces.

Class A foam is documented as being three to five times more effective than plain water. This results in saving water supplies.

It is important to understand the expansion ratio of Class A foams. **Expansion** is the resultant increase in volume of a solution as air is introduced to it. It is a ratio of the volume of the foam in its aerated state to the original volume of the nonaerated foam solution. A ten to one expanded foam, it is said, expands 200 gallons of water to 2,000 gallons of water. The expansion ratio is divided into three classes, related to how much foam is generated. Low-expansion foams are wetter and are used for direct fireline applications, while high-expansion foams have more air, are drier, and stick better to vertical surfaces. They generally are used for structure protection. The three classes are as follows:

- **Low-expansion foam**—expands up to 20 times (1:1 to 20:1)
- **Medium-expansion foam**—expands up to 200 times (21:1 to 200:1)
- **High-expansion foam**—expands up to 1,000 times (201:1 to 1,000:1)

Figure 11-3 shows the four types of Class A foams that are produced.

low-expansion foam
foam with an expansion between 1:1 and 20:1

medium-expansion foam
foam with an expansion between 21:1 and 200:1

high-expansion foam
foam with an expansion between 201:1 and 1,000:1

How Foam Is Generated—The Hardware

Class A foam can be batch mixed in an engine's water tank, or pumped into the discharge side of the pump tank through a proportioner. Air can also be added to the foam after the proportioner to make a compressed air foam.

Low-energy systems (Figure 11-4) use the energy of the water pump only to educt air into the class A foam solution. A good example is an aspirating foam nozzle that introduces air into the foam solution.

A high-energy system (Figure 11-5) is a foam-generating device that combines the energy of air to the energy already produced by the water pump in the low-energy system. A good example of this is a compressed air foam system (CAFS).

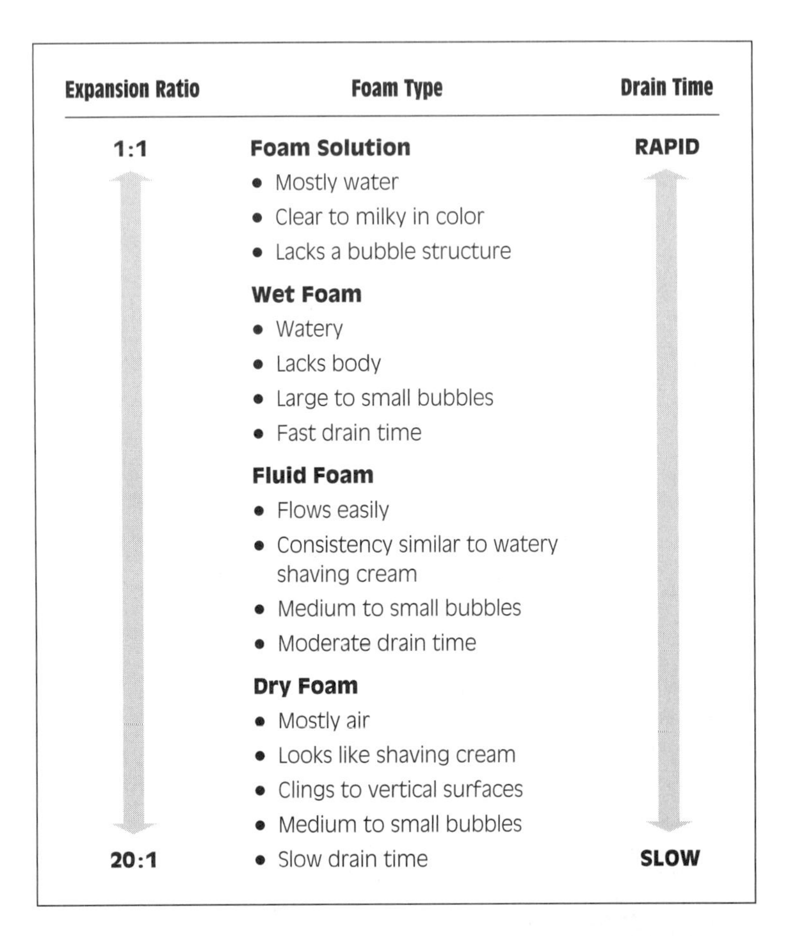

Figure 11-3 *Types of Class A foams.*

■ Note

Low-energy systems use the energy of the water pump only to educt air into the class A foam solution.

Figure 11-4 *A low-energy system with a medium-expansion nozzle. Photo courtesy National Interagency Fire Center.*

■ Note
A high-energy system is a foam-generating device that combines the energy of air to the energy already produced by the water pump in the low-energy system.

Figure 11-5 *A high-energy system. Compressed air is added to the foam solution. Photo courtesy National Interagency Fire Center.*

batch mixing
manually adding and mixing a concentrated chemical into solution in a tank or container

Batch mixing, directly dumping foam solution into an engine's tank, can be done but is not a preferred method of application. The advantage to batch mixing is that no hardware is required so the costs are minimal. The disadvantages outweigh the cost savings. Some of the disadvantages are as follows:

- Possible corrosion to the tank
- Foam forming in the tank
- Excessive foam concentrate use
- Cleaning of the lubricants in the pumping system
- Excess foam in the tanks, causing the pump to cavitate
- Water refill difficulties as a result of foam bubbles in the water tank

proportioner
device that adds a predetermined amount of foam concentrate to water to form foam solution

Proportioners add foam concentrate to the water supply, thus creating finished foam solution. Foam concentrate is generally added to the water supply on the discharge side of the pump. There are two basic types of proportioning systems: manually regulated systems and automatic regulating systems. Automatic regulating systems that inject foam concentrate directly into the discharge side of the water pump give the most desirable results.

A foam concentrate proportioner should ideally have the following characteristics:

- Be proportional over the entire operating pressure and flow capacities of the water pump.
- Be unaffected by changes in engine pressure, changes in hose length and size, changes in water flow, changes in nozzle adjustment, size, or elevation.
- Be suitable for use with CAFS.
- Ingest the foam concentrate directly into the discharge side of the pump's water stream. It should be proportioned correctly to achieve the right ratio to foam concentrate to water. The foam solution should then flow directly through the engine's piping system into the hose lines. There should be no chance of the foam solution recirculating back into the tank, pump, or engine piping system.
- Be highly reliable, simple in design, and easy to repair.
- Be able to proportion either Class A or Class B foams and proportion down to 0.1% or less.
- Have an indicator showing how much foam concentrate is left in the system for use.
- Have enough foam concentrate to treat a full tank of water on the engine.
- Not cause a water pressure loss as water flows through the proportioning unit.
- Be able to operate without any external power source. Should the proportioner require the use of the engine's 12-volt electrical system to power it, then it should draw less than 30 amperes.
- Have corrosion-resistant storage tanks and be compatible with foam.
- Be easy to routinely flush.

Manually regulated proportioning systems require manual adjustment to the mix ratio when there is a change in pressure or flow through the proportioner. There are five different types of manually regulated proportioners, as follows:

- Suction-side proportioning systems
- In-line eductor proportioning systems
- Bypass eductor proportioning systems
- Around-the-pump proportioning systems
- Direct injection, manually regulated proportioning systems

Automatic regulating systems automatically adjust the flow of foam concentrate into the water stream, which automatically maintains the proper ratio of concentrate to water. There are three types of automatic regulating systems, as follows:

- Balanced pressure, venturi proportioning systems
  - Pump systems
  - Pressure tank systems
- Water-motor meter proportioning systems
- Direct ingestion, automatic regulating proportioning systems

With an automatic regulating system, once the mix ratio is set and the proportioner is operated, there should be no further need to adjust the proportioning unit.

COMPRESSED AIR FOAM SYSTEMS

■ Note
The compressed air foam system (CAFS) originated in the Texas Forest Service.

The compressed air foam system (CAFS) originated in the Texas Forest Service. In 1972 the Texas Forest Service used pine soap as the foaming agent and thus came about the phrase, "the Texas snow job."

■ Note
In a CAF system, air or gas is injected into the water stream after the proportioner.

In a CAFS, air or gas is injected into the water stream after the proportioner (Figure 11-6). It is a high-energy foam system because pneumatic power (compressed air) is being added to the water stream.

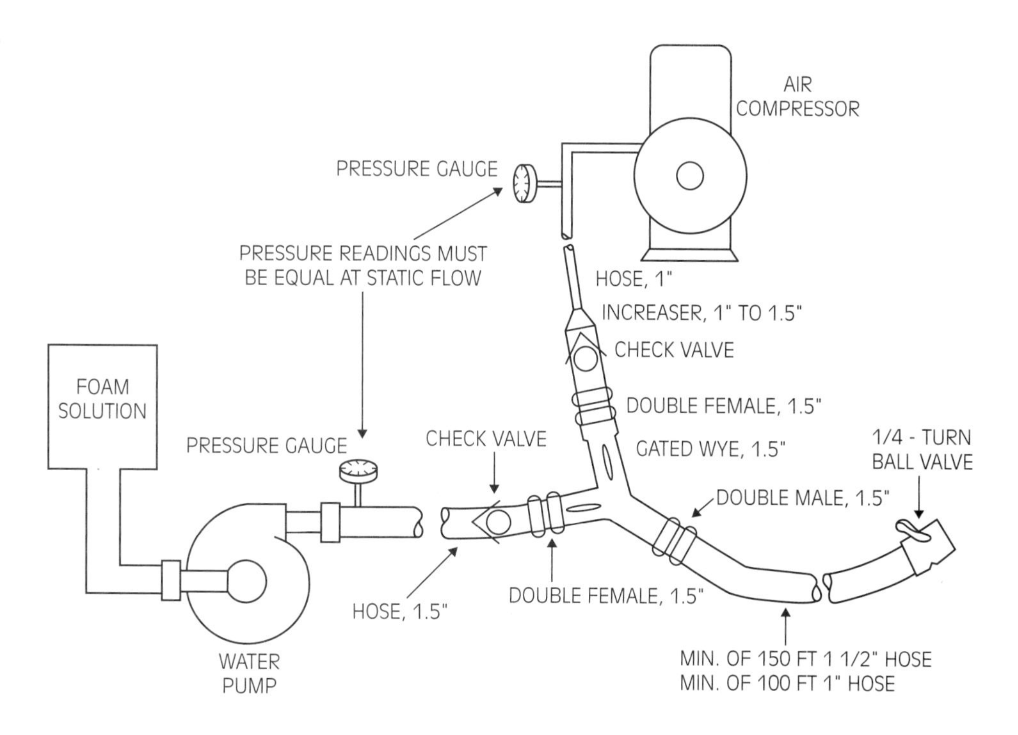

Figure 11-6 *A compressed air foam system.*

■ Note
After combining the foam concentrate and air, foam solution flows through the hose line, where it is agitated by the inside of the hose lining.

scrubbing
agitating foam solution and air within a confined space (usually a hose) that produces tiny, uniform bubbles

static pressure
water pressure head available at a specific location when no water is being used so that no friction loss is being encountered

slug flow
discharge of distinct pockets of water and air due to the insufficient mixing of foam concentrate, water, and air in a compressed system

centrifugal water pump
pump that expels water by centrifugal force through the ports of a circular impeller rotating at high speed

After combining the foam concentrate and air, foam solution flows through the hose line where it is agitated by the inside of the hose lining. This action is called **scrubbing.** Scrubbing produces tiny uniform bubbles within the foam solution. It takes approximately 100 feet of 1½-inch hose to effectively scrub the foam solution. Scrubbing can also be done with the use of a device called a *motionless mixer.*

It is important in a CAFS for the **static pressures** of the air and foam concentrate to be the same. It is also important to remember that when there is not enough foam solution to mix with the air, inadequate mixing occurs. This situation is called **slug flow** because pockets of water (or plugs) and air move through the nozzle, causing a hammer effect as these plugs pass through the hose and exit the nozzle. To remedy this problem, return to the use of straight water until you can correct the foam solution to air ratio.

There are three general CAFS configurations found on engines. They are as follows:

- The fire truck engine drives a **centrifugal pump** and an air compressor through load sense hydraulic drive systems.
- A single auxiliary engine mechanically drives both the centrifugal pump and the air compressor.
- The fire truck engine drives a centrifugal pump and an air compressor through a mechanical drive system.

A CAFS usually has the following components:

- Centrifugal pump
- Air compressor
- Foam concentrate proportioner system
- Drive or power system
- Control and instrument systems

There are seven main advantages to a CAFS:

1. Foam can be projected greater distances.
2. Foam can be pumped great distances in hose lines, as far as 5,000 feet.
3. They produce less head pressure when pumping up a slope.
4. They create drier foam, which is advantageous for structure protection.
5. They create very light hose lays—half the hose is filled with air. A 1½-inch hose line will carry 50 gallons per minute of foam solution and 50 cubic feet per minute of air.
6. Less foam concentrate is used.
7. Foam with smaller bubbles is more uniform and lasts longer.

Some of the drawbacks of a CAFS are as follows:

- A CAFS is very expensive.

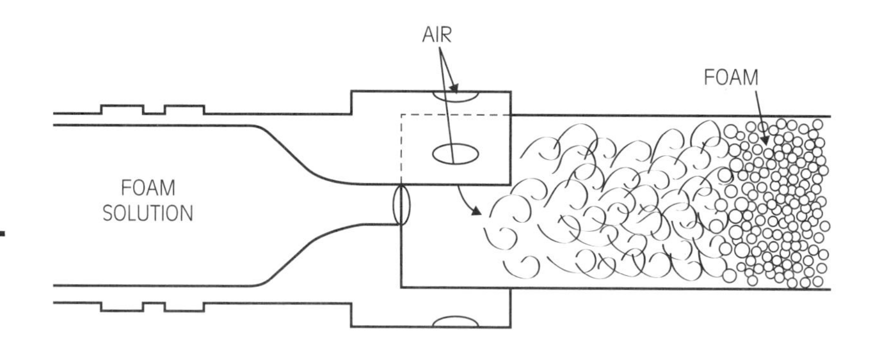

Figure 11-7 *How an aspirating foam nozzle works.*

Note
The nozzle draws air into the water stream, usually by venturi action, to create and mix the solution.

You must wear your personal protective equipment when working with Class A foam.

Class A foam concentrate spilled on the ground can be slippery, so watch your footing if a spill occurs.

Note
Do not mix Class A and Class B foams together.

- It requires more extensive training.
- The complexity of the system affects the reliability.

Aspirating Foam Nozzles

Proportioning systems add foam solution to the water supply to create a foam solution. To produce foam, air has to be introduced. In a low-energy system, this is accomplished by an aspirating foam nozzle (Figure 11-7). The nozzle draws air into the water stream, usually by venturi action, to create and mix the solution. The expansion chamber, in the nozzle, strengthens the bubbles before they are discharged. Aspirating nozzles are the least expensive and most reliable way to produce foam.

Generally speaking, aspirating nozzles that have long reaches only produce wet, frothy foam. The foam expansion ratio will be less than 10:1.

SAFETY PRACTICES USING CLASS A FOAM

Class A foam is like a strong household detergent. You must wear your personal protective equipment including safety goggles when working with Class A foam, because foam solution irritates the eyes. Class A foam concentrate spilled on the ground can be slippery, so watch your footing if a spill occurs.

Class A foam is environmentally safe, when used at the manufacturer's recommended mixing ratios, which are usually only up to 1% foam solution.

Foam concentrate leaks near a waterway or storm drain may need to be diked to prevent the product from mixing with the water and causing nuisance foaming. Keep foam solutions out of ponds, lakes, and other water sources. It can inhibit oxygen exchange and eventually kill the fish.

Do not mix Class A and Class B foams together. They are not compatible and combining both types of foam will cause the solutions to gel. This can result in damage to the equipment.

If foam solution is ingested, seek medical attention as soon as possible.

If foam solution is ingested, seek medical attention as soon as possible. Large oral doses could produce narcosis.

TACTICAL APPLICATION

Class A foam has been implemented and its tactical applications thoroughly tested in fire departments. Class A foam is three to five more times effective than water. (A southern California department performed an initial attack on a small grass fire with two Type III engines.) One engine took the right flank of the fire, with class A foam, and the other used plain water in its attack. The one using plain water ran out of water while the engine equipped with a Class A foam proportioner was able to complete the attack on its flank of the fire. In addition, the engine using Class A foam was able to extinguish the head of the fire and eventually complete the knockdown of the fire on the other engine's flank of the fire.

Basically there are five tactical applications for Class A foam: direct attack applications on the fire edge, indirect applications, use in mop-up operations, barrier protection, and structure protective applications.

Note
Foam solution (water and foam concentrate mixed) should be applied to the base of the flames.

Direct Attack Applications

Consult your manufacturer's recommendations for mix ratio of foam concentrate with water.

Foam solution (water and foam concentrate mixed) should be applied to the base of the flames. Also, while working the fire's edge, some of the foam solution can be directed toward the unburned fuels on the fire's edge.

Note
Leave a foam blanket where there are hotter fuels that will take longer to extinguish.

Where there are hotter fuels that will take longer to extinguish, leave a foam blanket. This will help smother the fuel and continue to wet it. The concentrate ratio amount may have to be increased by the engine operator. Creating a foam blanket requires a 1% concentrate level and an aspirating nozzle. A dual function nozzle on a direct attack hoseline works as a regular nozzle when using water solely as a wetting agent. If foam is desired, a slide can be moved forward thus exposing small teeth that cause the nozzle to then work as an aspirating nozzle. A wet foam solution is then produced, provided the concentrate level is adjusted.

Note
What makes Class A foam so effective is its ability to wet and cool fuels long after its initial application.

Mobile pumping from engines works well with Class A foam. The key is not to move too rapidly, risking not extinguishing the fire completely.

What makes Class A foam so effective is its ability to wet and cool fuels long after its initial application. This allows the firefighter to move to a new area, because the foam continues to work where it was initially applied.

Note
Class A foam can be used as a wet line from which to start a firing operation.

Indirect Attack Applications

Class A foam can be used as a wet line from which to start a firing operation (Figure 11-8). A 40 gallons per minute medium-expansion nozzle can be used for this purpose. It produces a fluid foam solution that is wet enough to penetrate into the

Figure 11-8 *Type IV engine putting down a wet line from which a firing operation will be started. Photo courtesy Duncan Todd, North Tree Fire.*

ground and surface fuels. This is important, as foams that are too dry tend to hang on surface fuels and allow the fire to creep under the foam line.

The wet line is put down on the ground by an engine crew mobile pumping Class A foam through a medium-expansion nozzle. The foam line is generally 2½ times as wide as the flame lengths. Be sure to coat all sides of the fuel and also make sure that it has penetrated the ground and surface fuels. The igniters involved in the firing operation follow along after the foam line is constructed. They fire from this line. In areas where it is not suitable to mobile pump, hose lines can be extended for this purpose. Also consider ladder fuels, as it may be necessary to supply foam to low hanging limbs, brush, tree trunks, or canopy to help provide an insulting barrier.

Mop-up Operation

■ Note
Class A foam penetrates fuels well and puts out deep-seated fires.

Class A foam works well during mop-up operations. It penetrates fuels well and puts out deep-seated fires. Wet or fluid foam can be applied to heavier fuels. The blanket smothers the fire, holds water in solution longer so as to release it slower, and helps hold down residual smoke levels. Where there are pockets of deep-seated fire, such as duff or litter, cover the area with a foam blanket. Class A foam also works well for deep-seated fires in log decks (stacked logs ready for the lumber mill) and stumps.

In mop-up operations, start mopping up along the edge of the burn and work in, concentrating on hot spots. As a general rule, mop up 100 feet in from the edge of the burn. At times of expected increase in wind activity or if close to an interface area, 100% mop-up of the area can be done. Follow your departmental policy and closely examine the fireline situation that exists.

Note
Residual pockets of deep-seated fire can be seen by steam releasing through the foam blanket.

Residual pockets of deep-seated fire can be seen by steam releasing through the foam blanket. Give these areas more attention.

Consider the use of medium expansion nozzles where there are wide areas of smoldering surface heat. They will get the job done quickly and provide a foam blanket to smother fire, eliminate smoke, and cause a slow release of water into the fuel.

Barrier Protection

Note
Class A foam works well for barrier protection because it insulates the fuel from the fire.

Class A foam works well for barrier protection, because it insulates the fuel from the fire. In addition, it is white and reflects radiant heat. It can be applied to brush and grass, timber stands, low hanging limbs to interrupt spread of fire to crown fuels, log decks, and used to protect environmentally sensitive areas.

The length of time the barrier lasts depends on the fuel loading, moisture content, wind factor, relative humidity, and air temperature. Fluid foams provide for good blanketing and have good wetting properties. Compressed air foams remain longer than foam produced by aspiration nozzles. They produce a strong blanket and are very slow to wet the fuel.

Note
Application of foam should take place just prior to ignition, as it is a short-term treatment option.

Application of foam should take place just prior to ignition, as it is a short-term treatment option. Always evaluate the durability of the foam you are using. Wet foam provides a wet blanket that allows for rapid wetting. The opposite extreme is dry foam, such as compressed air foam, which insulates well because of the strong foam blanket produced and has a very slow drain time.

The width of the foam line will be dependent on the fuel model present and the fire behavior indicators observed. Always remember to apply foam to all sides of the fuel if possible.

Structure Protection

Note
Fluid foams and dry foams are best used in structure protection.

Fluid foams and dry foams are best used in structure protection. They adhere well to vertical surfaces and the underside of deck and roof projections on buildings.

Again, apply Class A foam just prior to the arrival of the flaming front. In hot weather, generally speaking, foam produced by aspirating nozzles lasts about 30 minutes. CAFS foam will last approximately 1 hour in hot weather.

Apply foam to all roof surfaces, outside walls, deck and roof eave projections, and supporting columns of the structure. LPG tanks will also benefit from the application of Class A foam. In addition, survey any outbuildings and, if time permits, use foam to protect them.

FIRE-BLOCKING GEL

Note
One of the main differences between Class A foams and Fire-Blocking Gels is that the bubbles are water filled in Fire Gel solutions.

A class of super absorbent polymers, Fire-Blocking Gels were successfully used in 1998 at the Slave Lake fire in Canada and in Flagler and Putnam counties in Florida.

One of the main differences between Class A foams and Fire-Blocking Gels is that the bubbles are water filled in fire gel solutions. Class A foam has air-filled bubbles that burst at a much lower temperature when heat is applied. These water-filled bubbles, found in gels, stack on each other and provide an excellent barrier against radiant heat. Fire Gel also has an extremely long evaporation time, which allows it to be applied long before the arrival of the flaming front. As a layer of gel starts to burn off, more water-filled bubbles slide down to cover the exposed area.

■ Note

Better atomization of the water stream takes place as a result of the addition of the polymer to the hose stream.

Better atomization of the water stream takes place as a result of the addition of the polymer to the hose stream. This exposes greater wetting surface areas, which in turn provide for a greater cooling effect. Thinner gel solutions are therefore used when greater extinguishment properties are desired. More gel concentrate is added when a thicker gel is desired for insulating value. Gels can be applied as thick as 1 inch to exposed vertical surfaces.

There are two different types of fire-blocking gel products. One is batch mixed and the other is applied at the end of a hose stream with a backpack application (Figure 11-9). The friction loss in a gel-filled hose line is many times that of water; therefore, backpack applications seem to provide greater promise. Batch mixing of product is more economical but results in limitations when longer hose lays are desired. Backpack applications seem to provide better insulating values.

Figure 11-9
Barricade backpack applicator.

Tactical Applications

Fire-Blocking Gels will inevitably change the way structure protection is performed. They can be applied up to 24 hours before the flaming front arrives and still have good insulating values.

■ Note
Firefighters now have the opportunity to coat a structure with gel solution and then retreat to a safe zone.

Firefighters now have the opportunity to coat a structure with gel solution and then retreat to a safe zone as a high-intensity fire moves through the area of structures. After the flaming front passes, engine companies can be redeployed to put out any residual fire left in the area. There have been tests on both exposed vegetative materials and the sides of structures where a 3,500° propane torch was applied to the gel solution for 10 minutes or longer. The plant materials did not burst into flames and the side of the structure did not burn. Certainly, these temperatures are much hotter than those produced in a wildland fire.

In situations where a lower intensity fire is running through the area, gel solution can be applied to the structure, and firefighters can then turn their attention to fighting the fire.

Fire Blocking Gels[1] can also be used to provide barrier protection, as a wet line in firing operations, and in mop up operations.

This technology was only tested in research situations until its use in the recent 1998 Florida and Canadian fires. It was used with great success at both fires. One product, Barricade, prevented extensive losses to log decks in the Canadian fires and successfully protected large groupings of homes in the Florida fires; exactly what test research had shown it would do. It has also been used at prescription burns in both northern California and Florida.

Part of the reason for the burn in northern California was to test the effectiveness of Barricade. The top control line on a midslope road burn was only about 8 feet wide. The slope was steep and the fuels were a model 12 (slash model) on the burn site. The aspect was receiving maximum solar radiation because of the time of day and the wind was in alignment with the slope.

The top edge of the control line was pretreated with Barricade by helicopter application using the Baker Aerial Snow System. The burning operation started and the fire moved uphill as expected and bumped the upper control line with great intensity.

Because of the intensity levels, holding crews had to be moved to safety zones. The fire did not jump the control line or spot into the unburned area. The product possesses very good fire-stopping power.

Fire-Blocking Gels work. Consider adding them to your wildland toolbox and pull them out when they are needed. They are biodegradable, nontoxic, and environmentally safe.

---

[1] The Fire-Blocking Gel technology is still in its infancy. Contact John Bartlett of Barricade at 1-800-201-3927 for the latest information on the use and application of these products.

Summary

Class A foam is an inexpensive tool for use on any wildland fire. Used at 0.5%, it costs about 6 cents per gallon.

Both low- and high-energy systems are used to disburse foam solutions. Aspirating nozzles are used in low-energy systems to make foam solution with more body and durability. CAFS produces dry foam that adheres well to the sides of structures.

Class A foam is nontoxic when used at its recommended concentrations. It absorbs heat, acts as a blanket to separate oxygen from the fuel, suppresses smoke, has insulating properties, reflects radiant heat, and interrupts the chemical chain reaction of fire.

Fire-blocking gels provide the best opportunities for favorable outcomes when doing structure protection. These products adhere better to vertical surfaces than Class A foams and protect the structure from heat exposure for dramatically longer times than Class A foam.

They will change the way interface fires are fought.

Review Questions

1. What is a surfactant?
2. What is drain time as it applies to Class A foam?
3. Describe what medium expansion foam is.
4. What is a high-energy foam system?
5. Can Class A and Class B foam solutions be mixed in a proportioner?
6. What does the term *slug flow* mean?
7. What is Fire-Blocking Gel?
8. Do fire-blocking gels experience drain time problems?

References

Franklin, Scott E., "Foam Applications for Prescribed Fire," *American Fire Journal,* February 1990.

National Wildfire Coordinating Group, *Foam vs. Fire-Class A Foam for Wildland Fires* (Boise, ID: National Interagency Fire Center, 1992).

National Wildfire Coordinating Group, *Proportioners for Use on Wildland Fire Applications* (Boise, ID: National Interagency Fire Center, 1992).

National Wildfire Coordinating Group, *Compressed Air Foam Systems For Use in Wildland Fire Applications* (Boise, ID: National Interagency Fire Center, 1992).

National Wildfire Coordinating Group, *Foam Applications for Wildland and Urban Fire Management,* Vol. 1, No. 1 (Boise, ID: National Interagency Fire Center, 1998).

National Wildfire Coordinating Group, *Foam Applications for Wildland and Urban Fire Management,* Vol. 1, No. 3 (Boise, ID: National Interagency Fire Center, 1988).

National Wildfire Coordinating Group, *Foam Applications for Wildland and Urban Fire Management,* Vol. 2, No. 1 (Boise, ID: National Interagency Fire Center, 1989).

National Wildfire Coordinating Group, *Foam Applications for Wildland and Urban Fire Management,* Vol. 2, No. 2 (Boise, ID: National Interagency Fire Center, 1989).

National Wildfire Coordinating Group, *Foam Applications for Wildland and Urban Fire Management,* Vol. 3, No. 1 (Boise, ID: National Interagency Fire Center, 1990).

National Wildfire Coordinating Group, *Foam Applications for Wildland and Urban Fire Management,* Vol. 4, No. 1 (Boise, ID: National Interagency Fire Center, 1991).

Use of the Global Positioning System and Maps

Learning Objectives

Upon completion of this chapter, you should be able to:

- Explain what the Global Positioning System (GPS) is.
- Describe any limitations of the GPS.
- Identify the GPS reference lines on a United States Geological Survey (USGS) topographic map.
- Describe how to read a USGS topographic map.
- Explain the Universal Transverse Mercator Grid System used in GPS.
- Describe latitude and longitude.
- Use a coordinate ruler to find your position on a map.

INTRODUCTION

This chapter discusses the use of the Global Positioning System (GPS) in fire department operations. It also covers topographic map reading and the use of GPS with the topographic map.

THE GLOBAL POSITIONING SYSTEM

Global Positioning System (GPS)
system of navigational satellites operated by the U.S. Department of Defense that can track objects anywhere in the world with an accuracy of approximately 40 feet

The NAVSTAR GPS is an acronym that stands for Navigation Systems with Time and Ranging Global Positioning System. The **Global Positioning System (GPS)** is funded and controlled by the United States Department of Defense. It is operated by the United States Air Force. Stanford University Professor Bradford Parkinson, now retired from the U.S. Air Force, is considered the father of the GPS. The GPS was declared fully operational in January of 1994.

■ Note
The space segment of GPS involves 24 satellites that orbit the Earth every 12 hours at a height of about 12,000 miles above the Earth.

The space segment of GPS involves 24 satellites that orbit the Earth every 12 hours at a height of about 12,000 miles above the Earth. At times there may be more than 24 operational satellites, as new ones are launched when older satellites need replacing. These satellites transmit microwave carrier signals back to Earth, which enables the user, with a receiver, on the ground to plot his or her location and obtain the current time, date, and elevation. It takes three satellites over a receiver to obtain a horizontal location on the Earth's surface. Adding a fourth satellite enables you to determine your altitude as well.

■ Note
It takes three satellites over the receiver to obtain a horizontal location on the Earth's surface. Adding the fourth enables you to determine your altitude.

The heart of the GPS master control system is located at Schriever Air Force Base in Colorado. At the master control facility, the satellites are constantly monitored. Corrections are made from the control facility to the satellite clock and orbital data is also uploaded.

The user segment of the system is made up of GPS receivers that are hand carried or installed in aircraft, aboard ships, or on ground vehicles.

Accuracy

The GPS was originally developed for military use and was successfully used by U.S. forces in the Gulf War to guide carrier missiles to specified targets deep inside Iraq. It was also used to give aircraft and planes their exact location.

The U.S. military, its allies, and certain U.S. government agencies, have access to the most accurate positioning available from the system. The U.S. government also approves selected civilian users to have the same access.

meter
basic unit of length in the metric system; 1 meter is equal to 39.37 inches

The most accurate positioning system is called the Precise Positioning Service. Authorized users of the service have access to accuracy rates of 22 **meters** horizontal accuracy, 27.7 meters vertical accuracy, and 100 nanoseconds time accuracy. A nanosecond is one billionth of a second.

Civilian users have access to the Standard Positioning Service. This service is not as accurate. A comprehensive national policy on the management of GPS

Note
Civilian users can expect accuracy to 100 meters horizontally, 156 meters vertically, and 340 nanoseconds of time accuracy.

was enacted on March 29, 1996. Part of that policy allowed the Department of Defense, in the name of national security, to introduce deliberate inaccuracies into positioning information being transmitted by the GPS satellite. This is done by the use of selective availability, which degrades satellite signals by a time-varying bias. Civilian users can expect accuracy to 100 meters horizontally, 156 meters vertically, and 340 nanoseconds of time accuracy.

GEOGRAPHIC COORDINATES FOUND ON A TOPOGRAPHIC MAP

latitude
angular distance, measured in degrees, creating imaginary lines circling the Earth's globe. The lines extend in an easterly and westerly direction, parallel to the equator

longitude
angular distance, measured in degrees, creating imaginary lines extending from the north pole to the south pole, which identify geographical positions on the Earth's globe

Universal Transverse Mercator
a rectangular coordinate system of determining location on the Earth's surface, similar to latitude and longitude, but is defined in meters rather than degrees, minutes, and seconds

prime meridian
an imaginary line on the ground running north and south that is accurately laid out to serve as the reference meridian in land survey

In order to be able to locate ourselves on the Earth's surface, a coordinate system with positional values was developed. Reference lines were projected onto maps that represented the Earth's surface. These lines are called parallels of **latitude** and meridians of **longitude.** We discuss two of these systems: latitude and longitude and the **Universal Transverse Mercator** (UTM) Grid system.

A network of reference lines is formed around the Earth. Horizontal lines that run east and west around the Earth are called parallels. Vertical lines that run north and south are called meridians. These lines converge at the poles. With these reference lines in place, we have a coordinate system that can be used to locate a position on the Earth's surface.

Latitude is a point north or south of the equator, the equator being 0° latitude. To find a location's latitude, you simply count the values of the lines either north or south of the equator. On a topographic map, the degrees, minutes, and seconds of latitude can be found in the right and left margins (Figure 12-1). These horizontal values represent latitude.

Longitude is a point on the Earth's surface that is east or west of what we call the prime meridian. The prime meridian, defined as 0° longitude, runs through Greenwich, England. To find a longitudinal point of the Earth's surface, simply count the values of the lines either east or west of the **prime meridian.** On a topographic map, these values can be found in the top and bottom margins (Figure 12-2). With this system of reference lines in place, you can now see how easy it is to locate a position on the map.

Before learning how to locate your position on a topographic map using longitude and latitude, two things must be defined. They are the smaller unit measurements used with latitude and longitude, and what a topographic map is and a description of its features.

Each degree of latitude and longitude is broken down into 60 minutes and each minute is broken down into 60 seconds. It must be emphasized that these measurements have nothing to do with time; they are angular measurements. One minute of latitude is equal to 1.15 miles. One minute of longitude at the equator is also equal to 1.15 miles. One second is equal to 33.82 yards for both latitude and longitude. These figures change as the lines of longitude converge on the poles. A GPS handheld receiver is capable of reporting degrees, minutes, and seconds, if it is set on that feature.

■ Note
Horizontal lines that run east and west around the Earth are called *parallels*.

■ Note
Vertical lines that run north and south are called *meridians*.

■ Note
Latitude is a point north or south of the equator, the equator being 0° latitude.

■ Note
Longitude is a point on the Earth's surface that is east or west of what we call the prime meridian.

■ Note
These measurements have nothing to do with time; they are angular measurements.

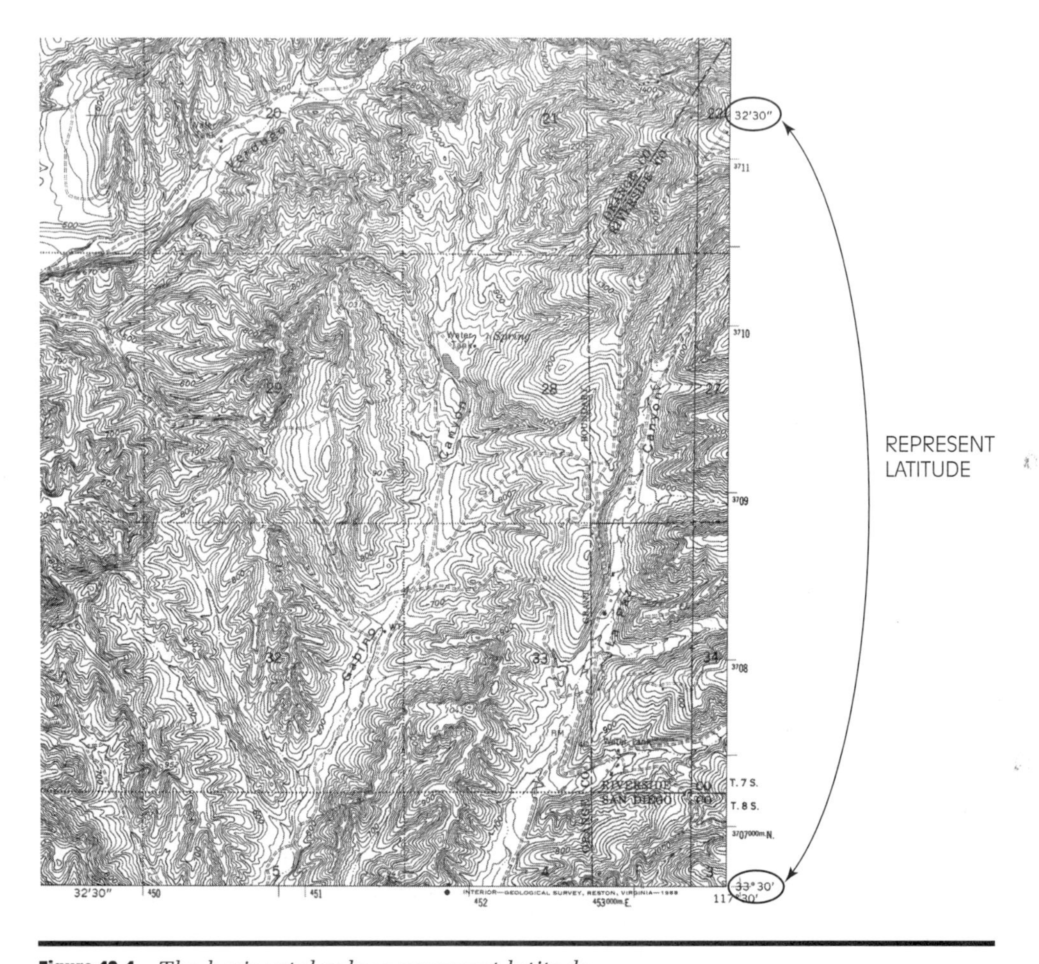

Figure 12-1 *The horizontal values represent latitude.*

UNDERSTANDING TOPOGRAPHIC MAPS

■ Note
Topographic maps are a relief of the Earth's surface.

Topographic maps are a relief of the Earth's surface. They show both the horizontal and vertical positions of the Earth's features. Topographic maps are also called *quadrangle maps*.

There are basically two different types of topographic maps: contour maps or shaded relief maps. The one that is used most commonly on wildland fires is the contour relief map.

A contour relief map is a common method of representing the shape and elevation of the land. A contour line is a line of equal elevation on the ground and

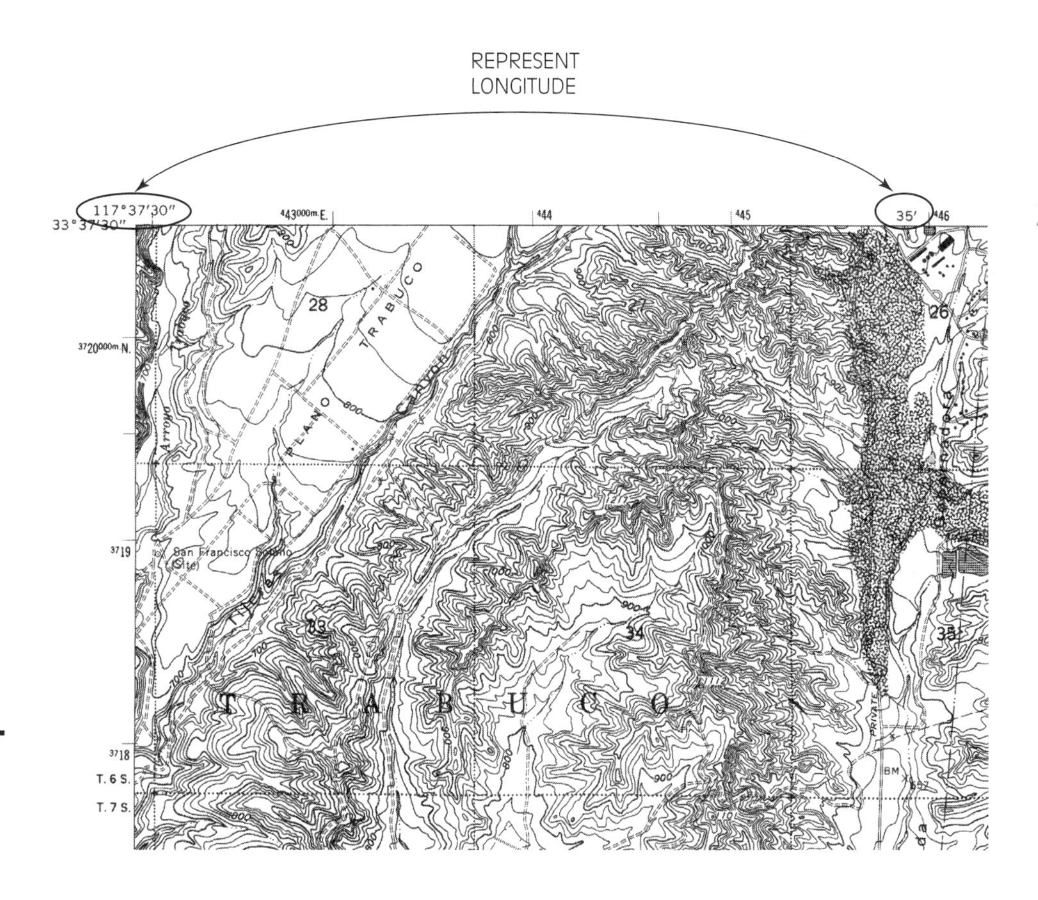

Figure 12-2 *Longitude values are represented by vertical lines.*

■ Note

A contour map is a common method of representing shape and elevation of the land.

delineates the same elevation above sea level. It is referenced from the average sea level of the closest ocean.

A common map used on a wildland fire is a 7.5 minutes (refers to minutes in latitude or longitude), 1:24,000 scale topographic map (Figure 12-3). It is sometimes called a *quad map*.

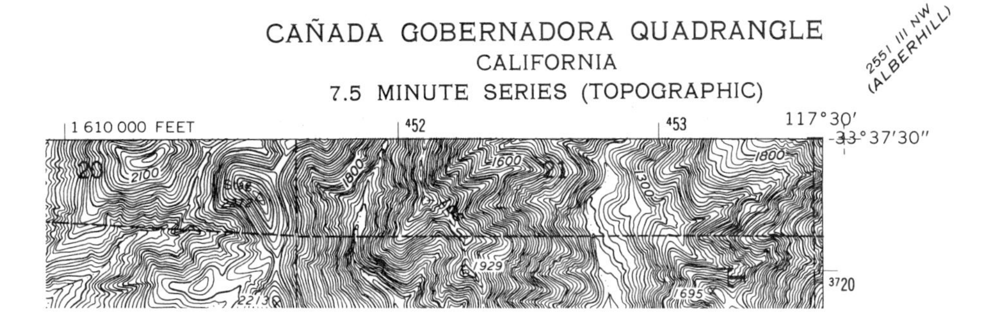

Figure 12-3 *This notation is usually found in the upper right hand corner of a USGS 7.5 minute series map.*

map's legend
a key accompanying a map that shows information needed to interpret that map

The 1:24,000 is the map's representative fraction. One inch on the map equals 24,000 inches on the ground. Another way to think of this is that 1 inch on the map equals 2,000 feet on the ground. The representative fraction, for a map, is found in the bottom margin area and is part of the **map's legend** (Figure 12-4).

A topographic map is a representation of the Earth's surface with mountains, man-made features, water features, and so forth. It is represented on a flat piece of paper, therefore, there needs to be a way to show these terrain features. This is done by the use of contour lines (Figure 12-5). Again, a contour line is a line of equal elevation on the ground and delineates the same elevation above sea level. Any point on that contour line is the same elevation.

■ **Note**

To find the distances between contour lines, look in the map's legend.

To find the distances between contour lines, you can look in the map's legend (Figure 12-6). The contour interval is found there.

There are several different types of contour lines. To make it easier to read a topographic map, every fourth or fifth line is printed darker. These lines are called *index contours* and somewhere along an index contour can be found an elevation (Figure 12-7). The lighter contour lines in between the index contours are called *intermediate contours* (Figure 12-8).

If the contour interval is unknown, then find two index contours adjacent to each other. Read the elevations and subtract one from the other. Then divide the difference by 4, if the number of spaces between index contours is 4. Divide by 5, if the number of spaces is 5. This gives the unknown contour interval. You may be required to figure the contour interval on the map displayed in an incident action plan.

A depression contour line (Figure 12-9) represents areas either man-made or natural that are lower than the immediate surrounding terrain. A good example would be a gravel pit (Figure 12-10) or a natural depression. Depressions also contain no outlets for water drainage. Note the profile in Figure 12-11.

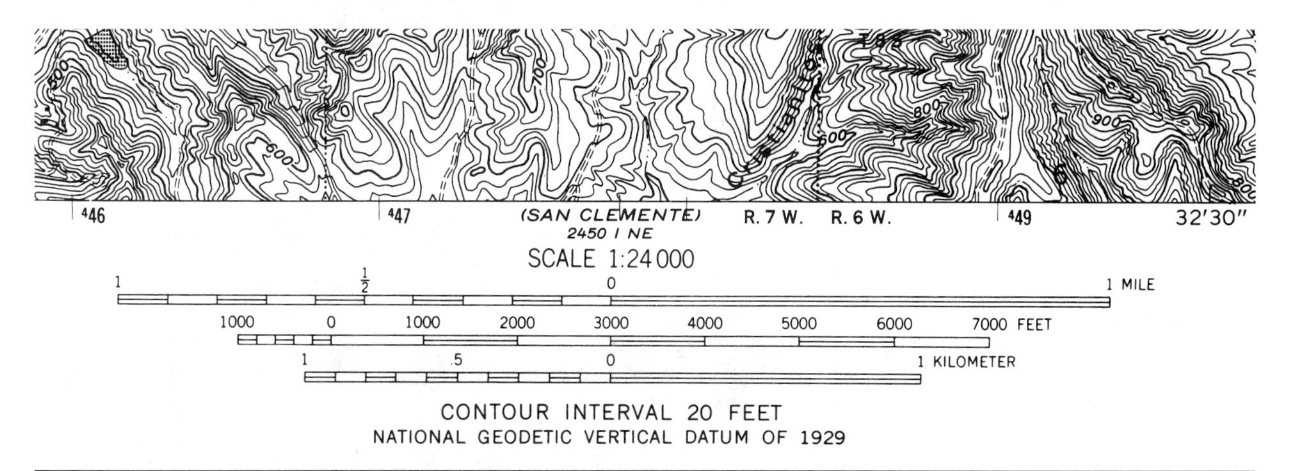

Figure 12-4 *The map's representative fraction is found in the bottom margin of the map.*

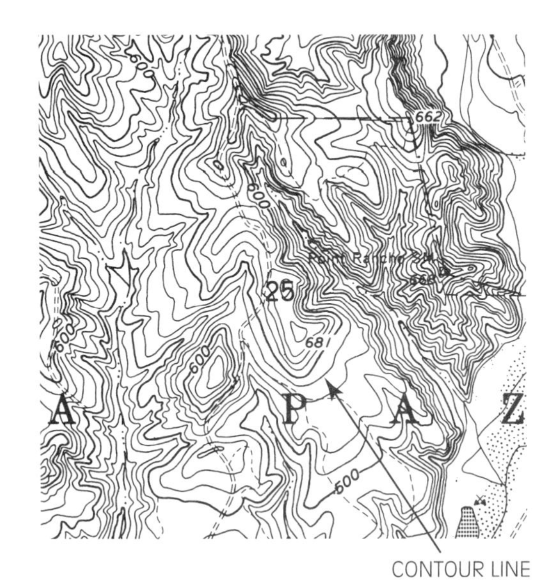

Figure 12-5 *A contour line is a line of equal elevation on the ground and delineates the same elevation above sea level.*

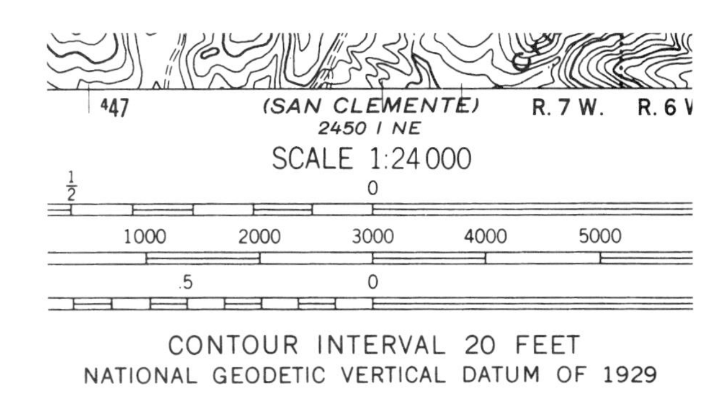

Figure 12-6 *The contour interval is found in the lower margin.*

Figure 12-7 *The darker contour lines with an elevation marking are index contour lines.*

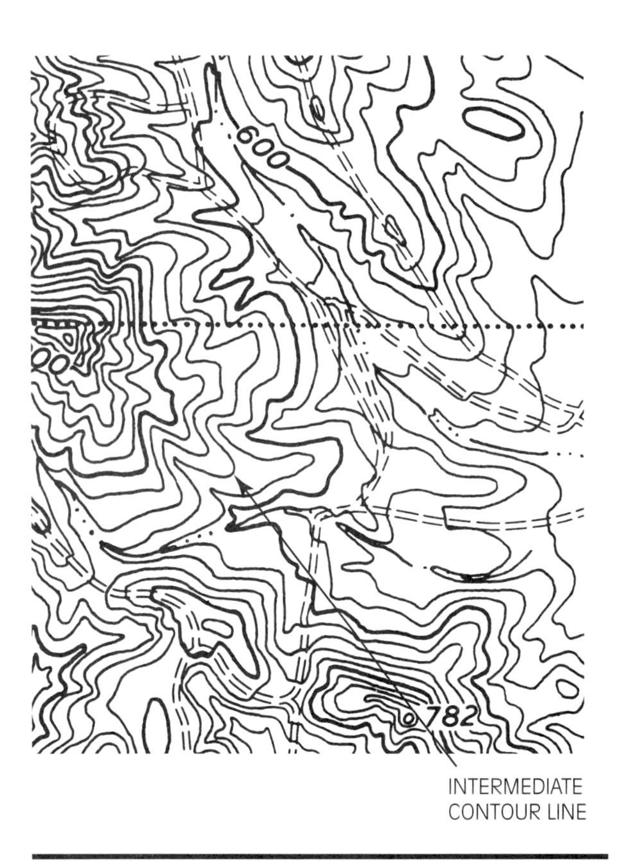

Figure 12-8 *The lighter lines between the contour index lines are called* intermediate contour lines.

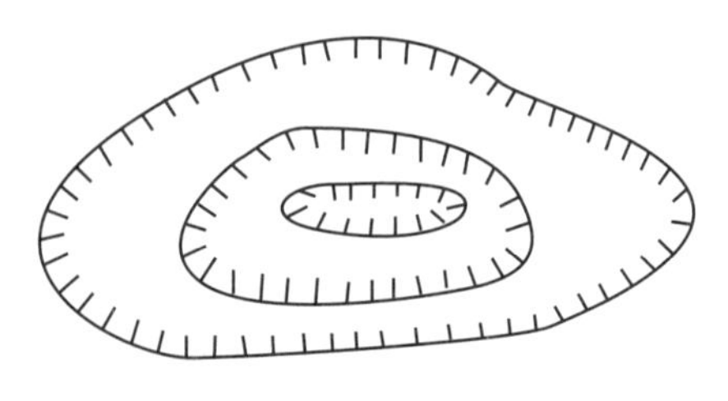

Figure 12-9 *Depression contour lines indicate an area that is lower than the surrounding areas.*

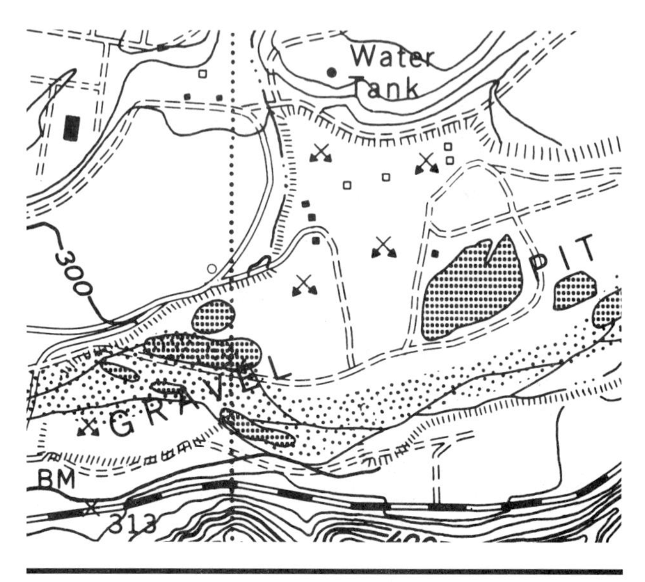

Figure 12-10 *Gravel pit.*

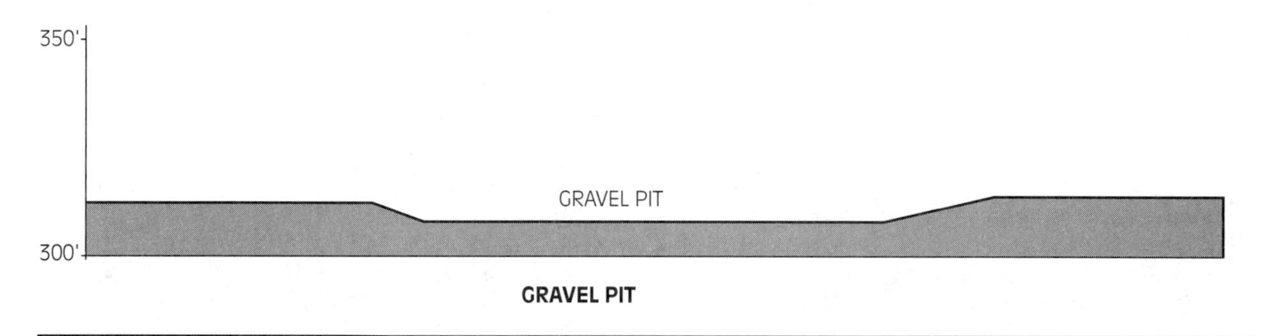

Figure 12-11 *Gravel pit profile.*

■ Note

The closer the contour lines are together, the steeper the slope.

The closer the contour lines are together, the steeper the slope (Figure 12-12) will be. Figure 12-13 is a profile of the steep slope in Figure 12-12. Other terrain features examined are a hill, a ridge, a saddle, a valley, a spur, a draw, and a depression.

A hill (Figure 12-14) is a land mass that usually tends to be round and falls off in all directions. See a profile of the hilltop in Figure 12-15. Its crest is usually lower than the surrounding mountains.

A ridge (Figure 12-16) is a long elevation of land that is often located on a mountainside. It usually has three downward slopes and one slope that projects upward. The ridge is profiled in Figure 12-17.

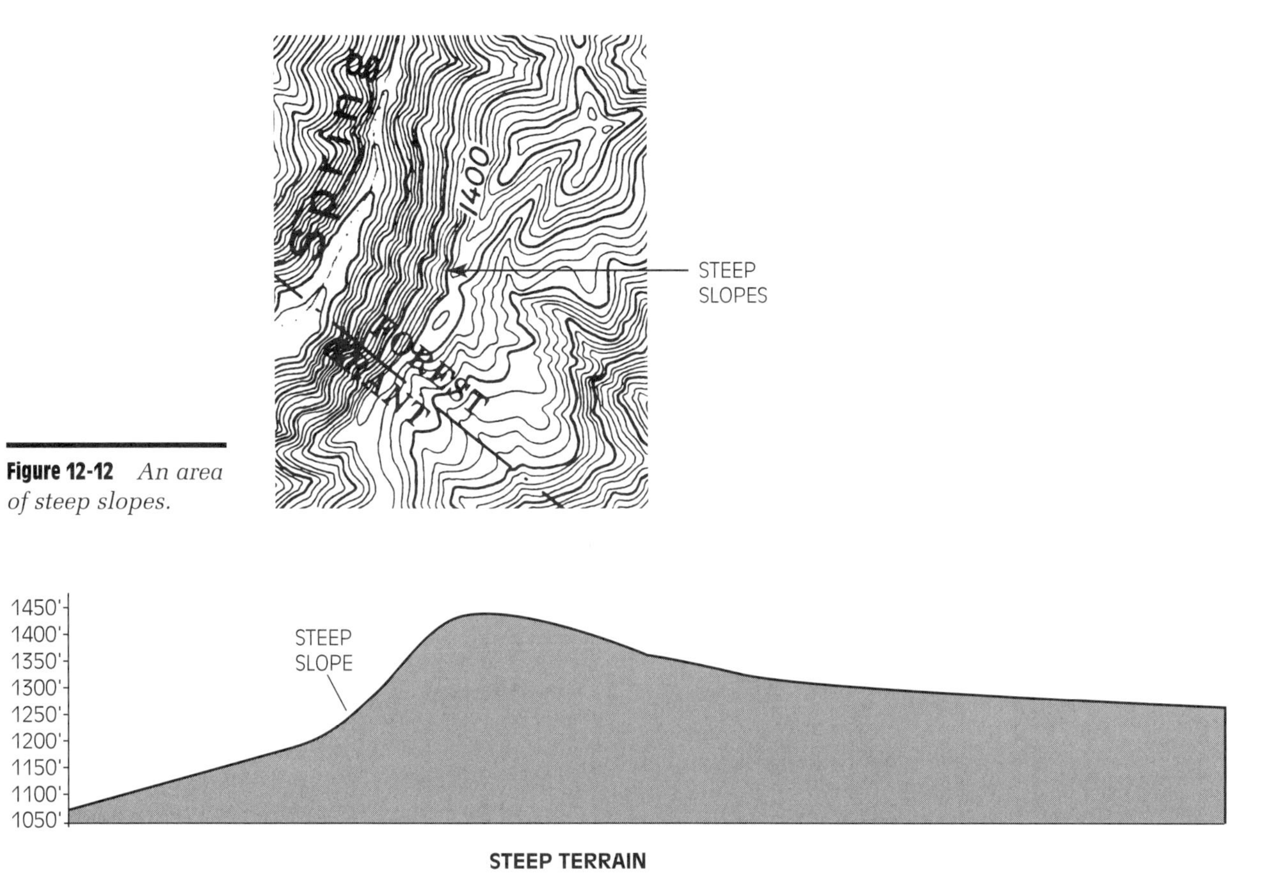

Figure 12-12 *An area of steep slopes.*

Figure 12-13 *Steep terrain profile.*

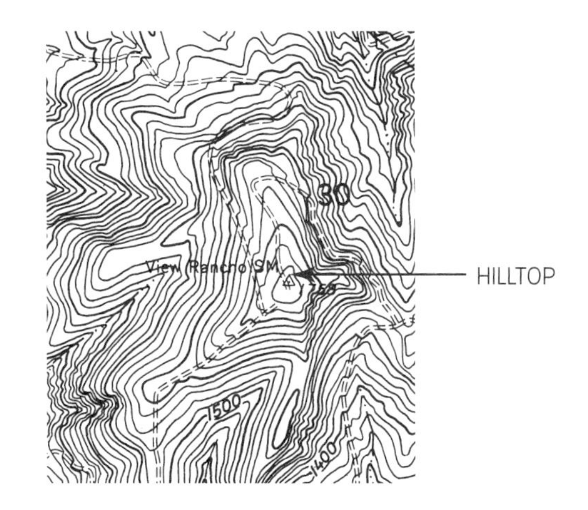

Figure 12-14 *A hilltop.*

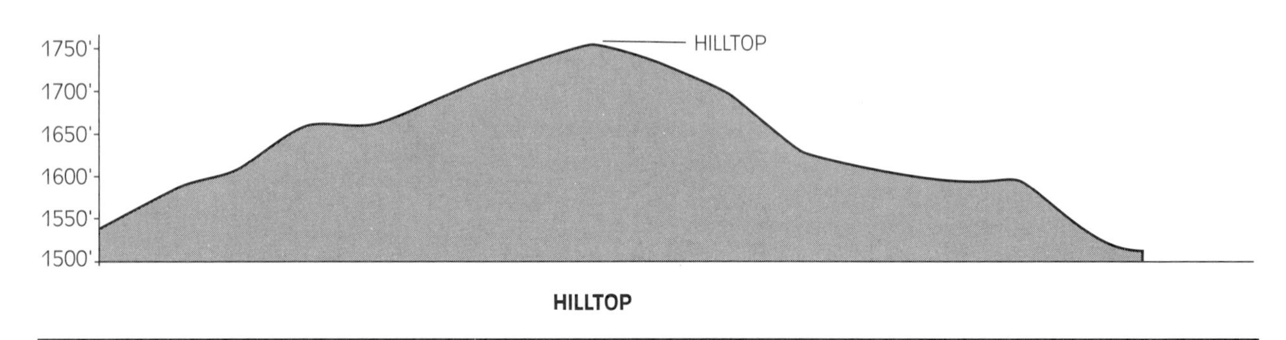

Figure 12-15 *Hilltop profile.*

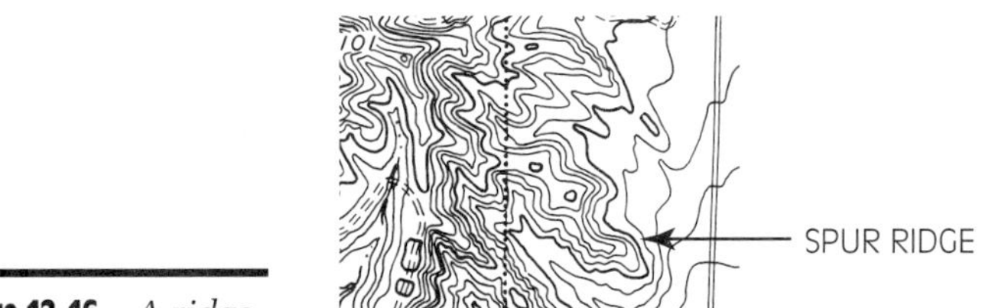

Figure 12-16 *A ridge.*

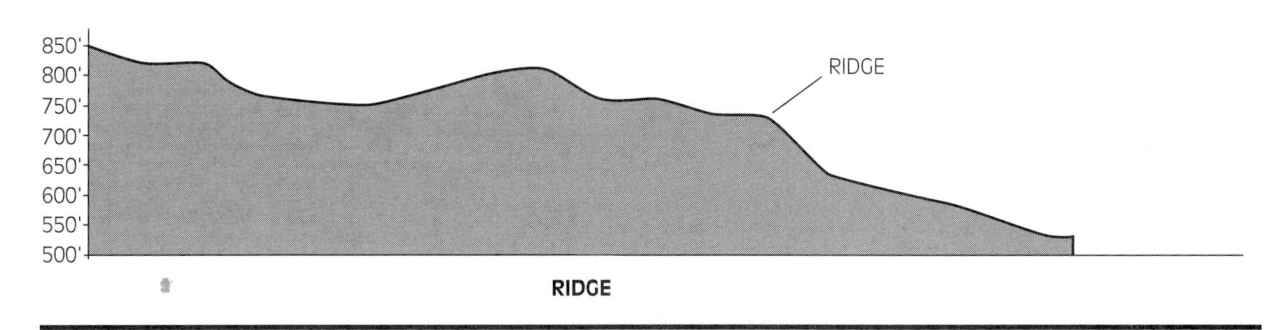

Figure 12-17 *Ridge profile.*

A saddle (Figure 12-18) is a low topographic point between two hills or summits. A saddle can be very deep or shallow; it can also be wide or narrow. As noted in Chapter 2, they have a channeling effect on a wildland fire. Expect greater rates of spread through a saddle. Note the profile of the saddle (Figure 12-19).

A valley (Figure 12-20) is an area of low land between two hills or mountains. The valley floor sometimes has a stream or river in it. The sides of the valley floor slope upward toward the adjacent hills or mountains. Note in the profile in Figure 12-21 that the valley floor also slopes both upward and downward.

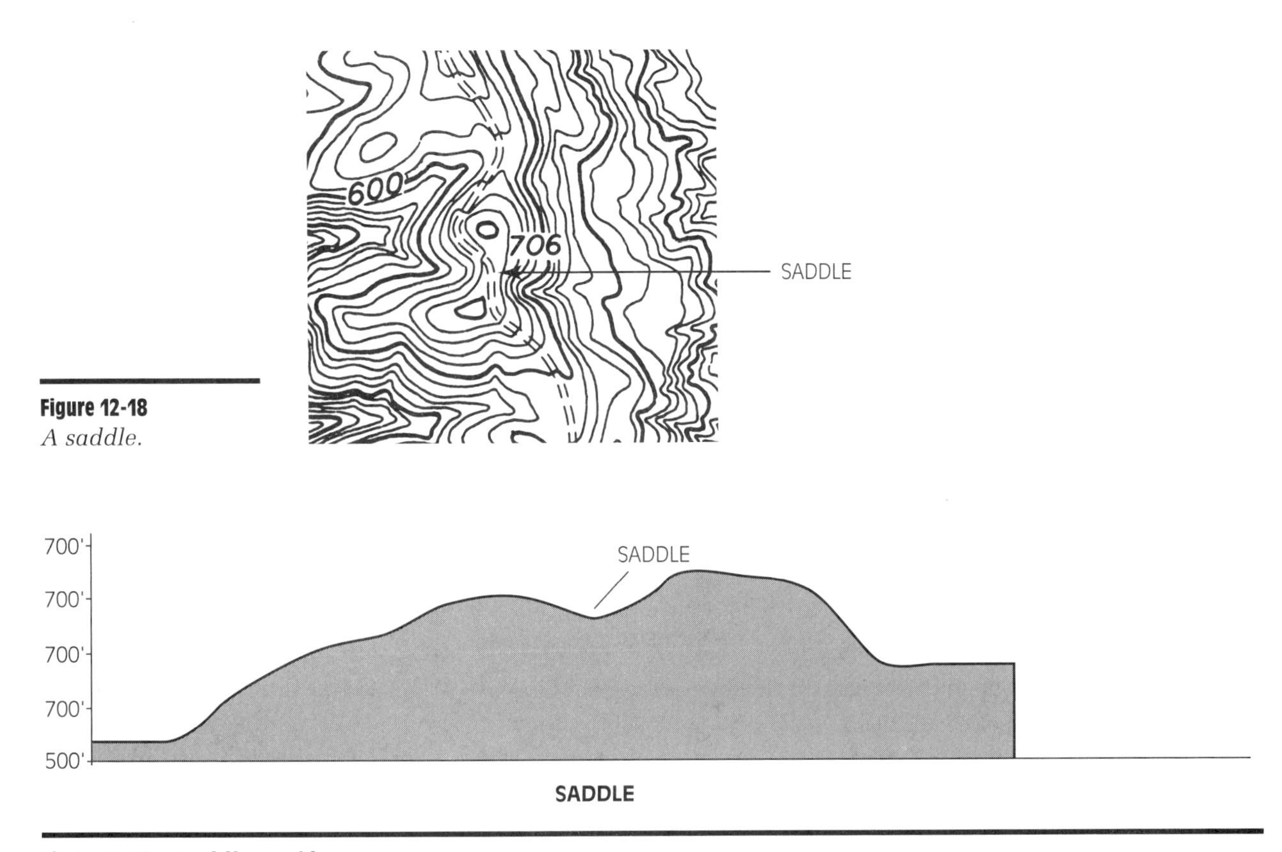

Figure 12-18
A saddle.

Figure 12-19 *Saddle profile.*

Figure 12-20
A valley.

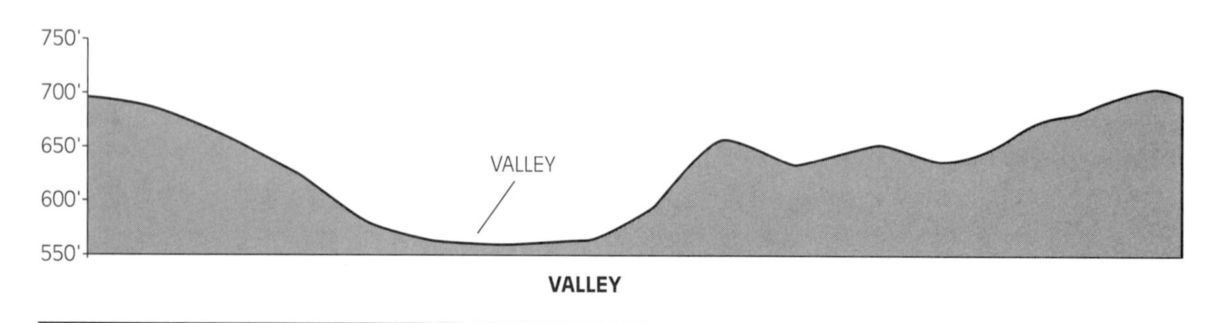

Figure 12-21 *Valley profile.*

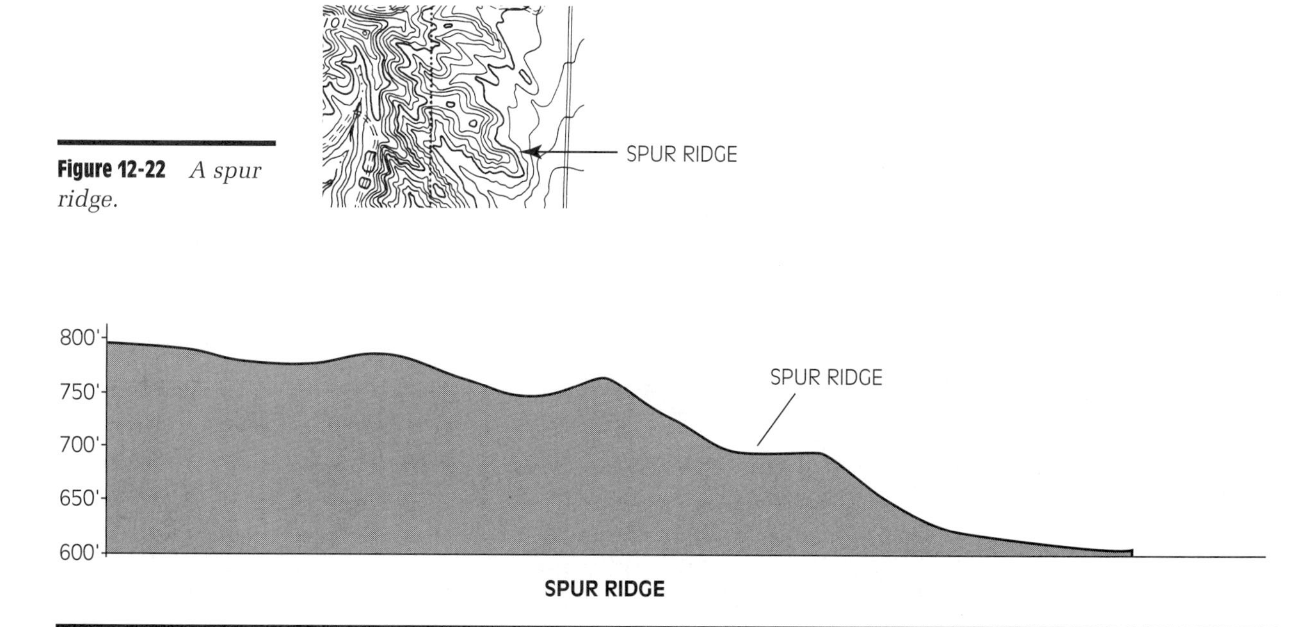

Figure 12-22 *A spur ridge.*

Figure 12-23 *Spur ridge profile.*

■ Note
Topographic map legends also contain details that describe the map symbols.

Spurs (Figure 12-22) are small ridges off the main mountain ridge. Note the profile of the spur ridge in Figure 12-23.

Topographic map legends also contain details that describe the map symbols (Figure 12-24). Topographic maps are printed in color which helps to interpret the map symbols more easily. Different colors mean different things. A list of colors commonly found on a topographic map follows:

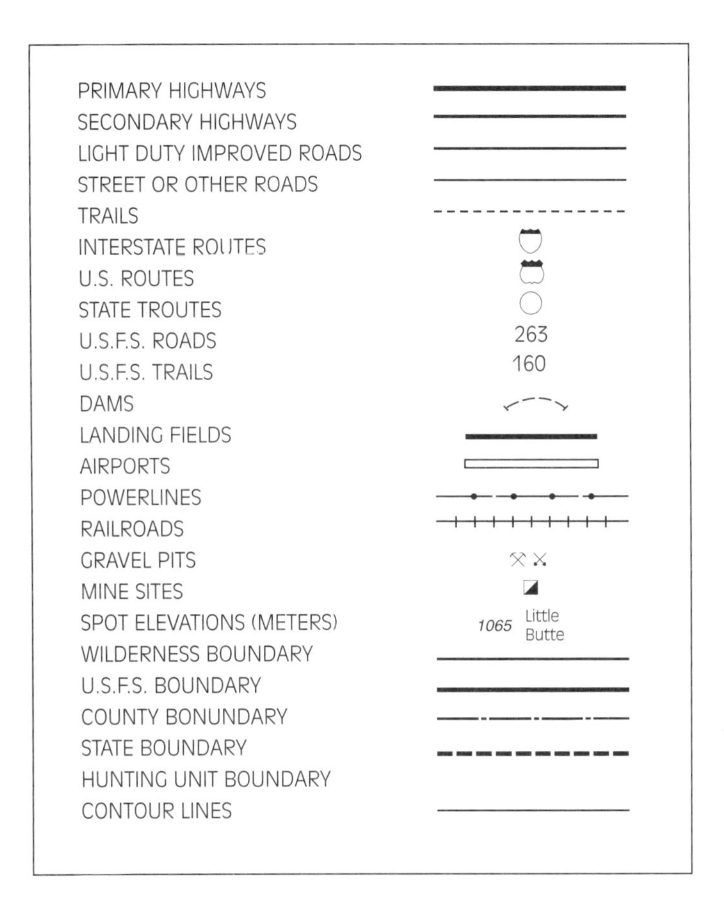

Figure 12-24 *Map legend common details.*

Black—Most cultural or human-made features.

Blue—Water features, such as lakes, rivers, or swamp areas.

Green—Vegetation features, such as woods, vineyards, or orchards.

Brown—All relief features, contours, cuts, and fills.

Red—Main roads, built up areas, boundaries, and special features.

Other colors—Special purposes; their key is found in the legend of the map.

UNIVERSAL TRANSVERSE MERCATOR (UTM) GRID

Mapmakers have always had to introduce distortion errors into their maps because of the mathematical treatment required to portray a curved surface on a flat map surface (Figure 12-25). To best illustrate this, take a grapefruit and draw sixty 6° wide vertical lines (zones) from top to bottom. Now, with a knife, cut off the top

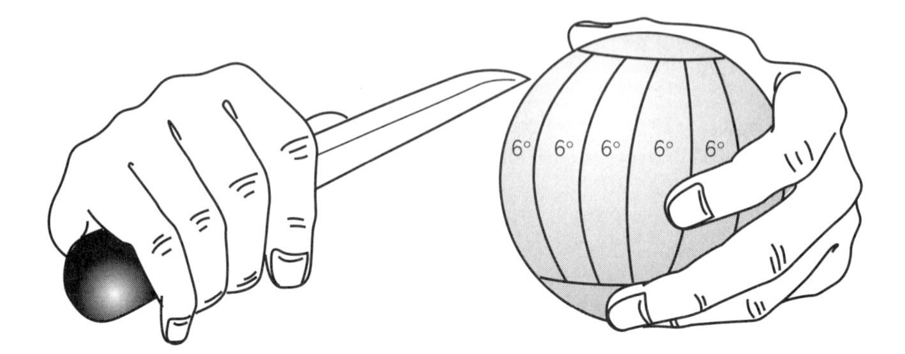

Figure 12-25 *Each section of the grapefruit, once peeled and flattened, illustrates a UTM zone.*

and bottom rind of the grapefruit (areas to be excluded from the map). Next, cut all the lines that are drawn and peel the rind from the grapefruit. Each section represents a Universal Transverse Mercator (UTM) zone once it is placed on a flat surface. This best illustrates the term *map projection,* which is a system used to project the Earth on a flat map surface.

To make maps simpler to use and to avoid the inconvenience of pinpointing locations on curved reference lines, a rectangular grid surface was designed. The grid system consists of two sets of parallel lines, each uniformly spread and perpendicular to each other that are superimposed on the map. With the UTM grid system, any point on the map can be designated by its latitude and longitude or by the grid coordinates. A reference in one system can be converted into a reference in another system.

■ Note
UTM is a special grid system adopted by the Defense Mapping Agency for military use throughout the world.

UTM is a special grid system that was adopted by the Defense Mapping Agency for military use throughout the world. It consists of sixty north-south zones that are only 6° of longitude wide. Each zone is so narrow that, when it is peeled from the globe and flattened, only a minimal distortion results. The areas above north latitude 84° and below south latitude 80° are excluded from maps due to the minimal distortion that takes place as that portion of the globe is flattened (Figure 12-26).

■ Note
It consists of 60 north-south zones that are only 6° of longitude wide.

All 60 UTM grid zones are numbered from west to east (left to right) beginning at the international date line which is 180° longitude. Zone 1 is a vertical area 6° wide running between the meridians located at 180° west and 174° west longitude.

The zones are numbered progressively in an eastward direction and end at zone 60, which is located between 174° and 180° east longitude. The conterminous United States are covered by 10 zones from zone 10 on the West Coast through zone 19 in New England.

In each of the UTM zones, all the meridians and parallels experienced a slight distortion when the segment was flattened (map projection). The central meridian, which runs through the center of each 6° zone, and the equator are not

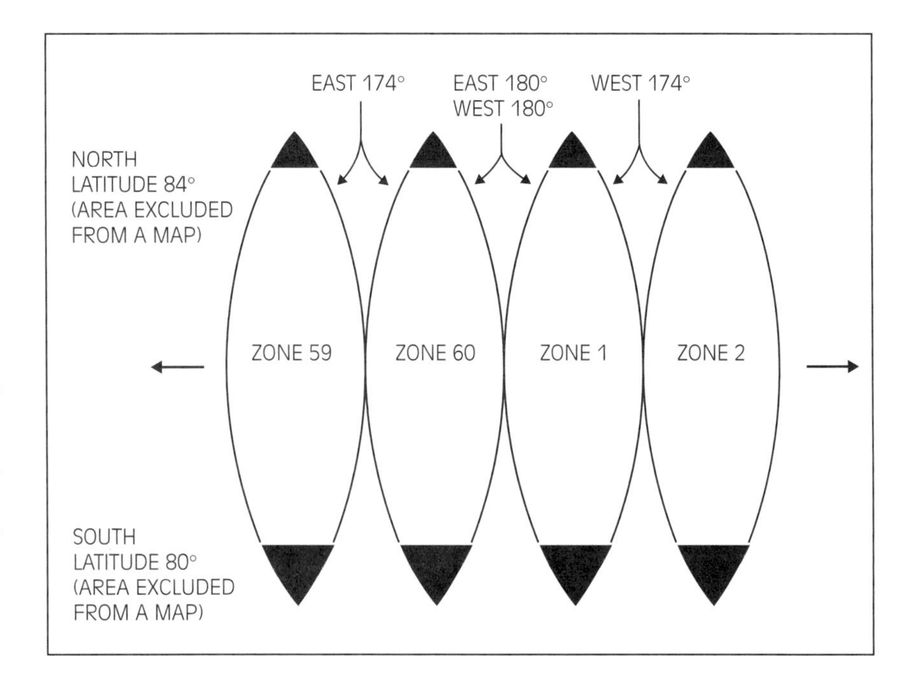

Figure 12-26 *UTM consists of 60 north-south zones. Each is 6° wide. Areas above north latitude 84° and below south latitude 80° are excluded from maps.*

easting
the distance in meters of the position east of the zone line

northing
the distance in meters of the position north of the equator, measured along a zone line

Note
To find UTM coordinates on a topographic map (7.5 minute series, scale 1:24,000) look at the margins of the map.

distorted and act as perpendicular UTM grid lines. The central meridian and equator also serve concurrently as lines of latitude and longitude (Figure 12-27).

Each central meridian that runs through a 6° grid zone is assigned an **easting** value of 500000 ME (Meters East). This helps you determine if you are east or west of the central meridian. Grid values to the west of the central meridian are less than 500000 ME and as we move east away from the central meridian the values are more than 500000 ME.

The equator serves as the grid line for **northing** values. It is either assigned a value of 0000000 MN or 10000000 MN, depending whether you are in the northern hemisphere or southern hemisphere. When your location is north of the equator, the true value assigned is 0000000 MN. Grid lines will increase in value as you move north of the equator. The value assigned for the southern hemisphere is 10000000 MN. However, remember that it is still referenced to the equator. Grid lines decrease in value as you move southward from the equator.

UTM coordinates, then, consist of an easting (in meters), a northing (in meters), and a zone number. To express the coordinates, read right and up.

To find UTM coordinates on a topographic map (7.5 minute series, scale 1:24,000), look at the margins of the map. Blue tick marks will be found that denote 1,000 meter grid lines. The marks on the top and bottom are called eastings and are used to locate an east-west location. The blue tick marks on the left and right margin of the map are called northings and are used to locate a north-south location.

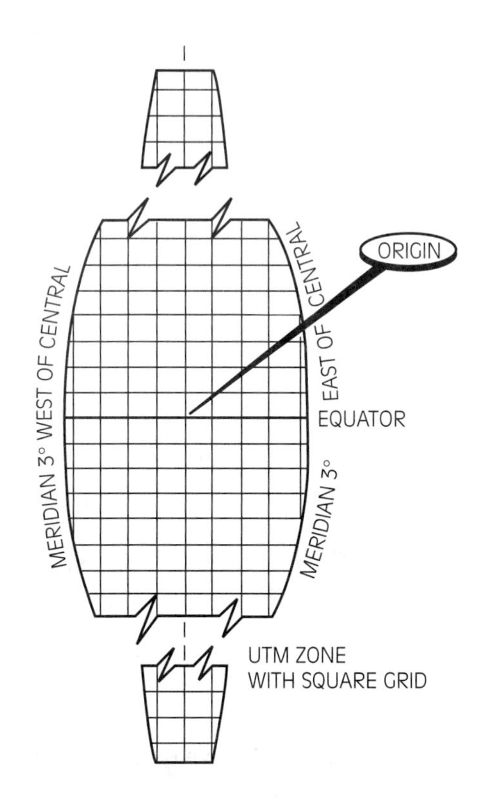

Figure 12-27 *Flat representation from the Earth's surface. Note the reference lines. Photo courtesy National Wildfire Coordinating Group.*

Many of the newer topographic maps are printed with the UTM grids already in place. When using an older topographic map, take a large straight edge and draw lines at each 1,000-meter blue tick mark, from top to bottom on the map and from left margin to right margin. This preparation will create UTM grids.

■ Note **If the easting numbers are increasing, you are headed in an easterly direction.**

Here is an example of an easting coordinate 265000 ME. The distance between 365000 ME and 366000 ME is 1,000 meters. If you are traveling anywhere on the same topographic map, then you should be concerned with only the two large numerals. They are an abbreviation of grid lines commonly printed 1,000 meters apart. The last three numbers stand for meters. A northing value would look like this: 464000 MN. Again, the distance between 464000 MN and 465000 MN is 1,000 meters.

■ Note **Larger northing numbers indicate movement in a northerly direction.**

If easting numbers are increasing, you are headed in an easterly direction. Larger northing numbers indicate movement in a northerly direction.

The advantage in using the UTM grid system is that you will always be working with grid lines that are 1,000 meters apart or .62 miles apart. This makes it easy to determine your location on the map.

HOW TO USE A COORDINATE RULER

A coordinate ruler will help you quickly locate your position on a topographic map. The one demonstrated in this text is called the TopoTool Coordinate Ruler (Figure 12-28). It is made for use on a 7.5 minute topographic map with a representative fraction of 1:24,000.[1]

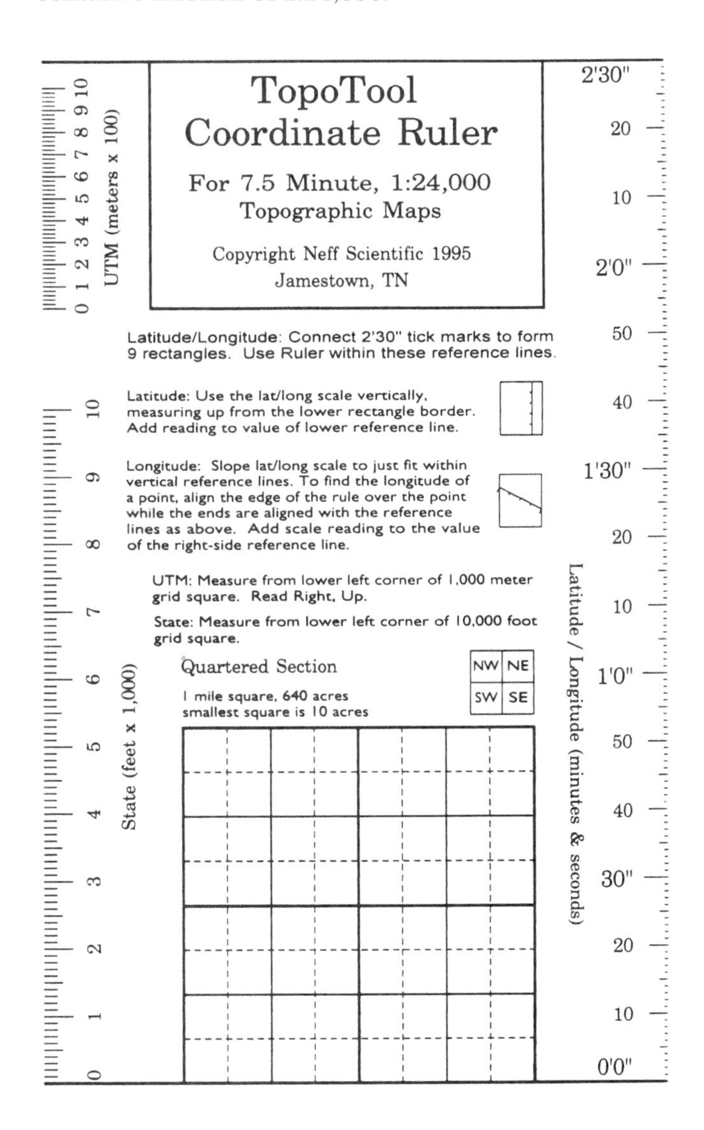

Figure 12-28 *A coordinate ruler.*

[1] Instructions for using the TopoTool Coordinate Ruler are furnished by NEFF Scientific, 103 Cottle Lane, Jamestown, Tennessee 38556. Contact the company for additional information. It is a simple and accurate tool to use.

Use

> **Note**
> The coordinate ruler is a set of scales printed on clear plastic.

The coordinate ruler is a set of scales printed on clear plastic. These scales are used like a ruler to measure in different coordinate systems. The scales are sized to be used with the United States Geological Survey (USGS) system of 7.5 minute, 1:24,000 topographic maps we use in the continental United States. The scales can be used to measure distance in feet and to determine slope. The Public Land Survey Quartered Section is divided into 10-acre squares, which can be used to estimate acreage.

Coverage

The UTM, feet, and Public Land Survey scales are useable on any 1:24,000 map. The latitude/longitude scale is useable over the continental United States.

Additional Information

Conversions Dealing with units like minutes and seconds can be a little difficult. The first three graphic scales in Figure 12-29 are used to help with minute-second ad-

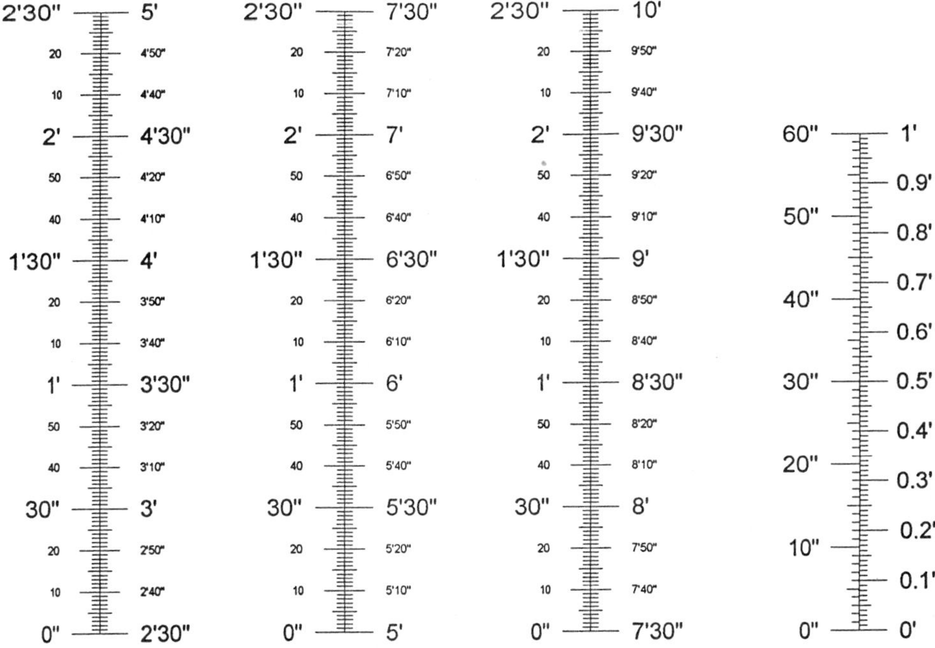

Figure 12-29
Conversion scale.

dition when finding or plotting coordinates. The left side of each scale shows the ruler's numbers. The map coordinates are on the right. Note the top and bottom numbers correspond to the minutes and seconds of the border lines (the tens of minutes are left off). Suppose you locate a point with the ruler. The scale reads 1′43″. The right-side reference line is 17′30″. Use the scale starting with 7′30″. Go up the left side of the scale to the ruler reading 1′43″. The right side reads 9′13″ here. Since the tens digit is 1, the point is at 19′13″.

Decimal Minutes Loran C and GPS receivers often give locations with latitude/longitude in degrees and decimal minutes. The graphic scale on the right converts decimal minutes to and from seconds. Suppose a Loran displayed a latitude of 35° 14.67″. Finding 0.67 minutes on the right side of the aid, this corresponds to about 40″ on the left side. Thus, the coordinate could be expressed as 35° 14′40″. (Figure 12-29).

Notes The coordinate ruler is silk screen-printed on clear plastic. The material is durable, but it gets superficial scratches. These are not noticeable when the ruler is placed on a map. The printing on the ruler is quite durable as well, but it can be damaged. Use care to avoid unnecessary abrasion.

The ruler's scales (except the Section) can be used in a pinch on 1:25,000 maps by using the sloping scale technique for both north–south and east–west coordinates.

Using the Coordinate Ruler to Find Latitude or Longitude

The coordinate ruler is sized to fit the common USGS 7.5 minutes, 1:24,000 topographic maps as used in the continental United States.

Latitude and Longitude on 7.5 Minute Maps A 7.5 minute topographic map is 7.5 minutes high and 7.5 minutes wide. The lower right corner has the smallest values of degrees, minutes, and seconds. There are latitude numbers along both sides of the map, showing the angular distance from the equator, and longitude numbers along the top and bottom margins, showing the angular distance west of the prime meridian.

There are two small lines or tick marks showing the location of additional latitude or longitude lines along each margin of the map. These intermediate marks are labeled with minutes and seconds only. Also, there are small crosses inside the map where these lines would intersect. Using a sharp pencil, draw lines connecting equal values of latitude and longitude. By using the crosses, the long edges of the ruler enables you to draw these lines. This divides the map into 9 rectangles, each 2.5 minutes (2′30″) on all sides (Figure 12-30). These lines, along with the map margins, are used as reference lines of known latitude or longitude.

■ Note

There are two small lines or tick marks showing the location of additional latitude or longitude lines along each margin of the map.

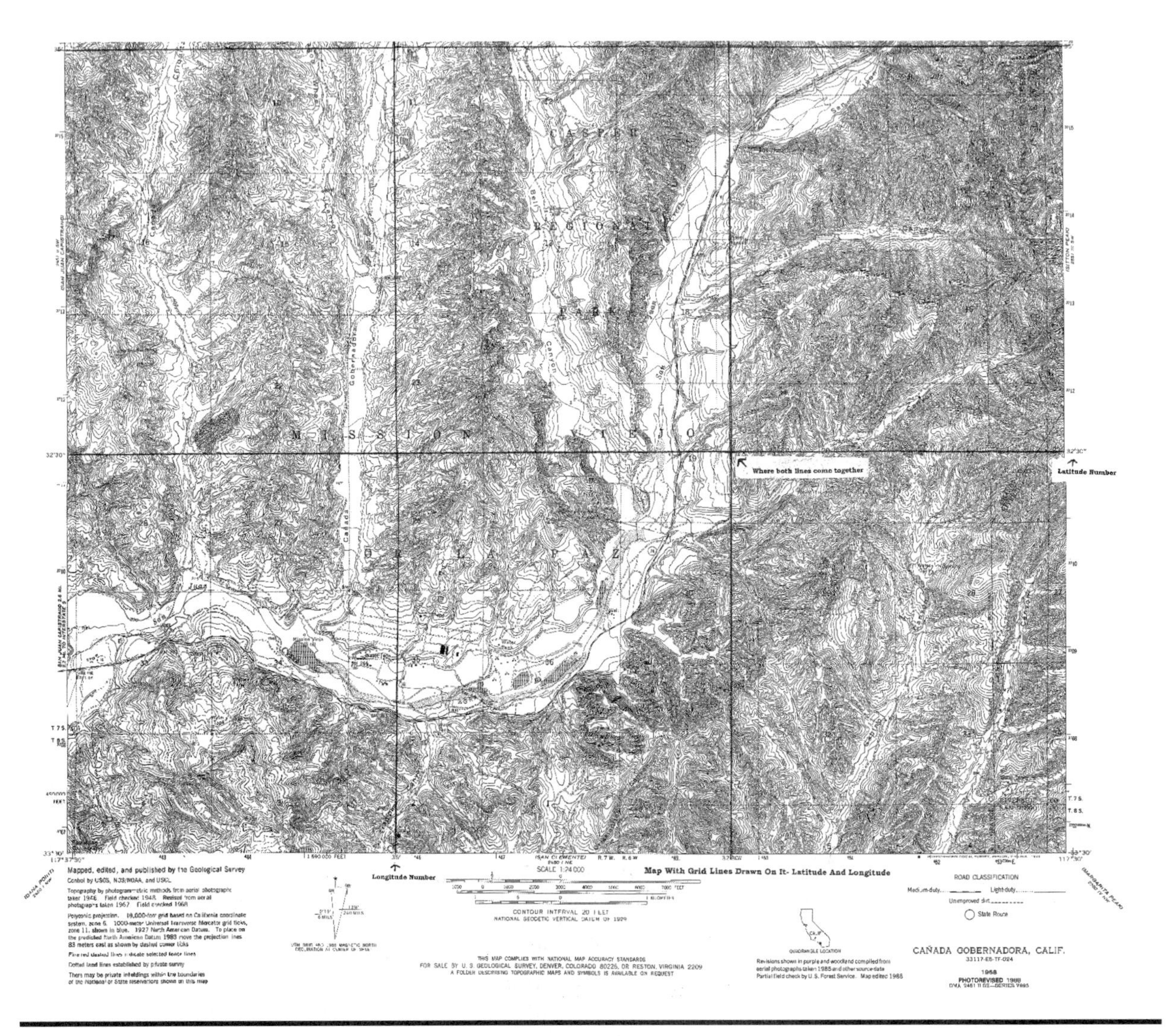

Figure 12-30 *The grid lines on the map have been drawn. Note that this is only the lower two-thirds of the example map. You should end up with nine squares. The interior crosses can be found where the arrow points to the comment "where both lines come together."*

Now, imagine a series of horizontal and vertical lines forming a grid, sort of like a piece of window screen, laid on the map. Spaced 1 second apart, there would be 150 horizontal latitude lines and 150 vertical longitude lines in each 2.5 minutes rectangle. If these lines were visible, the latitude of a point could be determined by counting the number of lines from a labeled reference line up to the point. Add the number of lines (seconds of latitude) to the value of the reference latitude line to find the latitude of the point.

Finding Latitude Latitude on a topographic map is found in just this way. The latitude/longitude scale is used just like a ruler to measure from a known latitude line to a point. Each ruler division shows the location of an imaginary latitude line. Measure from the lower labeled latitude (horizontal) line of the rectangle containing the point. Read the minutes and seconds from the ruler and add this to the value of the reference line. This is the latitude of the point.

Latitude-Finding Hints Paper maps vary somewhat in size with age and humidity. Rectangles typically do not measure exactly 2′30″ high. For demanding applications, the scale can be shifted slightly to minimize this inaccuracy when finding (or plotting) latitude. Align the lower line of the ruler on the lower rectangle border line if the point is in the lower quarter of the rectangle and align the upper line of the ruler with the upper rectangle border line if the point is in the upper quarter. Center the ruler over the rectangle borders if the point is in the middle. Add the rule reading to the lower reference line (Figure 12-31).

Adding the ruler's minutes and seconds of latitude to the latitude of the lower border line of the rectangle is treated like time addition. Using a scrap piece of paper to add the coordinates is a good idea, as it is easy to make a mistake with mental addition. See the foregoing Additional Information section for instructions and an example.

Finding Longitude The longitude width of a 2.5 minute rectangle is not a constant distance, since the longitude lines converge toward the poles. To compensate for this, the ruler uses a technique called sloping-scale. If you are trained in drafting, you are probably familiar with this technique.

The latitude/longitude scale is too long to fit the width of a 2′30″ rectangle. However, if the scale is tilted, it can be adjusted so that the two end lines of the scale lie directly over the side borders of the rectangle. The location of each imaginary longitude line is then at the edge of the rule.

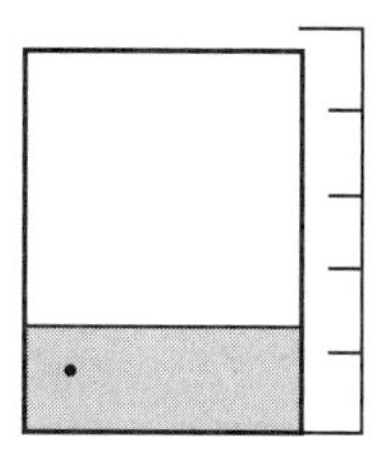
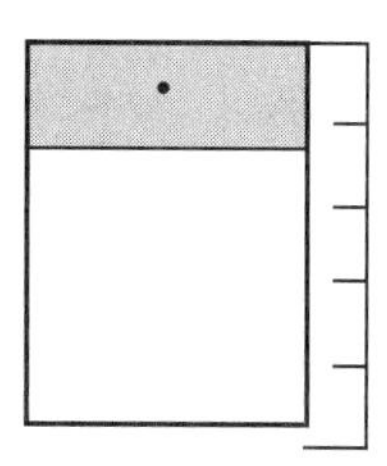
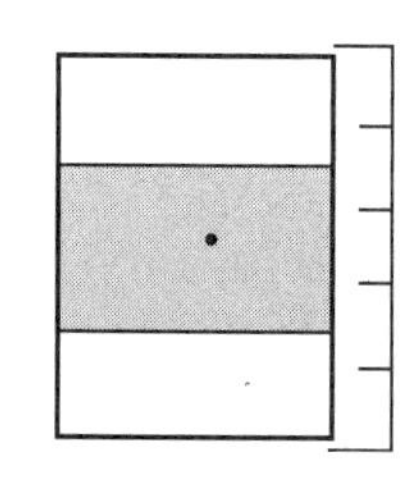

Figure 12-31 *Shifting the scales to minimize latitude inaccuracies. Ruler scale is shown to the right of the rectangle for clarity.*

To find the longitude of a point, align the edge of the ruler just over the point while the ends of the scale are precisely aligned with the vertical border lines. Correctly aligning the scale is a little time consuming, but it is quite accurate. With some practice, it becomes easier. When finding longitude, remember that the longitude lines are all vertical; they do not follow the slope of the ruler graduations. The longitude lines lie just under the ruler marks at the very edge of the rule. The rule numbers must increase to the left. Add the ruler reading at the point to the right-side reference line (Figure 12-32).

Longitude-Finding Hints Use care in adding the rule reading to the value of the longitude reference line. Remember that angular degrees, minutes, and seconds are added just like time's hours, minutes, and seconds. Use the graphic scales mentioned previously to aid your arithmetic. Note that since the borders of the rectangle are always multiples of 2′30″, the units digit for seconds will be the same. Using the example in the "Some Additional Information" section, the ruler reads 1′43″. The right-side reference line is 17′30″. Thus, it is not necessary to count each small graduation on the graphic aid—you know the coordinate will end with a 3. Just glance at the aid to see that 1′40″ corresponds to _9′10″ (the underline is for the missing tens-of-minutes digit). So, the ruler reading of 1′43″ added to the border line gives the point's minutes and seconds of longitude: 19′13″.

Since the ruler must be tilted or sloped to just fit the rectangle borders, it is helpful to extend the vertical reference lines out past the map proper onto the margins. This allows the tilted ruler to still find the longitude of points near the upper and lower borders. However, small triangular-shaped areas at the top margin of the map may not be accessible to the sloped rule. If the rule starts going off the paper because the point is too far up, just use the ruler like a draftsman's triangle to project a small line down into the accessible region of the rectangle. Align the top line of the ruler with the top margin line of the map and slide the edge of the scale up to the point. Now mark the longitude of the desired point along the edge of the ruler further down in the rectangle, where it can be measured.

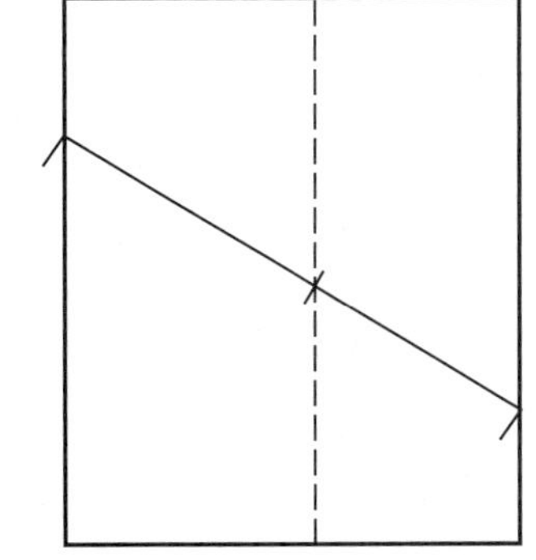

Figure 12-32 *Tilt scale to fit just within rectangle borders. Longitude line of point is shown dashed.*

Plotting Plotting coordinates is the reverse of finding coordinates and is actually a little easier. Determine longitude first. By inspection of the coordinates and the reference lines, decide which rectangle contains the point. Then align the ends of the ruler over the vertical rectangle borders (slope the ruler). Measure left the required number of minutes and seconds, and make a small vertical mark to indicate the position of the desired longitude line. Then, use the ruler like a draftsman's triangle to measure up at a right angle from the horizontal latitude reference line the required number of minutes and seconds. By aligning the ruler's lower line over the lower reference line and carefully sliding the ruler sideways to the longitude mark made above, the edge of the ruler is lying directly along the longitude line of the point. Again, just measure up the required number of minutes and seconds of latitude and mark the location of the point. Erase the first longitude reference mark (Figure 12-33).

The latitude/longitude scale is being used to find the longitude of an eastern Kentucky oil well. The ruler is used to connect the tick marks to the cross to form a rectangle with border lines of known latitude/longitude.

Note the position of the coordinate ruler's upper and lower lines. This indicates how to tilt the ruler to fit a 2′30″ longitude distance. Add the 35″ on the scale to the value of the right-side reference line, 86°07′30″W to get the longitude of the well, 86°08′05″W. The coordinates of this oil well are 36°45′47″ North, 86°08′05″ West.

Using the Coordinate Ruler to Find a UTM Grid

Finding UTM The UTM system is very simple to use. The coordinates increase to the north and to the east (up and to the right). Measurements within a 1,000-meter grid square are taken from the lower left corner and are always additive. To find a UTM coordinate, measure from the west (left) border of the grid square to the point. This is the easting. Then measure from the south (lower) border of the grid square to the point. This is the northing. The zone number is listed in the map margin.

Plotting UTM Plot UTM coordinates on a map by first locating (by inspection) the 1,000-meter grid square containing the coordinate. Using the UTM scale, measure from the lower left corner the proper number of meters such that the sum of the left grid line and the ruler UTM scale equals the easting coordinate. Make a reference mark on the lower border and then measure from this mark up the proper number of meters. Mark the point.

Advantages One advantage of UTM coordinates is that, due to the small size of the squares, folding the map does not affect coordinate accuracy as much as other systems (Figure 12-34).

As an example of UTM coordinate determination, suppose we want to find the UTM coordinates of the Fort Knox Gold Bullion Depository. The UTM grid

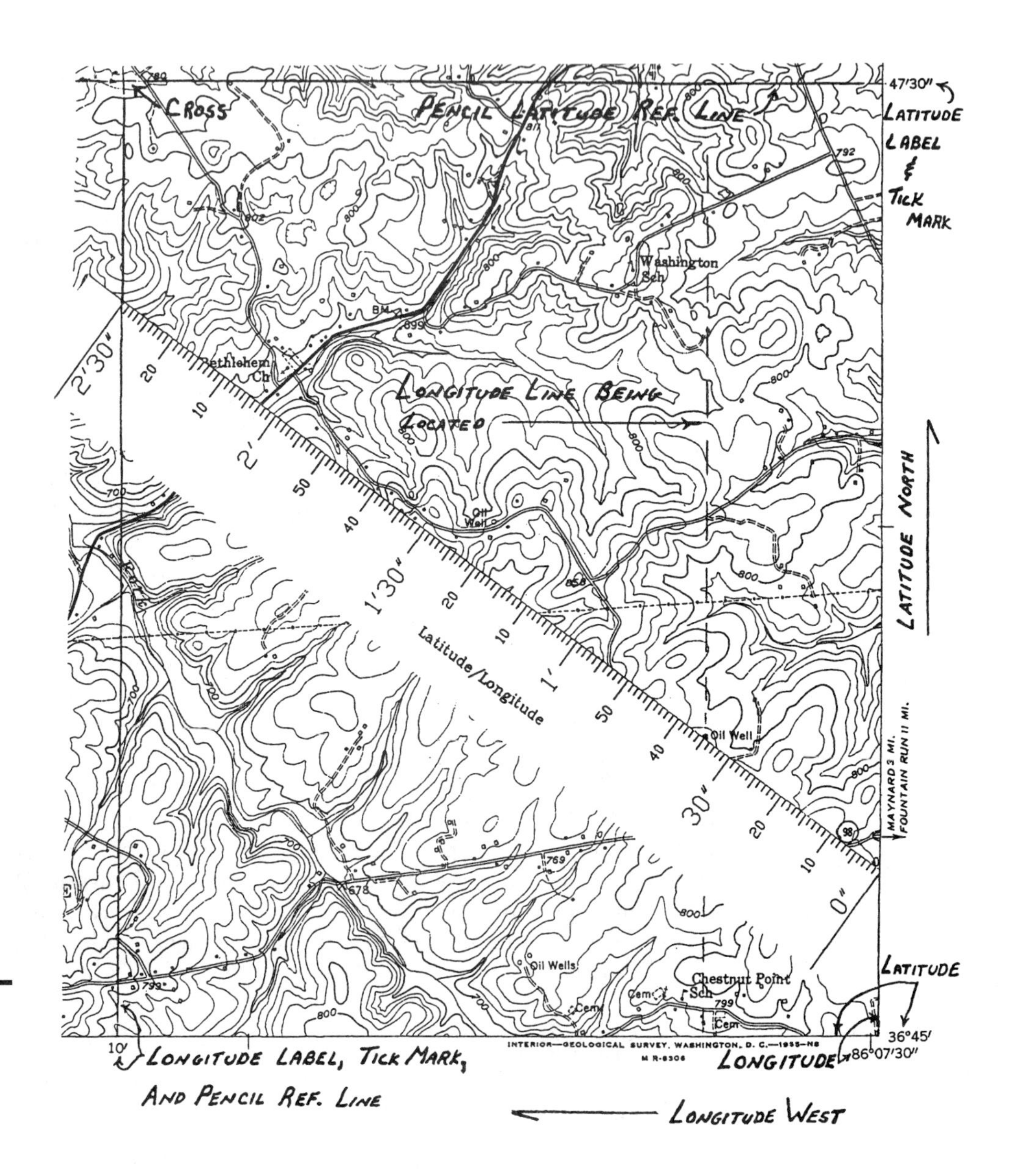

Figure 12-33
Example of how to use the coordinate ruler to find longitude.

is not preprinted, but a note in the lower left margin tells us the 1,000 meter grid ticks are shown in blue (on the original topographic map, not this copy). The zone number is 6.

Examining the map borders, we note the UTM numbers and construct our grid by connecting matching tick marks on the opposite borders of the map. To locate one point, only four lines need to be drawn around the point. For conve-

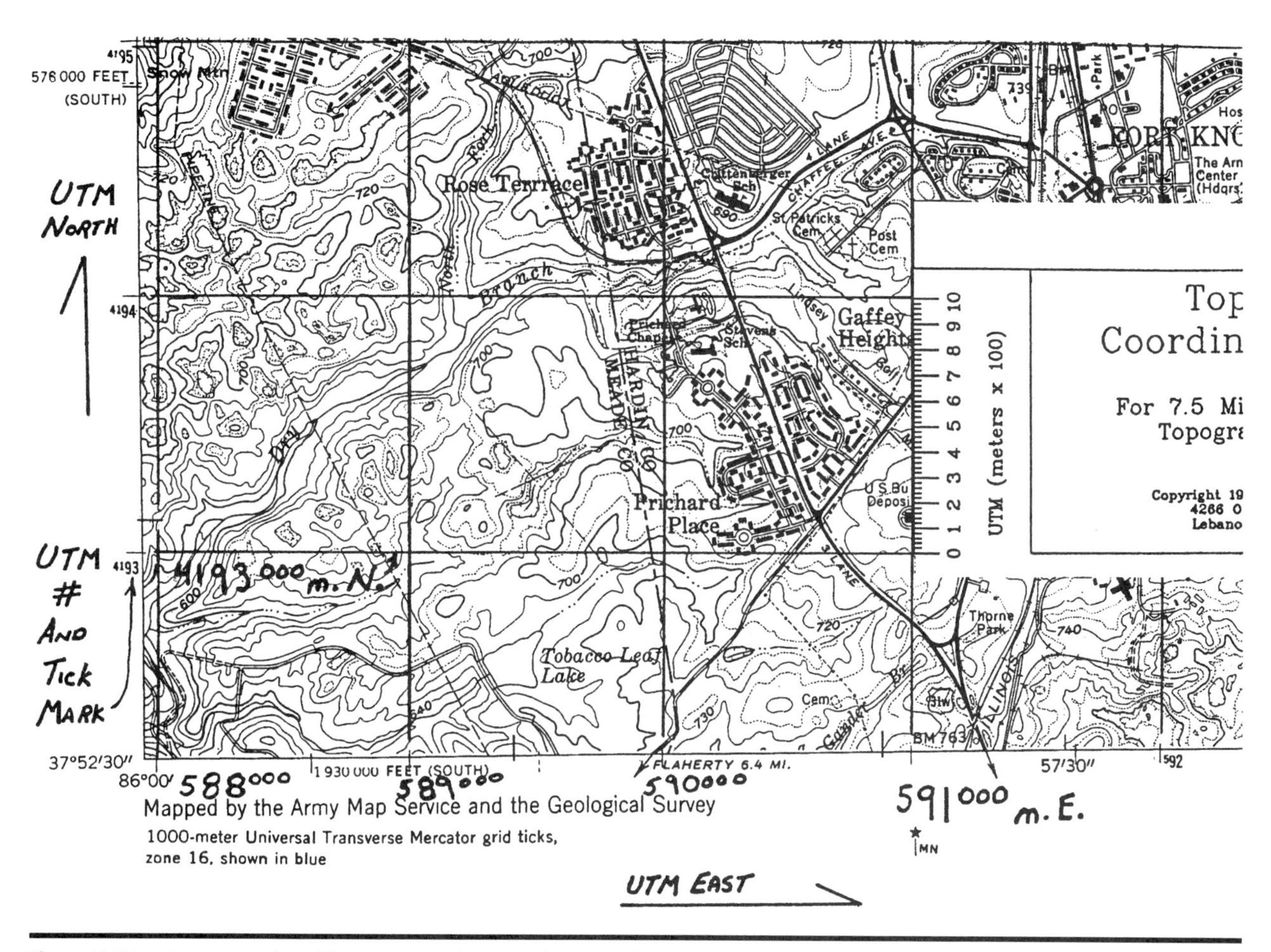

Figure 12-34 *An example of how to use the coordinate ruler to find UTM coordinates.*

nience, expand a couple of the grid numbers to full UTM coordinates. Remember, the two larger principal digits are in ten thousands and thousands of meters. Number any grid line that does not have a printed value. Note that the coordinates should increase from left to right and up.

The depository lies directly on the 591000 ME line, so this is its East coordinate. Use the coordinate ruler's UTM scale to measure from the lower border line of the square containing the point up to the point itself. The North UTM coordinate is 4193140 MN, found by adding the scale reading of 140 meters to the lower grid line number of 4193000 MN. The full UTM coordinates of the U.S. Bullion Depository are 591000ME; 4193140MN; Zone 16. Remember, the order of the coordinates is "Read Right, Up."

Summary

The Global Positioning System was developed under a project funded by the United States Department of Defense. It first became operational in January of 1994. The space segment involves 24 satellites that orbit the Earth every 12 hours.

In order to be able to locate ourselves on the Earth's surface, a coordinate system with positional values was developed. Reference lines were projected onto maps that represented the Earth's surface. These lines are called *parallels of latitude* and *meridians of longitude.*

Parallels are horizontal lines that run east and west around the Earth. Meridians are vertical lines that run north and south. Latitude is a point on the Earth's surface, represented by horizontal lines, that is north or south of the equator. Longitude, represented by vertical lines or meridian, is a point on the Earth's surface that is east or west of the prime meridian. The prime meridian, defined as 0° longitude, runs through Greenwich, England.

Topographic maps are a relief of the Earth's surface. A 7.5 minute (refers to minute in latitude or longitude) topographic map is generally used. The representative fraction on these maps is 1:24,000 or 1 map inch representing 24,000 actual ground inches. Topographic maps also contain details that describe the map symbols. A topographic map is of great use to wildland firefighters. It allows fire behavior analysts to use the map and do fire projections. The map allows fire crew members who are working in rural areas to find their way around or to describe their location. Topographic maps have multiple uses and are a must for anyone working in wildland fire suppression.

Since the advent of GPS, finding one's location is now easier and more accurate. Handheld receivers are light, relatively inexpensive, and easy to use. Handheld units use UTM or latitude/longitude to locate position. A coordinate ruler helps speed the process.

The information provided will assist in the use of a GPS unit with a topographic map. Learning to use a compass with your topographic map is suggested.[2]

---

[2] An excellent resource that contains practical exercises is *Be an Expert with a Map and Compass* by Bjorn Kjellstrom, published by Macmillan Publishing Company, 866 Third Avenue, New York, NY 10022. It can be found in most mountaineering shops.

The National Wildlife Coordinating Group also just published an excellent self-paced study guide entitled *Map and Compass Skills.* The address is National Interagency Fire Center, Attn: Supply, 3833 S. Development Ave., Boise, ID 83705.

Review Questions

1. GPS involves how many satellites?
2. Where is the GPS master control system located?
3. What does the term *latitude* mean?
4. What does the term *longitude* mean?
5. What is the Universal Transverse Mercator grid system?
6. Describe the term *easting*.
7. Describe the term *northing*.
8. What does the term *7.5 minutes topographic map* mean?

References

Kjellstrom, Bjorn, *Be an Expert with a Map and Compass* (New York: Macmillan, 1994).

National Wildfire Coordinating Group, *Map and Compass Skills* (Boise, ID: National Interagency Fire Center, 1998).

The Incident Command System and ICS Forms

Learning Objectives

Upon completion of this chapter, you should be able to:

- Explain how the ICS began.
- Describe the five major functional units in the ICS.
- Describe the ICS forms used in the ICS.
- Identify the forms used in an incident action plan.

INTRODUCTION

This chapter presents an overview of the Incident Command System (ICS) and gives ideas on how best to use it. Also discussed is how the ICS builds from an initial attack fire to a major fire situation. Various incident-related ICS forms are described so the reader will be familiar with a document used on larger fires, called an incident action plan.

WHY THE ICS BEGAN

In the early 1970s, fire managers realized that a new approach for managing rapidly escalating wildfires was necessary. Larger and more complex fires exposed weaknesses in the organizational command structure. More homes were being built in the interface, and rapidly spreading, high-intensity fires often taxed Incident Commanders.

Without a better system, Incident Commanders were often faced with the following problems:

- Too many people were reporting to only one supervisor. As a result, it was realized that a supervisor could optimally manage up to five people reporting to him or her with seven people being the maximum number. This is called **span of control.**

span of control
supervisory ratio of from three to seven individuals with five to one being established as optimum

- Fire agencies were often working with different organizational structures, which created confusion when agencies had to work together.
- There was no common language between fire agencies. Terminology differences existed.
- Unclear lines of authority developed. Personnel were unclear as to who they worked for or with.
- Incident objectives were often unspecified or unclear.
- Communications were a problem because radio systems were often incompatible.
- Information did not flow well within the organization, resulting in a lack of reliable incident information.
- There was no organizational framework coordinate planning in uniform command situations.

Note
To mitigate the adverse effects of these early command structures, a system needed to be developed.

To mitigate the adverse effects of these early command structures, a system needed to be developed. Finally through a cooperative effort of local, state, and federal agencies, the Incident Command System was developed. This interagency development effort was called FIRESCOPE (Firefighting Resources of California Organized for Potential Emergencies). In 1980, FIRESCOPE transitioned into a national program called The National Interagency Incident Management System (NIIMS). When this happened, the ICS was accepted nationally in all federal wildland fire agencies.

■ **Note**
The ICS is a modular system that is both flexible and functional.

Both flexible and functional, the ICS is an all-risk system that can be used on any type or size of incident or planned event. The system is modular and builds from the top down, with the Incident Commander initially being that point at the top.

The ICS (Figure 13-1) also provides for common terminology, a common organizational framework, and procedures that are understood by all those using it. The system is simple to use and offers low operational costs.

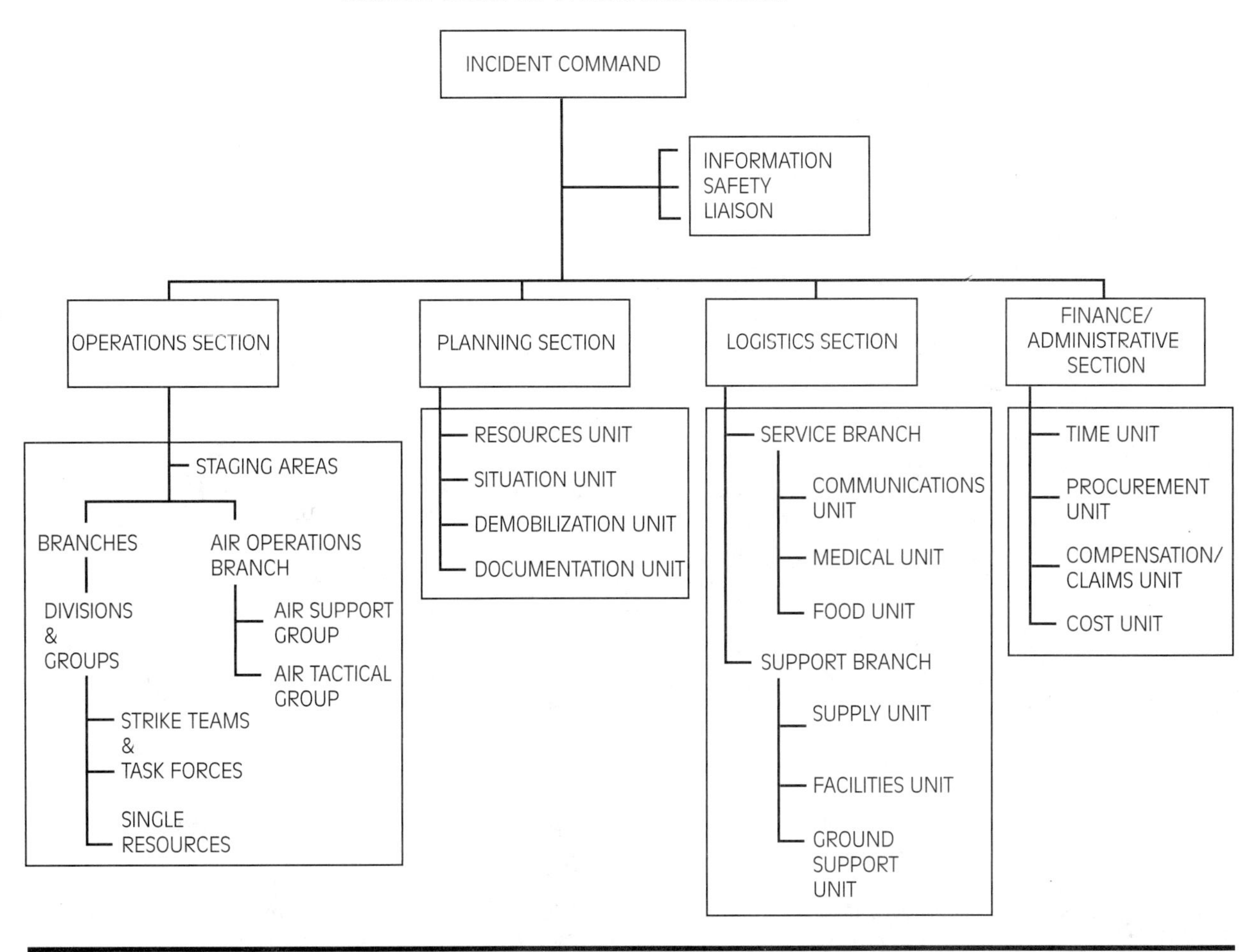

Figure 13-1 *Major functional units of the ICS broken down into sublevels.*

It is easier to understand if we first view all the major functional units of the ICS and then expand our look into its sublevels.

THE INCIDENT COMMAND SYSTEM

■ Note There are five major functional units in the ICS.

■ Note One of the major advantages of the ICS is that only those parts of the organization that are necessary are filled.

There are five major functional units in the ICS. On smaller incidents, the incident may be run by one individual, such as a fire captain of an engine company. He or she is then responsible for all the five major functional units as the Incident Commander. As the incident becomes larger and one or more of the other functional units requires independent management, then an individual is named to be responsible for that area. One of the major advantages of the ICS is that only those parts of the organization that are necessary are filled. Illustrated in Figure 13-2, the five major functional units of the ICS are as follows:

- Command
- Operations
- Planning
- Logistics
- Finance/Administration

Command

Command has overall responsibility for the management of the incident. The Incident Commander on any incident, regardless of the size, has the ultimate responsibility for the execution of each of these five functions. Once one of these functional units requires independent management, the Incident Commander may delegate the authority to manage certain functions within that area of responsibility. The Incident Commander's role is the first major component of the ICS—command.

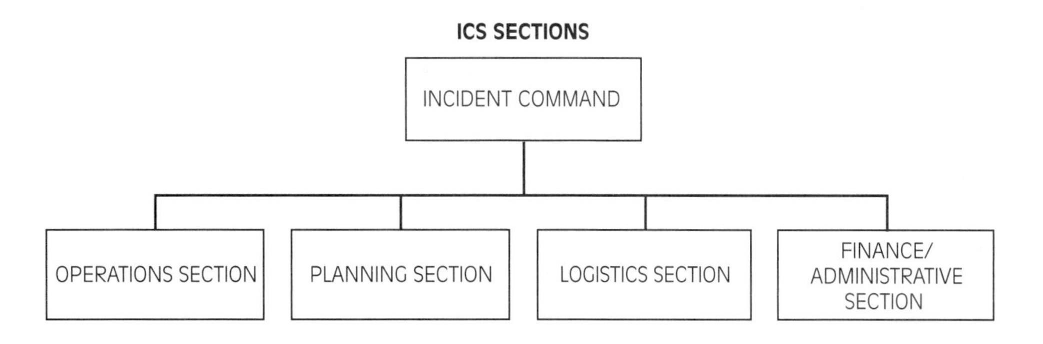

Figure 13-2 *Five major components of the ICS.*

■ Note
Command of an incident occurs in one of two ways: either as a single command under one Incident Commander or as a unified command.

Command of an incident occurs in one of two ways: either as a single command under one Incident Commander or as a unified command. Unified command is used on multijurisdictional incidents or multiagency planned events. It is a team effort in which all the stakeholders in an incident establish a common set of strategies to meet the incident objectives.

On a single command incident, the first-arriving officer usually assumes the role of Incident Commander and fills that position until there is a formal transfer of command. This transfer usually takes place upon the arrival of a higher ranking officer; however, upon arrival, the higher ranking officer may also choose to let the original Incident Commander retain command of the incident. If the higher ranking officer does take command of the incident, he or she then reassigns the original Incident Commander to a new position within the organization being set in place.

Unified command is usually established on multijurisdictional incidents. In unified command, the incident objectives are defined by the agency-designated officials working at the scene of the incident as a team. A single planning process takes place and an **Incident Action Plan** is developed for the incident.

Incident Action Plan (IAP)
a document that contains objectives reflecting the overall incident strategy and specific tactical actions and supporting information for the next operational period; the plan may be written or oral

Resource requirement operations at the incident are determined at the planning meeting. Incident tactical resources are managed by the Operations Section Chief under unified command.

It is highly recommended that you take I 400-Advanced ICS, which is part of the national training curriculum for the Incident Command System developed by the National Wildfire Coordinating Group.

The Incident Commander is the person in charge of an incident and has a wide variety of responsibilities; therefore, he or she must be fully qualified to do the job. Some of the major responsibilities include:

- Establishing the incident command post. The Incident Commander works from that location.
- Upon arrival on scene, the IC must establish the immediate priorities for the incident with safety of personnel being the first priority. The second priority is stabilization of the incident.
- Setting the incident objectives. This is a statement of intent that defines the ultimate aims.
- Developing appropriate strategies and tactics to meet the incident objectives.
- Monitoring safety on the incident. On small incidents the IC may do this; however, as the incident grows in size and/or complexity, this task is usually done by an assigned **Safety Officer.**
- Establishing the incident organization. As the incident grows in size and taxes his or her span of control, the Incident Commander sets up and mon-

Safety Officer
member of the Command Staff responsible to the Incident Commander for monitoring and assessing hazardous and unsafe situations and for developing measures for assessing personnel safety

Public Information Officer
a member of the Command Staff who acts as the central point of contact for the media

■ Note
An extended attack fire is a fire in which the initial attack resources must be supported by substantial additional numbers of personnel and equipment.

burning period
that part of each 24-hour period in which fires spread most rapidly, typically from 10:00 A.M. to sundown

General Staff
group of incident management personnel reporting to the Incident Commander, consisting of the Operations Section Chief, Planning Section Chief, Logistics Section Chief, and Financial/Administration Chief

itors a management organization. The IC coordinates the activities of both the Command and General Staff.

- Managing the planning meeting and approval of the Incident Action Plan.
- Approving requests for additional resources or the release of incident personnel or equipment.
- Authorizing and approving news media information. On larger incidents, the incident commander uses a **Public Information Officer.**

As you can see, the Incident Commander has a lot of responsibility. To help, he or she may appoint one or more Deputy Incident Commanders. They must have the same qualifications as the Incident Commander and can take over when relief is needed. This may be necessary on large fires where the Incident Commander has worked both operational periods. A deputy can also help the Incident Commander with specific tasks, lessening his work load.

Figure 13-3 and Figure 13-4 show an initial attack fire organization and an extended attack fire organization. An initial attack fire is one in which the fire is controlled by the first dispatched resources. An extended attack fire is a fire in which the initial attack resources must be supported by substantial additional numbers of personnel and equipment. The fire is controlled during the first **burning period.**

Note in the initial attack fire organization (Figure 13-3) that all resources are managed by the initial attack Incident Commander. That same Incident Commander also fills all the roles of both the Command and **General Staff.**

Figure 13-4 is an example of an ICS organization for an extended attack fire. The fire was not controlled by the initial attack companies and required additional resources and engines. It will, however, be controlled in the first burning period. Note that the Incident Commander is using a Logistics Sections Chief, who is a member of the General Staff.

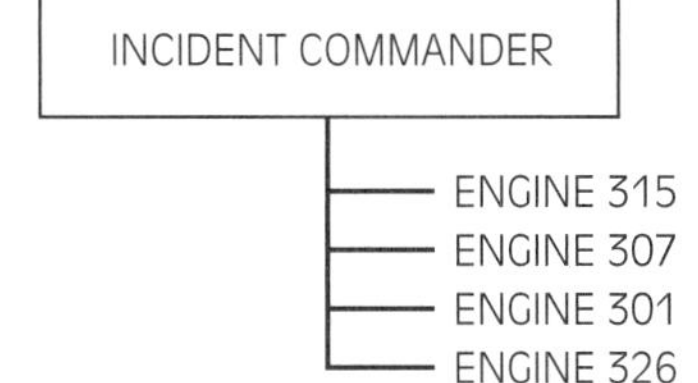

Figure 13-3 *Initial attack fire organization.*

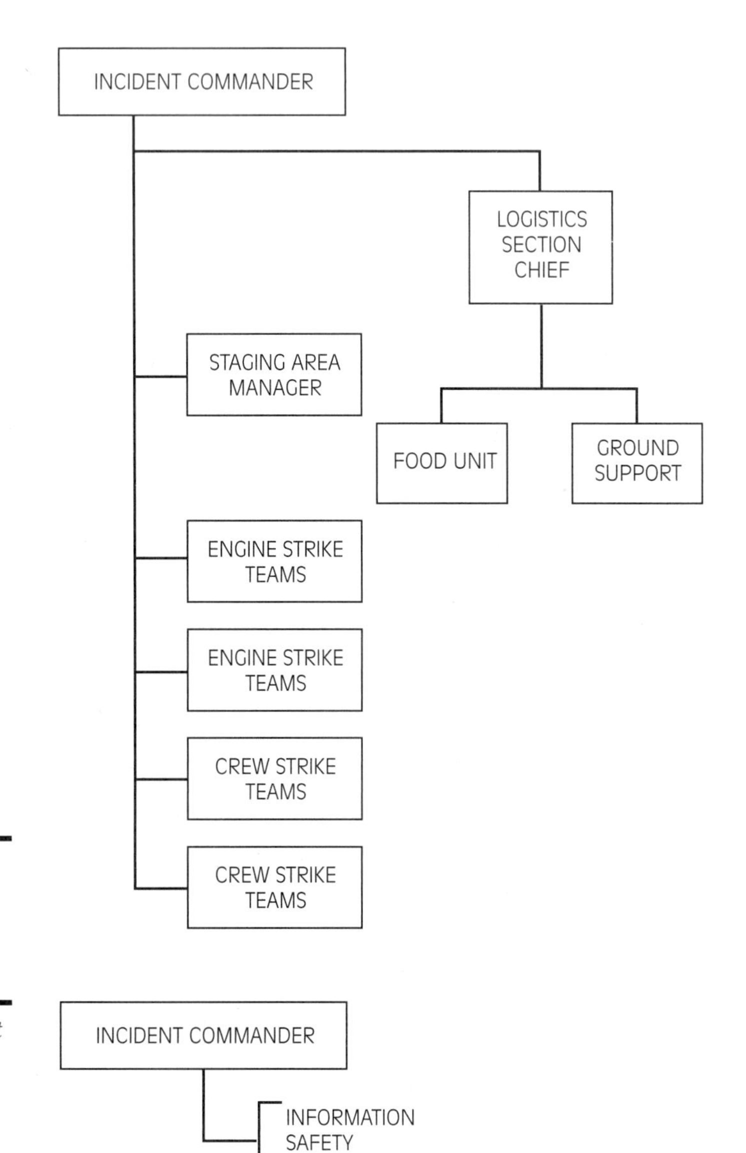

Figure 13-4 *Extended attack organization.*

Figure 13-5 *Incident Command Staff organization.*

■ Note

The Information Officer is the point of contact for the media or other appropriate agencies or organizations.

Incident Commander's Staff Command Staff is shown in Figure 13-5.

Persons who fill these positions are called officers. The Command Staff consists then of the Incident Commander, the Information Officer, the Safety Officer, and the Liaison Officer. These officers report directly to the Incident Commander. They can have assistants; they have no deputies. The Information Officer is the point of contact for the media or other appropriate agencies or organizations. He enables the Incident Commander to do the job and not be distracted by the numerous requests for information that are received on a large incident or event. The Safety Officer is the individual who monitors and assesses the incident or event for unsafe or hazardous conditions. He or she also develops measures that ensure personnel safety on the incident. The Safety Officer corrects unsafe acts through the regular chain of command. He or she has the emergency authority to stop or prevent an unsafe act if immediate action is required. Large incidents or events may have a Safety Officer with many assistants.

On large incidents or planned events where different agencies or jurisdictions are involved, a **Liaison Officer** is assigned. Each agency usually assigns an agency representative to represent its interest. The Liaison Officer is their primary point of contact.

Liaison Officer
member of the Command Staff responsible for coordinating with agency representatives from assisting and cooperating agencies

General Staff Larger incidents or events that tax the Incident Commander's span of control require the use of people to fill positions in Operations, Logistics, Planning, and Finance and Administration. They make up the IC's General Staff as illustrated in Figure 13-6, and fill the four major functions stated previously.

■ Note
General Staff may have deputies. All deputies must be as qualified as the person for whom they work.

The person in charge of one of these four major functions is designated as a chief:

- Operations Section Chief
- Planning Section Chief
- Logistics Section Chief
- Finance/Administration Section Chief

General Staff may have deputies. Each **deputy** must be as qualified as the person for whom he or she works. There may be more than one deputy assigned to a position. A deputy can work in a relief capacity, be assigned a specific task, or work with the designated chief assigned to one of the four major areas of the General Staff.

Deputy
qualified individual who could be delegated the authority to manage a functional operation or perform a specific task

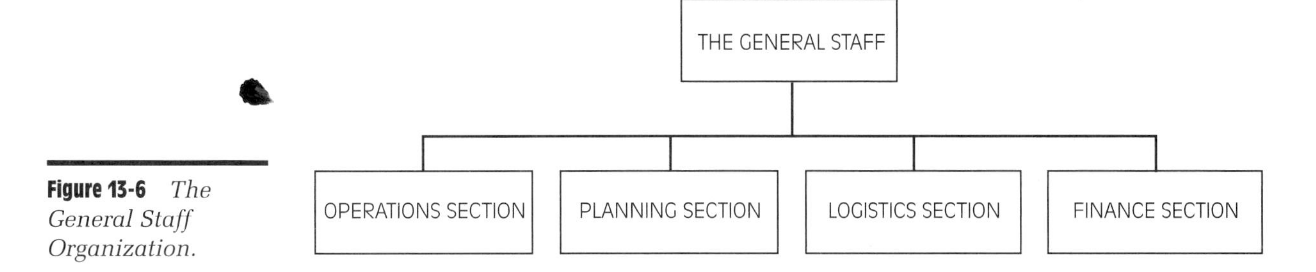

Figure 13-6 *The General Staff Organization.*

Operations Section
section responsible for all tactical operations at the incident

■ **Note**

The Operations Section Chief supervises and directs the suppression and rescue elements of the Incident Action Plan.

Division/Group Supervisor
an Operations Supervisor responsible for all suppression activities on a specific division of a fire; a group is also managed by a Division/Group Supervisor

Operations Branch Director
a person under the direction of the Operations Section Chief who is responsible for implementing that portion of the Incident Action Plan appropriate to the branch

Operations Section

All tactical operations on an incident are coordinated by the **Operations Section** (Figure 13-7). The Operations Section Chief's position is usually activated when the Incident Commander commits additional resources to the incident or event and when the number of people reporting to him or her is too great.

The Operations Section Chief supervises and directs the suppression and rescue elements of the Incident Action Plan. For each operational period on an incident or event, there is only one Operations Section Chief. He or she can use deputies to assist.

Subsections within the Operations Sections organization must be addressed: divisions, groups, and branches.

Divisions Divisions are managed by the **Division/Group Supervisor.** His or her job is to implement the control objectives stated in the Incident Action Plan for the division. He uses the resources assigned to the division to meet those control objectives. The Division Supervisor reports directly to the Operations Section Chief or an **Operations Branch Director** when this position is activated.

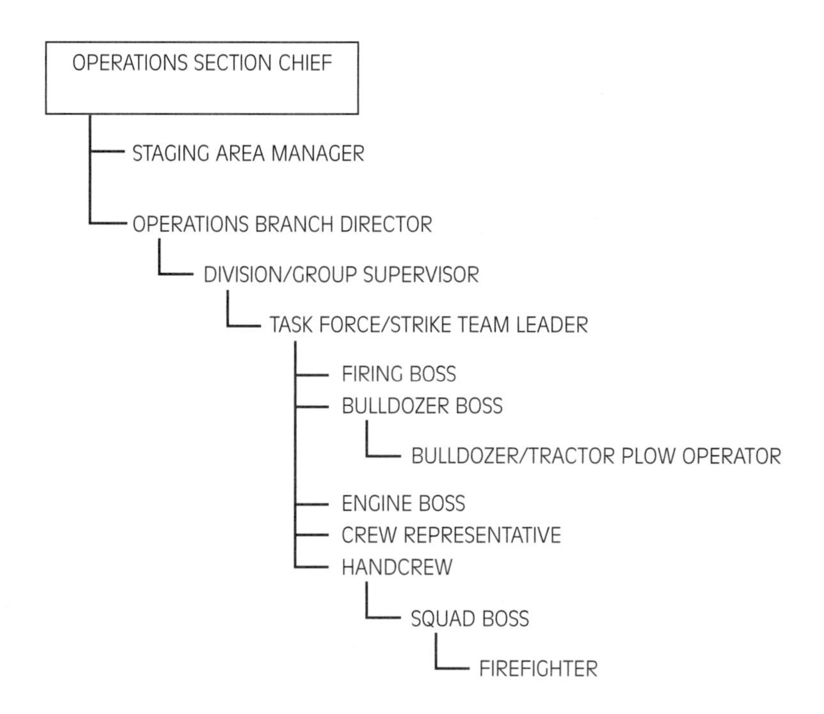

Figure 13-7 *Sample organizational chart for Operations Section.*

■ **Note**
Setting up divisions allows the incident to be split up geographically.

Setting up divisions allows the incident to be split up geographically. On wildland incidents, divisions are usually labeled by letters of the alphabet. Incidents that involve buildings, such as a high-rise, usually use the number of the floor as a division designator.

groups
established to divide the incident into functional areas of operation

Groups To define functional areas of operation, **groups** are established. A good example of this on a wildland interface fire is the establishment of a Structure Protection Group. The incident needs dictate the kind of group that should be established.

■ **Note**
Divisions and groups can be used together on the same incident.

Divisions and groups can be used together on the same incident. Again, on a large interface fire, divisions may be assigned to fight the wildland fire while the Structure Protection Group protects a group of houses. In this case, the Structure Protection Group Supervisor works directly for the Operations Section Chief or Operations Branch Director if the incident is large enough.

On small incidents, where the Operations Section Chief position has not been activated, the Group Supervisor reports directly to the Incident Commander.

Task Force
any combination of single resources assembled for a particular tactical need, with common communications and a leader

Strike Team
specified combinations of the same kind and type of resources, with common communications and a leader

Task Forces and Strike Teams As an incident grows in size, additional fire equipment needs to be brought in. These additional resources can either be classified as a **Task Force** or a **Strike Team.**

A task force is any combination of single resources assembled for a particular tactical need, with common communications and a leader. The leader is called a Task Force Leader. A task force may include a bulldozer, an engine, and a handcrew.

A Strike Team is a specified combination of the same kind and type of resource with common communications and a leader. The leader is called the Strike Team Leader. There are three primary types of strike teams: engines, handcrews, or bulldozers.

Depending on the incident size and needs, these tactical resources can report to:

- Incident Commander or Operations Section Chief
- Division/Group Supervisor
- Operations Branch Director

■ **Note**
On a large incident where there are numerous divisions and groups, an Operations Branch Director may be assigned.

staging areas
locations set up at an incident where resources can be placed while awaiting a tactical assignment on a 3-minute available basis

Branches On a large incident where there are numerous divisions and groups, an Operations Branch Director may be assigned. He or she is responsible for those objectives identified in the Incident Action Plan in his branch and for those units within his branch. He reports directly to the Operations Section Chief. An Operation Branch Director is used when the span of control limit is reached by the Operations Section Chief. The ICS branch structure can be established to represent geographic or functional areas. Geographic branches can either be delineated areas on the ground or can be set up by jurisdiction.

Note
The Air Operations Branch Director is responsible for preparation of the air operations portion of the Incident Action Plan.

Note
The Planning Section collects, evaluates, and disseminates information of the incident.

Note
The planning section chief is a member of the General Staff.

Field Observers persons responsible to the situation unit leader for collecting and reporting information about an incident obtained from personal observations and interviews

Situation Unit Leader the person responsible for the collection and organization of incident status and situation information and the evaluation, analysis, and display of that information for use by ICS personnel and agency dispatchers

Two additional areas of the Operations Section for which an understanding is necessary are staging areas and air operations.

Staging Areas To temporarily locate available resources that are assigned to the incident, **staging areas** are established. All resources assigned to staging are under the control of the Operations Section Chief.

Air Operations On larger incidents that have complex needs for aircraft use, a new branch level is activated by the Operations Section Chief. This position is ground based. The Air Operations Branch Director is responsible for preparation of the air operations portion of the Incident Action Plan. Upon approval of the plan, the Air Operations Branch Director is responsible for implementing its strategic aspects and all logistical operations.

Planning Section

The Planning Section (Figure 13-8) collects, evaluates, and disseminates information about the incident. The section also develops the incident action plan for each of the operational periods. This major functional unit of the ICS also maintains information on the status of equipment and personnel resources, maintains all incident documentation, and is responsible for preparing the Demobilization Plan.

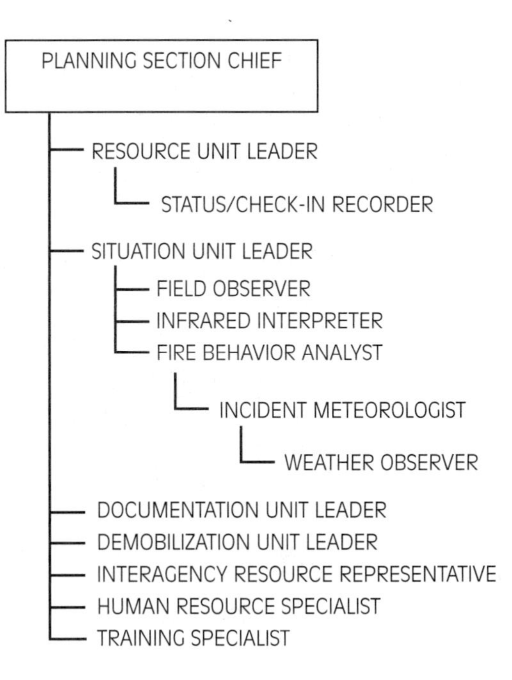

Figure 13-8 *Sample organizational chart for Planning Section.*

Fire Behavior Analyst person responsible to the Planning Section Chief for establishing a weather data collection system and for developing fire behavior predictions based on fire history, fuel, weather, and topography

Incident Meteorologist specially trained meteorologist who provides site-specific weather forecasts and information to an incident

The Planning Section Chief is a member of the General Staff; therefore, he or she can have deputies. The Planning Section Chief reports directly to the Incident Commander.

The Planning Section Chief conducts planning meetings where information is needed to understand the current incident situation and information is shared on the probable course of incident events. The information gathered to help with these predictions comes from several sources including **Field Observers** who are sent out by the **Situation Unit Leader** to gather site-specific incident information. They act as the eyes and ears for that section.

The **Fire Behavior Analyst** develops projections on the fireline intensity levels and flame lengths, predicts how large the fire will grow, develops rate of spread projection, and predicts where the fire will spread. The **Incident Meteorologist** and Weather Observer give the Fire Behavior Analyst information needed to complete fire behavior predictions. Infrared Interpreters also provide interpretations on the results of infrared aerial reconnaissance flights. The information gathered shows where active fire exists.

The Planning Section can further organize into four unit level positions, as follows:

■ Note The resource unit is responsible for checking in all incoming resources and personnel on the incident.

Resource Unit Leader the person responsible for checking in and tracking of all incident resources

Resource Unit The **Resource Unit Leader** is responsible for all activities within this section. The unit is responsible for checking in all incoming resources and personnel on the incident. Displays, charts, and lists reflect the current status and location of all incident resources.

The Status/Check-In Recorder, provided at each incident check-in location to account for incoming resources, works for the Resource Unit Leader.

■ Note The situation unit collects and processes information on the current incident status.

Situation Unit This section is run by the Situation Unit Leader. This unit collects and processes information on the current incident status. Once this information is evaluated and analyzed, it is disseminated in the form of summaries, maps, and projections.

■ Note The documentation unit is responsible for maintaining complete and accurate incident files.

Documentation Unit This unit is important as it is responsible for maintaining complete and accurate incident files. Duplication services are also provided by this section. Incident files are stored and packed by the section for legal, historical, and analytical purposes.

Demobilization Unit This unit is responsible for developing an Incident Demobilization Plan. The plan ensures that incident personnel and equipment are released in an orderly, safe, and cost-effective method when they are no longer needed.

Technical Specialists Depending upon the requirements of the Planning Section Chief or incident needs, **Technical Specialists** may be assigned to the Planning Section or may function within the unit, such as a Fire Behavior Analyst working for the

■ Note
The demobilization unit is responsible for developing an incident demobilization plan.

Technical Specialists personnel with special skills who are activated only when needed

Logistics Section Chief the person in charge of the Logistics Section who is responsible for providing facilities, services, and materials for the incident

■ Note
The Logistics Section provides all services and support needs to the incident.

■ Note
The Service Branch Chief is responsible for all incident-related communications, medical, and food-related needs.

■ Note
The Communications Unit distributes and maintains all forms of communication equipment on the incident.

Situation Unit Leader. They may also form a separate unit within the Planning Section. Technical Specialists can also be reassigned to other parts of the organization if needed. Technical Specialists provide the incident with personnel that have special skills. Some examples of Technical Specialists include:

- Meteorologist
- Fire Behavior Analyst
- Structural Engineer
- Environmental Impact Specialist
- Flood Control Specialist

Logistics Section

A member of the General Staff, the **Logistics Section Chief** can have a deputy position assigned to him. Under his direction, units within the section provide all services and support needs related to the incident. The Logistics Section Chief activates the branches and six functional units, as necessary, on an incident.

It is important, as the incident grows in size, for the Incident Commander to recognize that a need may exist for logistical service and support to the incident. Once this has been determined, the Incident Commander should activate the Logistics Section Chief position. It is important that the need for a separate logistics function be recognized. This will save both time and money on the incident.

The Logistics Section comprises two Branch Directors and six additional functional units as shown in Figure 13-9.

If necessary, the Logistics Section Chief can establish a two-branch structure to help facilitate span of control. This decision is determined by the size of the incident, the length of time the incident is expected to last, and the complexity of

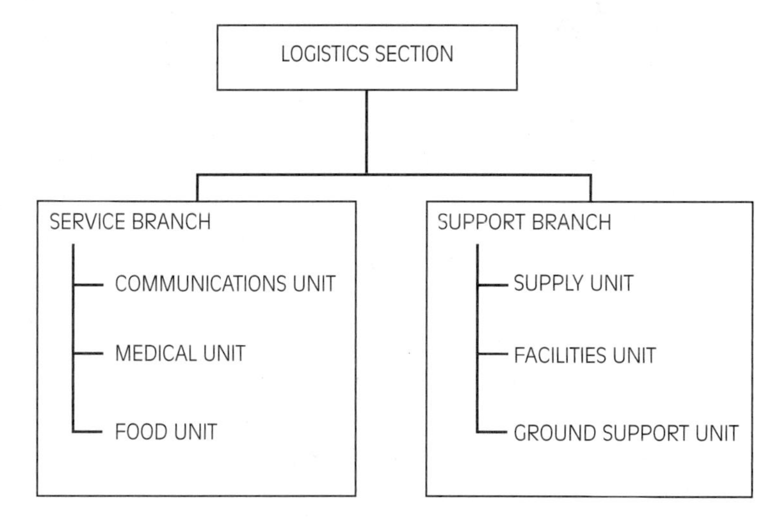

Figure 13-9 *The logistics two-section branch structure.*

■ Note
The Medical Unit provides and maintains first aid and light medical treatment stations for persons assigned to the incident.

Food Unit
functional unit within the logistics section responsible for providing meals for incident personnel

■ Note
The support branch is responsible for development and implementation of logistics plans in support of the Incident Action Plan.

Supply Unit
functional unit within the support branch of the logistics section responsible for ordering equipment and supplies required for incident operations

Facilities Unit
functional unit within the logistics section that provides the layout, activation, and management of all incident facilities

■ Note
The facilities unit establishes, sets up, and maintains all facilities required to support the incident.

the support needs. On smaller incidents, only a portion of the six separate units are activated. On a major fire, it is likely that all six functional units will be established.

Service Branch Director When activated, this position is under the supervision of the Logistics Section Chief, who is responsible for all incident-related communications, medical, and food-related needs.

Communications Unit This unit distributes and maintains all forms of communication equipment on the incident and is responsible for developing a plan, included in the Incident Action Plan, for the most effective use of the communication equipment. The Incident Communication Center is also managed by this unit.

Medical Unit This unit provides and maintains first aid and light medical treatment stations for persons assigned to the incident. Depending on the fire agency, these stations can be manned by Emergency Medical Technicians, Paramedics, or even a doctor.

The Medical Unit Leader is responsible for developing the Incident Medical Plan. Included in the Incident Action Plan, this document names the incident medical aid stations, addresses transportation issues, and gives information on hospitals.

Food Unit Responsibility for determining and providing all food and potable water needs to the incident rests with the **Food Unit.** This responsibility can be handled in a variety of ways. On large agency fires, portable kitchens are set up and staffed by incident personnel. On smaller fires or for agencies that do not have portable kitchen capabilities, a catering service may be brought in.

Support Branch Director When activated, this position is under the supervision of the Logistics Section Chief, who is responsible for development and implementation of logistics plans in support of the Incident Action Plan. This person supervises the **Supply Unit,** the **Ground Support Unit,** and the **Facilities Unit.**

Supply Unit This unit orders personnel, equipment, and supplies. In the ICS, all resource orders are placed through the Logistics Section's Supply Unit. If the Supply Unit is not activated, then the Logistics Section has the responsibility for all ordering requests. The unit also receives and stores all incident supplies, services all nonexpendable supplies and equipment, and maintains an inventory of supplies.

Facilities Unit The unit establishes, sets up, and maintains all facilities required to support the incident, including such items as sleeping facilities, security services, sanitation needs, lighting, trash removal and clean-up, and set up of incident command post trailer or other forms of shelter. The unit also provides base and camp managers.

Ground Support Unit
functional unit within the logistics section responsible for fueling, maintaining, and repairing vehicles and transporting personnel and supplies

Ground Support This unit is responsible for the maintenance and repairs of all vehicles assigned to the incident. It takes care of all the fueling of these vehicles and provides transportation services in support of incident operations. It does not provide for air services. The unit also implements the Incident Traffic Plan.

Finance/Administration Section

■ Note

The Finance/Administration Section is set up for incidents that require on-site financial management.

Set up for incidents that require on-site financial management, the **Finance/Administration Section** is under the supervisor of the Financial/Administration Section Chief, who reports to the Incident Commander. This section monitors all incident-related costs and establishes any necessary procurement contracts.

The Finance/Administration Section Chief is a member of the General Staff and is responsible for all financial and cost analysis aspects of the incident.

The section may establish four additional units as necessary, shown in Figure 13-10.

On smaller incidents, the section may not be activated. As the incident grows and a need is determined for the section, units are added.

Finance/Administration Section
section responsible for all incident costs and financial considerations; includes time, procurement, compensation/claims, and cost units

Time Unit The **Time Unit** is responsible for recording all personnel time on an incident or event. The unit also manages the commissary operations.

Time Unit
functional unit within the Finance/Administration Section responsible for recording time for incident personnel and hired equipment

Procurement Unit All financial contracts and the associated administrative paperwork related to vendors are administered by the **Procurement Unit.** The vendor contracts deal with incident equipment rentals and supply contracts. This section is also responsible for recording equipment time.

Procurement Unit
functional unit within the Finance/Administration Section responsible for financial matters involving vendor contracts

Compensation/Claims Unit This unit provides two important functions to the incident; it provides compensation for injury and claims. Compensation is responsible to see that all forms required for a workers' compensation claim are filled out correctly. Injuries or illness records for the incident are also maintained by this unit.

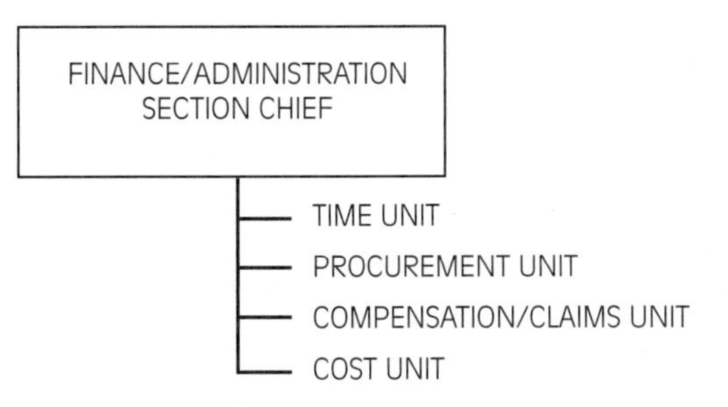

Figure 13-10 *Example of Finance/Administration Section organization.*

■ **Note** **The Compensation/Claims Unit provides compensation for injury and claims in the same unit.**

Cost Unit functional unit within the Finance/Administration Section responsible for tracking costs, analyzing cost data, making cost estimates, and recommending cost-saving measures

Cost Unit All cost data for the incident are collected by the **Cost Unit.** As incident costs are obtained and recorded, summaries are prepared. Cost estimates are shared with the Planning Section. The unit is also responsible for providing cost-saving recommendations to the Financial/Administration Section Chief.

MAJOR FIRE ORGANIZATION EXAMPLE

Figure 13-11 shows an example of a major fire organization. You can see that most of the ICS positions are filled.

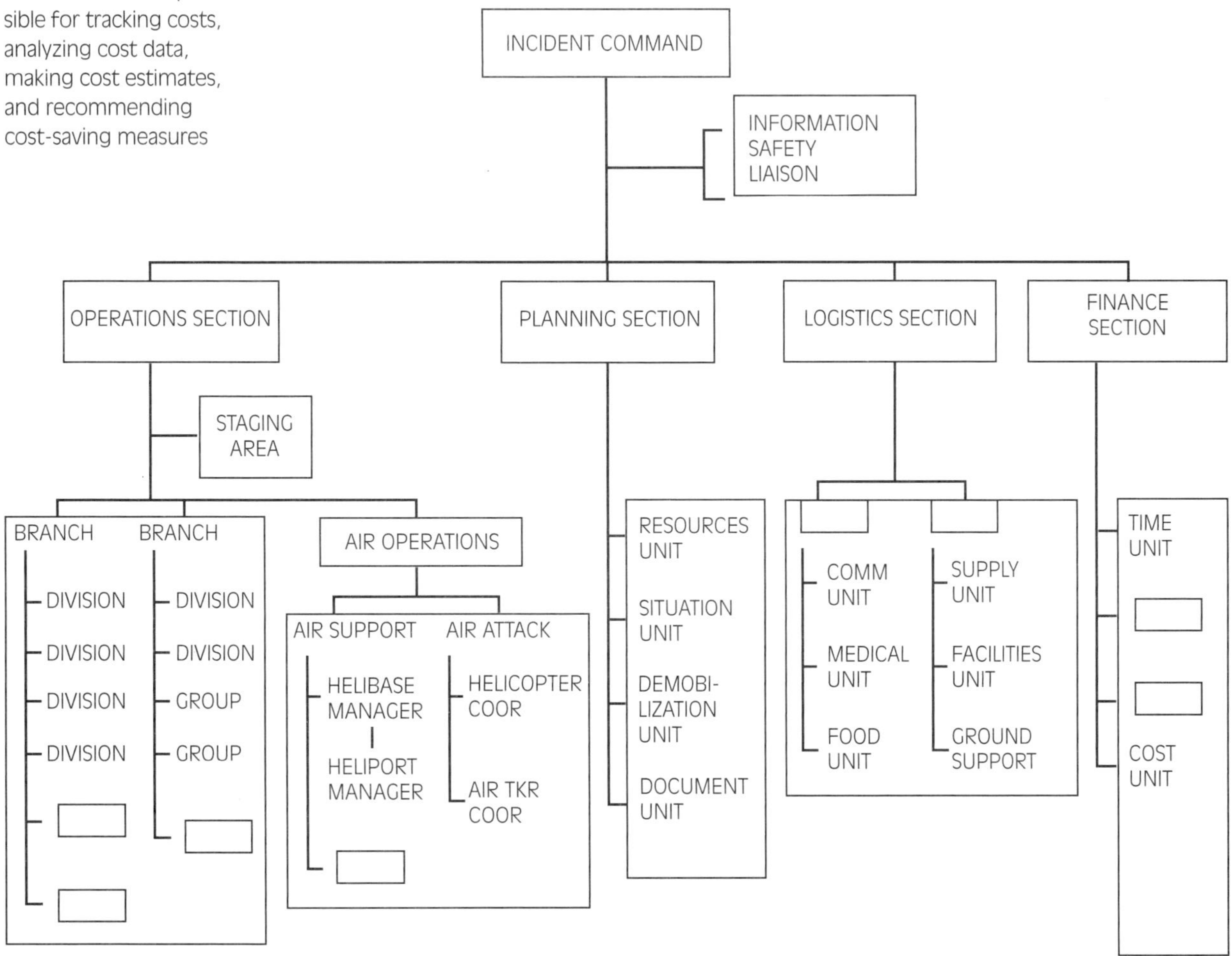

Figure 13-11 *Major fire organizational chart example.*

ICS FORMS

An Incident Action Plan, with various examples of the commonly used ICS forms, can be found in Appendix B. The Incident Action Plan represented is from a series of fires in northern California called the Feather River Complex.

Table 13-1 is a list of important ICS forms found in an Incident Action Plan that you should review.

Other forms found in the Incident Action Plan include:

Table 13-1 *Important ICS forms.*

ICS Form Number	Form Title	Description
202	Incident Objectives	Used to communicate the incident objectives and guide tactical implementations. It is developed by incident's management team.
203	Organizational Assignment List	Identifies people in primary overhead positions on the management team.
204	Division Assignment List	Identifies the work location, supervisor's name, and the resources assigned to the division; assigns tasks for the division (control operations); defines the operational period; gives radio frequencies; and gives any special instructions to the division.
205	Incident Radio Communications Plan	Lists the radio frequencies assigned to the functional areas and divisions or groups working within the functional areas. Check the operational period as radio frequency changes do occur. Also lists the radio frequency allocations for the incident. Check the remarks on the form as they may contain special considerations for radio use.
206	Medical Plan	Gives procedures to follow if medical attention is needed. Check the medical emergency procedures box as it differentiates the way medical emergencies shall be handled.
220	Air Operations Summary	Summarizes aircraft missions planned for the operational period; Defines key air operations section overhead personnel and the radio frequencies assigned to each; and identifies aircraft assigned to the incident as well as those available to it.

Form	*Description*
Safety Message	Gives a narrative of general and specific hazards facing personnel assigned to the incident. Remarks are not only restricted to fireline conditions; they address all incident safety concerns.
Fire Weather Forecast	Summarizes the weather influences for that operational period and gives forecasts for later operational periods.
Fire Behavior Forecast	Discusses past, present, and future fire behavior; identifies areas of potential concern.

You should become familiar with those forms found in an Incident Action Plan. Other routinely used forms include:

ICS Form Number	*Form Title*
201	Incident Briefing
207	Organizational Chart
209	Incident Status Summary
210	Status Change Summary
211	Check-In List
212	Vehicle Demobilization Form
213	General Message Form
214	Unit Log
215	Operational Planning Worksheet
216	Radio Requirements Worksheet
217	Radio Frequency Assignment
218	Support Vehicle Inventory
219	Resource Status Card
221	Demobilization Check-out
224	Crew Performance Rating
225	Incident Personnel Rating

Summary

The ICS is a modular system that is both flexible and functional. The system builds from the top down and can be used on any incident or event. It is built around five major functions: (1) Incident Command, (2) Operations, (3) Planning, (4) Logistics, and (5) Finance/Administration.

The incident organization should develop to meet the functional requirements of the incident or event. The Incident Commander determines the organization he needs and fills only those parts of the organization required. There is not one "best" way to organize an incident. This comes with a good working knowledge of the system and experience.

Lines of supervisory authority and formal reporting relationships are established by the ICS.

Remember that incidents or events are not always static and your organizational structure needs may change. The beauty of the system is that it can be expanded to meet virtually any need.

Review Questions

1. How did the ICS come about?
2. Name the five major functional units in the ICS.
3. Name the members of the Incident Commander's staff.
4. Who makes up the General Staff?
5. Who does a Division/Group Supervisor work directly for?
6. What is the function of the Planning Section?
7. What is the function of the Logistics Section?
8. What does the ground support unit do?

References

National Wildfire Coordinating Group, *S 200 Initial Attack Incident Commander* (Boise, ID: National Interagency Fire Center, 1989).

National Wildfire Coordinating Group, *S 300 Incident Commander Multiple Resources—Extended Attack* (Boise, ID: National Interagency Fire Center, 1986).

National Wildfire Coordinating Group, *ICS Forms Manual* (Boise ID: National Interagency Fire Center, 1987).

National Wildfire Coordinating Group, *Incident Command System,* National Training Curriculum Series (Boise, ID: National Interagency Fire Center, 1994).

National Wildfire Coordinating Group, *History of ICS* (Boise, ID: National Interagency Fire Center, 1994).

Fire Behavior Fuel Model Descriptions*

GRASS GROUP

Fuel Model 1 (1 foot deep) Fire spread is governed by the fine herbaceous fuels that have cured or are nearly cured. Fires are surface fires that move rapidly through cured grass and associated material. Very little shrub or timber is present, generally less than one-third of the area.

Grasslands and savanna are represented along with stubble, grass-tundra, and grass-shrub combinations that meet the foregoing area constraint. Annual and perennial grasses are included in this fuel model.

Fuel Model 2 (1 foot deep) Fire spread is primarily through the fine herbaceous fuels, either curing or dead. These are surface fires where the herbaceous material, besides litter and dead-down stemwood from the open shrub or timber overstory, contribute to the fire intensity. Open shrub lands and pine stands or scrub oak stands that cover one-third to two-thirds of the area may generally fit this model but may include clumps of fuels that generate higher intensities and may produce firebrands. Some pinyon-juniper may be in this model.

---

*Source: National Wildfire Coordinating Group, *Fireline Handbook 3* (Boise, ID: National Interagency Fire Center, 1998).

Fuel Model 3 (2.5 feet deep) Fires in this fuel are the most intense of the grass group and display high rates of spread under the influence of wind. The fire may be driven into the upper heights of the grass stand by the wind and cross over standing water. Stands are tall, averaging about 3 feet, but considerable variation may occur. Approximately one-third or more of the stand is considered dead or cured and maintains the fire.

SHRUB GROUP

Fuel Model 4 (6 feet deep) Fire intensity and fast spreading fires involve the foliage and live and dead fine woody materials in the crowns of a nearly continuous secondary overstory. Examples are stands of mature shrub, 6 or more feet tall, such as California mixed chaparral, the high pocosins along the East Coast, the pine barrens of New Jersey, or the closed, jack pine stands of the north-central states. Besides flammable foliage, there is dead woody material in the stand that significantly contributes to the fire intensity. Height of stands qualifying for this model vary with local conditions. There may also be a deep litter layer that confounds suppression efforts.

Fuel Model 5 (2 feet deep) Fire is generally carried in the surface fuels made up of litter cast by the shrubs and the grasses or forbs in the understory. Fires are generally not very intense as surface fuel loads are light, the shrubs are young with little dead material, and the foliage contains little volatile material. Shrubs are generally not tall, but nearly cover the entire area. Young, green stands with little or no deadwood such as laurel, vine maple, alder, or even chaparral, manzanita, or chamise are examples. As the shrub fuel moisture drops, consider using a Fuel Model 6.

Fuel Model 6 (2.5 feet deep) Fires carry through the shrub layer where the foliage is more flammable than Fuel Model 5, but require moderate winds (greater than 8 miles per hour) at midflame height. Fire will drop to the ground at low windspeeds or openings in the stand. Shrubs are older, but not as tall as shrub types of Model 4, nor do they contain as much fuel as Model 4. A broad range of shrub conditions is covered by this model. Typical examples include intermediate stands of chamise, chaparral, oak brush, low pocosin, Alaskan spruce taiga, and shrub tundra. Cured hardwood slash can be considered. Pinyon-juniper shrublands may fit, but may overpredict rate of spread except at high winds, for example, 20 miles per hour at the 20-foot level.

Fuel Model 7 (2.5 feet deep) Fire burns through the surface and shrub strata equally. Fire can occur at higher dead fuel moisture contents due to the flammable nature of live foliage. Shrubs are generally 2 to 6 feet high. Examples are palmetto-gallberry understory-pine overstory sites, low pocosins, and Alaska Black Spruce-shrub combinations.

TIMBER LITTER GROUP

Fuel Model 8 (0.2 foot deep) Slow burning ground fires with low flame heights are generally the case, although an occasional "jackpot" or heavy fuel concentration may cause a flare-up. Only under severe weather conditions do these fuels pose fire problems. Closed-canopy stands of short-needle conifers or hardwoods that have leafed out support fire in the compact litter layer. This layer is mainly needles, leaves, and some twigs since little undergrowth is present in the stand. Representative conifer types are white pine, lodgepole pine, spruce, true fin, and larches.

Fuel Model 9 (0.2 foot deep) Fires run through the surface litter faster than Model 8 and have higher flame height. Both long-needle conifer and hardwood stands, especially the oak-hickory types, are typical. Fall fires in hardwoods are representative, but high winds will actually cause higher rates of spread than predicted because of spotting caused by rolling and blowing leaves. Closed stands of long-needled pine like ponderosa, Jeffrey, and red pines or southern pine plantations are grouped in this model. Concentrations of dead-down woody material contribute to possible torching out of trees, spotting, and crowning activity.

Fuel Model 10 (1 foot deep) The fires burn in the surface and ground fuels with greater fire intensity than other timber litter models. Dead-down fuels include greater quantities of 3-inch or larger limb wood resulting from overmaturity or natural events that create a large load of dead material on the forest floor. Crowning out, spotting, and torching of individual trees are more frequent in this fuel situation leading to potential fire control difficulties. Any forest type may be considered when heavy down materials are present; examples are insect or diseased stands, wind-thrown stands, overmature situations with deadfall, and cured light thinning or partial-cut slash.

LOGGING SLASH GROUP

Fuel Model 11[1] (1 foot deep) Fires are fairly active in the slash and herbaceous material intermixed with the slash. The spacing of the rather light fuel load, shading from overstory, or the aging of the fine fuels can contribute to limiting the fire potential. Light partial cuts or thinning operations in mixed conifer stands, hardwood stands, and southern pine harvests are considered. Clear-cut operations generally produce more slash than represented here. The <3 inch material load is less than 12 tons per acre. The >3 inch material is represented by not more than 10 pieces, 4 inches in diameter along a 50-foot transect.

[1]When working in Fuel Model 11 or 12 with significant "red" needles attached to limbs, consider using the next heavier model. For example Fuel Model 11 with "red" needles, use Fuel Model 12.

Fuel Model 12[2] (2.3 feet deep) Rapidly spreading fires with high intensities capable of generating firebrands can occur. When fire starts, it is generally sustained until a fuel break or change in fuels is encountered. The visual impression is dominated by slash and much of it is less than 3 inches in diameter. These fuels total less than 35 tons per acre and seem well distributed. Heavily thinned conifer stands, clear-cuts and medium or heavy partial cuts are represented. The >3 inch material is represented by encountering 11 pieces, 6 inches in diameter, along a 50-foot transect.

Fuel Model 13[3] (3 feet deep) Fire is generally carried by a continuous layer of slash. Large quantities of >3 inch material are present. Fires spread quickly through the fine fuels and intensity builds up as the large fuels start burning. Active flaming is sustained for long periods and a wide variety of firebrands can be generated. These contribute to spotting problems as the weather conditions become more severe. Clear-cut and heavy partial-cuts in mature and overmature stands are depicted where the slash load is dominated by the >3 inch material. The total load may exceed 300 tons per acre, but the <3 inch fuel is generally only 10% of the total load. Situations where the slash still has "red" needles attached, but the total load is lighter like a Model 12, can be represented because of the earlier high intensity and faster rate of spread.

NOTE

Anderson, Hal E., *Aids to Determining Fuel Models for Estimating Fire Behavior.* Gen. Tech Report INT-122 (Missoula, MT: USDA Forest Service Technology and Development Program, 1982).

[2]See Note 1.
[3]If "red" needles are attached, consider using a Fuel Model 4.

Incident Action Plan

Incident Action Plan

Operational Period (Date / Time): **Night Operations 1800 hrs. 8/26/99 - 0600 hrs. 8/27/99**

Incident Name: **Feather River Complex**

Incident Number: **PNF 1066**

- Red Flag Warning -

Issued for the West Side of Sierras From 1200 hrs. to 2400 hrs. 8/26/99

National Weather Service

Approved by Incident Commander: Tom Hutchison, ICTI

Date / Time Prepared: ***8/26/99 4:23:15 PM***

ICS-202 Incident Objectives

Incident Name	Incident Number	Date / Time Prepared:
Feather River Complex	**PNF 1066**	**8/26/99 6:00:00 PM**

Operational Period

Night Operations 1800 hrs. 8/26/99 - 0600 hrs. 8/27/99

General Incident Objectives

- Through the use of the standard fire orders, 18 situations, LCES and evacuations provide for firefighter and public safety.

- Protect private property and structures using aggressive perimeter control.

- Keep Devil's Gap Fire, Division E, southeast of the South Fork Feather River, Northeast of Road 23N21, North of Road 21N16, West of Road 21N15.1

- Keep Hibbard Creek Fire, Division C, South and West of Road 23N11X, North of Road 23N31, East of the North Fork Feather River.

- Mosquito Ridge, China Gulch, and Stag Point Fires status will be monitored only. Div B, D, and Z

- Keep Bean Creek Fire, Division A, North and West of Middle Fork Feather River, East of Road 21N36D and South of Brush Creek

- With media releases and community meetings, keep the affected areas well informed on fire activities.

Weather Forcast for Operational Period

See attached

General / Safety Message

See attached

ICS-202 Incident Objectives:	☑	Safety Plan:	☑
ICS-203 Organization Assignment List	☑	Weather Forecast:	☑
ICS-204 Division Assignment:	☑	Incident Map:	☑
ICS-205 Radio Communications Plan:	☑	Incident Base Map:	☑
ICS-206 Medical Plan:	☑	Transportation Plan / Map:	☑
ICS-220 Air Operations Plan:	☐	Other **LCES**	■

Prepared by Plans Section Chief:	Approved by Incident Commander:
Tony Sarzotti, PSCI	**Tom Hutchison, ICTI**

ICS-203 Organization Assignment List

1. Incident Name	Feather River Complex
4. Ops. Period	NIGHT Ops. 1800 hrs 8/26/99 to 0600 8/27/99
2. Date/Time Prepared:	08/26/99 18:00:00

5. Incident Command and Staff

Position	Name
Incident Commander	TOM HUTCHISON
I.C. / Deputy	STEVE GAGE
I.C. / Deputy	
Safety Officer	TOM CABLE/RALPH TAYLOR
Information Officer	JACK HORNER
Liaison Officer	

Agency:	Rep Name:

7. Planning Section

Position	Name
Planning Sect. Chief	TONY SARZOTTI
Deputy PS	DAN GOSNELL
Resource Unit	MARK GEARY
Situation Unit	MIKE FERDIG
Documentation Unit	
Demobilization Unit	

Technical Specialist:

FBAN - LINDA ADAMS
IMET - MIKE SMITH

8. Logistic Section

Position	Name
Logistic Section Chief	A. KELLOGG
Deputy LSC	
Support Branch Director	
Supply Unit	ROSS PECKINPAH
Facilities Unit	WALLY GROGAN
Ground Support Unit	RALDON BROWN
Service Branch Director	
Communications Unit	JACK FROGGATT
Medical Unit	
Food Unit	GINGER SHAW
Prepared by Res. Unit:	MARK GEARY

Operations Section

Position	Name
Ops Sect. Chief	TERRY MOLZAHN
Ops Chief / Deputy	JERRY McGOWAN
Branch I Director	BR. I - JOE MOLHOEK
Branch I Deputy	
Division / Group	DIV. A - KARL SMITH
Division / Group	DIV. B - NOT STAFFED
Division / Group	DIV. C - NOT STAFFED
Division / Group	DIV. D - NOT STAFFED
Division / Group	
Branch I Director	BR. 2 - NOT STAFFED
Branch I Director	
Division / Group	DIV. E - GREG BLACK
Division / Group	DIV. G - NOT STAFFED
Division / Group	DIV. K - LALO GONZALEZ
Division / Group	DIV. Z - NOT STAFFED
Division / Group	POST RIDGE SECONDARY N/S
Branch I Director	
Branch I Director	
Division / Group	
Division / Group	
Division / Group	
Division / Group	
Division / Group	

Air Operations Branch

Position	Name
Air Ops Branch Dir	
Air Attack Grp Supv	
Air Support Grp Supv	
Helicopter Coord.	
Air Tanker Coord.	

Finance Section

Position	Name
Finance Sect. Chief	KATHY REMLEY
Deputy FSC	
Time Unit	DIANE LACEY
Procurement Unit:	DOUG HYDE-SADO
Comp / Claim Unit	MARIE GLETNE
Cost Unit	

ICS-204 Division Assignment List

1. BRANCH NUMBER	2. DIVISION / GROUP
BRANCH I	DIVISION A ** (P54393)**
3. INCIDENT NAME	OPERATIONAL PERIOD
Feather River Complex	NIGHT 1800 hrs 8/26/99 - 0600 hrs 8/27/99

5. OPERATIONS PERSONNEL

OPERATIONS CHIEF	JERRY McGOWAN	DIVISION / GROUP SUPVR	KARL SMITH
BRANCH DIRECTOR	JOE MOLHOEK	AIR ATTACK GRP SUPVR.	

6. RESOURCES ASSIGNED THIS PERIOD

Strike Team / Task Force / Resource Designator	Leader	Number Persons	Transport Needed	Drop Off Point/Time	Pick Up Point/Time
SCORPIONS #1,#2,#3		60	NO	Div A/1900HRS	
ENF E-63		5	NO	Div A/1900HRS	
ENF E-53		5	NO	Div A/1900HRS	

7. CONTROL OPERATIONS

Continue to hold and mop up 100 feet inside line.

Provide structure protection through aggressive perimeter control.
BURN OUT AS DIRECTED BY DIVS

8. SPECIAL INSTRUCTIONS

Known archaeological site on BLM land at South end.
Do not use dozers unless threat to life or property exists.
L.C.E.S.!!! DO NOT FALL SNAGS. FLAG ALL SNAGS FOR DAY SHIFT TO FALL.
BE AWARE OF ROLLING MATERIAL.
BE AWARE OF DOWNDRAFTS.

9. DIVISION/GROUP COMMUNICATION SUMMARY

		FREQ.	SYSTEM	CHANNEL			FREQ.	SYSTEM	CHANNEL
COMMAND	Local	171.425		1	SUPPORT	Local			
	Repeater	172.350		1		Repeater			
Div/Group TACTICAL		168.200		2	GROUND TO AIR		163.700		4

PREPARED BY RESOURCE UNIT LEADER	APPROVED BY PLANS SECTION CHIEF
MARK GEARY	TONY SARZOTTI

ICS-204 Division Assignment List

1. BRANCH	2. DIVISION / GROUP
BRANCH I	DIVISION B,C, and D
3. INCIDENT NAME	4. OPERATIONAL PERIOD
Feather River Complex	Night 1800 hrs 8/26/99- 0600 hrs 8/27/99

5. OPERATIONS PERSONNEL

OPERATIONS CHIEF	JERRY McGOWAN	DIVISION / GROUP SUPVR.	NOT STAFFED
BRANCH DIRECTO	JOE MOLHOEK	AIR ATTACK SUPERVISOR	

6. RESOURCES ASSIGNED THIS PERIOD

Strike Team / Task Force / Resource Designator	Leader	Number Persons	Transport Needed	Drop Off Point/Time	Pick Up Point/Time
NOT STAFFED					

7. CONTROL OPERATIONS

NOT STAFFED

8. SPECIAL INSTRUCTIONS

9. DIVISION/GROUP COMMUNICATION SUMMARY

		FREQ.	SYSTEM	CHANNEL			FREQ.	SYSTEM	CHANNEL
COMMAND	Local	171.475	King/NIFC	1	SUPPORT	Local			
	Repeater	172.350				Repeater			
	Div/Group TACTICAL	168.200	King/NIFC	2		GROUND TO AIR	163.700	King/NIFC	4

PREPARED BY RESOURCE UNIT LEADER	APPROVED BY PLANS SECTION CHIEF
MARK GEARY	TONY SARZOTTI

ICS-204 Division Assignment List

1. BRANCH NUMBER	2. DIVISION / GROUP
BRANCH II	DIVISION E
3. INCIDENT NAME	OPERATIONAL PERIOD
Feather River Complex	NIGHT 1800 hrs 8/26/99-0600 hrs 8/27/99

5. OPERATIONS PERSONNEL

OPERATIONS CHIEF	JERRY McGOWAN	DIVISION / GROUP SUPVR	GREG BLACK
BRANCH DIRECTOR	JOE MOLHOEK	AIR ATTACK GRP SUPVR.	

6. RESOURCES ASSIGNED THIS PERIOD

Strike Team / Task Force / Resource Designator	Leader	Number Persons	Transport Needed	Drop Off Point/Time	Pick Up Point/Time
SRV #29	RENDON	20	NO	DP 1 / 1900 HRS	
WATER TENDER "DIETZ" 3,800 GALS		1	NO	DP 1 / 1900 HRS	

7. CONTROL OPERATIONS

CONTINUE TO HOLD FIRELINE
MOP UP FIRE 100 FEET INSIDE FIRELINE.

8. SPECIAL INSTRUCTIONS

KEEP CLEAR OF HAZARD SNAGS. FLAG FOR DAY SHIFT TO FALL
HEADS UP FOR ROLLING MATERIALS
BE AWARE OF DOWNDRAFTS
L.C.E.S.!!!!!
MAINTAIN CONSTANT COMMUNICATIONS

9. DIVISION/GROUP COMMUNICATION SUMMARY

		FREQ.	SYSTEM	CHANNEL			FREQ.	SYSTEM	CHANNEL
COMMAND	Local	171.425		1	SUPPORT	Local			
	Repeater	172.350		1		Repeater			
	Div/Group TACTICAL	168.600		3		GROUND TO AIR	163.700		4

PREPARED BY RESOURCE UNIT LEADER	APPROVED BY PLANS SECTION CHIEF
MARK GEARY	TONY SARZOTTI

ICS-204 Division Assignment List

1. BRANCH NUMBER	2. DIVISION / GROUP
BRANCH II	DIVISION G and Z
3. INCIDENT NAME	OPERATIONAL PERIOD
Feather River Complex	Night 1800 hrs 8/26/99 - 0600 hrs 8/27/99

5. OPERATIONS PERSONNEL

OPERATIONS CHIEF	JERRY McGOWAN	DIVISION / GROUP SUPVR	NOT STAFFED
BRANCH DIRECTOR	JOE MOLHOEK	AIR ATTACK GRP SUPVR.	

6. RESOURCES ASSIGNED THIS PERIOD

Strike Team / Task Force / Resource Designator	Leader	Number Persons	Transport Needed	Drop Off Point/Time	Pick Up Point/Time
NOT STAFFED					

7. CONTROL OPERATIONS

NOT STAFFED

8. SPECIAL INSTRUCTIONS

9. DIVISION/GROUP COMMUNICATION SUMMARY

		FREQ.	SYSTEM	CHANNEL			FREQ.	SYSTEM	CHANNEL
COMMAND	Local	171.425	King/NIFC	1	SUPPORT	Local			
	Repeater	172.350				Repeater			
	Div/Group TACTICAL	168.600	King/NIFC	3		GROUND TO AIR	168.700	King/NIFC	4

PREPARED BY RESOURCE UNIT LEADER	APPROVED BY PLANS SECTION CHIEF
MARK GEARY	TONY SARZOTTI

ICS-204 Division Assignment List

1. BRANCH NUMBER	2. DIVISION / GROUP
BRANCH II	DIVISION K
3. INCIDENT NAME	OPERATIONAL PERIOD
Feather River Complex	NIGHT 1800 hrs 8/26/99-0600 hrs 8/27/99

5. OPERATIONS PERSONNEL

OPERATIONS CHIEF	JERRY McGOWAN	DIVISION / GROUP SUPVR	LALO GONZALEZ
BRANCH DIRECTOR	JOE MOLHOEK	AIR ATTACK GRP SUPVR.	

6. RESOURCES ASSIGNED THIS PERIOD

Strike Team / Task Force / Resource Designator	Leader	Number Persons	Transport Needed	Drop Off Point/Time	Pick Up Point/Time
2 ENGINES TBA		10	NO	DP 1 /1900HRS	
2 CREWS TBA		40	NO	DP 1 /1900HRS	

7. CONTROL OPERATIONS

CONTINUE TO HOLD LINE MOP UP 100' INSIDE LINE

8. SPECIAL INSTRUCTIONS

KEEP CLEAR OF HAZARD SNAGS. FLAG FOR DAY SHIFT TO FALL
HEADS UP FOR ROLLING MATERIALS
BE AWARE OF DOWNDRAFTS
L.C.E.S.!!!!!
MAINTAIN CONSTANT COMMUNICATIONS

9. DIVISION/GROUP COMMUNICATION SUMMARY

		FREQ.	SYSTEM	CHANNEL			FREQ.	SYSTEM	CHANNEL
COMMAND	Local	172.475		1	SUPPORT	Local			
	Repeater	171.350		1		Repeater			
Div/Group TACTICAL		168.600		3	GROUND TO AIR		163.700		4

PREPARED BY RESOURCE UNIT LEADER	APPROVED BY PLANS SECTION CHIEF
MARK GEARY	TONY SARZOTTI *aws*

ICS-204 Division Assignment List

1. BRANCH NUMBER	2. DIVISION / GROUP
BRANCH II	POST RIDGE SECONDARY
3. INCIDENT NAME	OPERATIONAL PERIOD
	Night 1800 hrs 8/26/99 - 0600 hrs 8/27/99

5. OPERATIONS PERSONNEL

OPERATIONS CHIEF	JERRY McGOWAN	DIVISION / GROUP SUPVR	NOT STAFFED
BRANCH DIRECTOR	JOE MOLHOEK	AIR ATTACK GRP SUPVR.	

6. RESOURCES ASSIGNED THIS PERIOD

Strike Team / Task Force / Resource Designator	Leader	Number Persons	Transport Needed	Drop Off Point/Time	Pick Up Point/Time
NOT STAFFED					

7. CONTROL OPERATIONS

NOT STAFFED

8. SPECIAL INSTRUCTIONS

9. DIVISION/GROUP COMMUNICATION SUMMARY

		FREQ.	SYSTEM	CHANNEL			FREQ.	SYSTEM	CHANNEL
COMMAND	Local	171.425	King/NIFC	1	SUPPORT	Local			
	Repeater	172.350				Repeater			
Div/Group TACTICAL		168.600	King/NIFC	3	GROUND TO AIR		168.700	King/NIFC	4

PREPARED BY RESOURCE UNIT LEADER	APPROVED BY PLANS SECTION CHIEF
MARK GEARY	TONY SARZOTTI

FEATHER RIVER COMPLEX
SAFETY MESSAGE
NIGHT SHIFT - 8/26/1999

ROSES ARE RED, VIOLETS ARE BLUE, MAKE SURE YOU TAKE CARE OF "YOU"!

* TAILGATE SAFETY SESSIONS ARE A MUST!

* ENSURE PROPER PPE IS WORN CORRECTLY, DRINK PLENTY OF LIQUIDS, PACE YOURSELF.

* DRIVE WITH YOUR LIGHTS ON. OBEY SPEED LIMITS. WATCH FOR THE OTHER DRIVER!

* BE AWARE OF THUNDERSTORMS AND ASSOCIATED WINDS. DOWNDRAFTS

* WATCH FOR ROLLING MATERIALS.

* POISON OAK. SEE THE MEDICAL UNIT FOR TECNU SOAP AND MEDICATION.

* MEAT BEES ARE VERY ACTIVE AND NEST IN THE GROUND. WATCH YOUR STEP AND FLAG NESTS.

* SNAG SAFETY. FELL OR FLAG HAZARD SNAGS. STAY AWAY FROM SNAG AREAS AT NIGHT.

RALPH TAYLOR - TOM CABLE SOFR'S

INCIDENT ACTION PLAN SAFETY ANALYSIS

1. Incident Name: FEATHER RIVER COMPLEX | AUGUST 26, 1999 | NIGHT SHIFT 8/26/99

	LCES Analysis of Tactical Applications Lookouts Communications Escape routes Safety zones											Other Risk Analysis						
Division/Group	Indirect Fireline	Downhill Fireline	Underslung Fireline	Mid-slope Fireline	Frontal Assault	Anchor points	Extreme Conditions (Spotting, Wind-driven)	Return Potential			LCES Mitigations	Hazard Materials	Transportation, 1HR +	Communications	Structure Protection			Other Risk Mitigations
	B	R	A	N	C	H	1											
A		X	X			X	X		X		LCES,FALL HAZ SNAGS, BURN OUT		X		X			DRIVE SPEED LIMITS, LIGHTS ON
B											NOT STAFFED							
C									X		NOT STAFFED							
D											NOT STAFFED							
	B	R	A	N	C	H	2											
E							X		X		KEEP CLEAR OF HAZARD SNAGS, HEADS UP FOR ROLLING MATERIALS, DOWNDRAFTS		X					DRIVE SAFELY W/LIGHTS ON
K		X				X	X	X	X		KEEP CLEAR OF HAZARD SNAGS, CURRENT WEATHER, WATCH ROLLING MATERIALS		X					SPEED LIMITS AND LIGHTS ON. CONSTANT COMMUNICATIONS.
G							X				KNOW CURRENT WEATHER INFO, HEADS UP FOR ROLLING MATERIALS						X	MAINTAIN CONSTANT COMMUNICATIONS
PR S											NOT STAFFED FOR NIGHT SHIFT							
Z											NOT STAFFED FOR NIGHT SHIFT							

Prepared by (Name and Position)
TOM CABLE - SOFR1 Ralph D. Taylor SOFR1

ICS 215A

ICS-205 Incident Radio Communication Plan

1. Incident Name	2. Date/Time Prepared:	3. Operational Period
Feather River Complex	8/26/99 2:33:26 PM	Night 1800 hrs 8/26 - 0600 hrs 8/27/99

4. Basic Radio Channel Utilization

System Cache	Channel	Function	Frequency	Assignment	Remarks
King/NIFC	1	Command	Rx 171.475 Tx 172.350	ALL DIVISIONS	PNF ADMIN NET
King/NIFC	2	TAC 2	168.200	DIV. A, B, C, AND Z	
King/NIFC	3	TAC 3	168.600	DIV. E, G, K, AND Z	
King/NIFC	4	Air to Ground	163.700	ALL DIVISIONS	

Prepared by Communication Unit Leader: JACK FROGGATT. COML

MEDICAL PLAN	1. Incident Name Feather River Complex	2. Date Prepared 8-26-99	3. Time Prepared 1500	4. Operational Period 1800-0600hrs Night

5. Incident Medical Aid Station

Medical Aid Stations	Location	Paramedics Yes	Paramedics No
Feather River Base			X
Brush Creek Camp			X

6. Transportation

A. Ambulance Services

Name	Address	Phone	Paramedics Yes	Paramedics No
Oroville Hospital	2767 Olive Highway	911 or	X	
		533-8500		

B. Incident Ambulances

Name	Location	Paramedics Yes	Paramedics No
None			

7. Hospitals

Name	Address	Travel Time Air	Travel Time Grnd	Phone	Helipad Yes	Helipad No	Burn Center Yes	Burn Center No
Oroville Hospital	2767 Olive Hwy			533-8500	X			X
Enloe Med Center	888 Lakeside Village Commo			332-6800	X		X	
Berry Creek Clinic	10 Townhill Way			589-2285		X		X

8. Medical Emergency Procedures

Operations will coordinate any incident within the incident. Notify DIVS. of any injury or medical problems. DIVS authorized to request a medevac directly through AOBD. Coordinate minor bee stings and poision oak through Base Medical Unit.

9. Prepared by (Medical Unit Leader) Paul Ringe	10. Reviewed by (Safety Officer) Tom [signature]

FIRE WEATHER FORECAST

FORECAST NO. 5

NAME OF FIRE: Feather River Complex **PREDICTION FOR:** Thursday Night **SHIFT**

UNIT: Fire Weather **SHIFT DATE:** August 26th /27th 1999

TIME AND DATE FORECAST ISSUED: 1400 August 25, 1999 **SIGNED:** Michael C. Smith
Incident Meteorologist

WEATHER DISCUSSION: Upper level low pressure system will drift slowly through central California during the night continuing the cloudiness and threat of light showers or thunderstorms. With very dry lower atmosphere, rainfall amounts will be very limited with any thunderstorms. The chances of convective activity will decrease after midnight with only some isolated showers possible through the early morning hours. Cloudiness should be clearing by midday Friday. Should see slight increase in Rh values for Friday with warming and drying period through the weekend.

WEATHER FORECAST

*** Red Flag warning this evening through around midnight for isolated thunderstorms with limited rainfall ***

WEATHER: Partly to mostly cloudy with isolated thunderstorms through around midnight with decreasing chances early morning hours. Lals 6 evening hours with lals 2/3 late night. Chance of wetting rain less than 10%.

TEMPERATURES: Minimum temperatures 60-65 lower to midslopes and around 70 ridges.

HUMIDITY: Maximum RH Recovery 60-70% lower to midslopes and 30-35% upper slopes and ridges.

20 FT Winds:

RIDGETOP - Light southerly through this evening 1-5 gusts to 8 mph becoming more westerly by morning except...gusty erratic winds near thunderstorms.

SLOPE/VALLEY - light upslope/upcanyon to 5 mph through early evening becoming light Downslope/Downcanyon 1-4 mph by midevening and through the night except...gusty and erratic winds near thunderstorms.

HAINES INDEX: 3-4

STABILITY/INVERSION: If no thunderstorms affect the area then nighttime inversions should begin to build in between 2100 and midnight.

EXTENDED FORECAST For Friday Day Shift:
Scattered clouds with a slight chance of an isolated thunderstorm early morning becoming partly cloudy to mostly sunny by midday. Max temps 92-97. Min RH values 19-25%. Light southwest winds to 12 mph on the ridges with upslope/upvalley winds to 7 mph.

OUTLOOK FOR Friday night and Saturday: Fair skies. Little change in max/min Temps with slightly lower Rh. Slightly higher southwest ridge winds.

FIRE BEHAVIOR FORECAST

FORECAST NO. 4

NAME OF FIRE: Feather River Complex PREDICTION FOR: Night SHIFT

UNIT: Plumas NF SHIFT DATE: 8-26-99

TIME AND DATE FORECAST ISSUED: 8-26-99 1330 SIGNED: J Adams SQF FIRE BEHAVIOR ANALYST

WEATHER DISCUSSION:

Possibility of thunderstorms until midnite & after.

FIRE BEHAVIOR

GENERAL:

SPECIFIC:

- Cloudiness has raised RH 5-10%, but poor RH recovery is still expected.
- If thunderstorms are near, expect increased fire activity with possible torching.
- Burnout maybe difficult, with RH generally over 30%, & the threat of thunderstorms.

AIR OPERATIONS:

SAFETY:

- Post Lookouts to monitor thunderstorms.

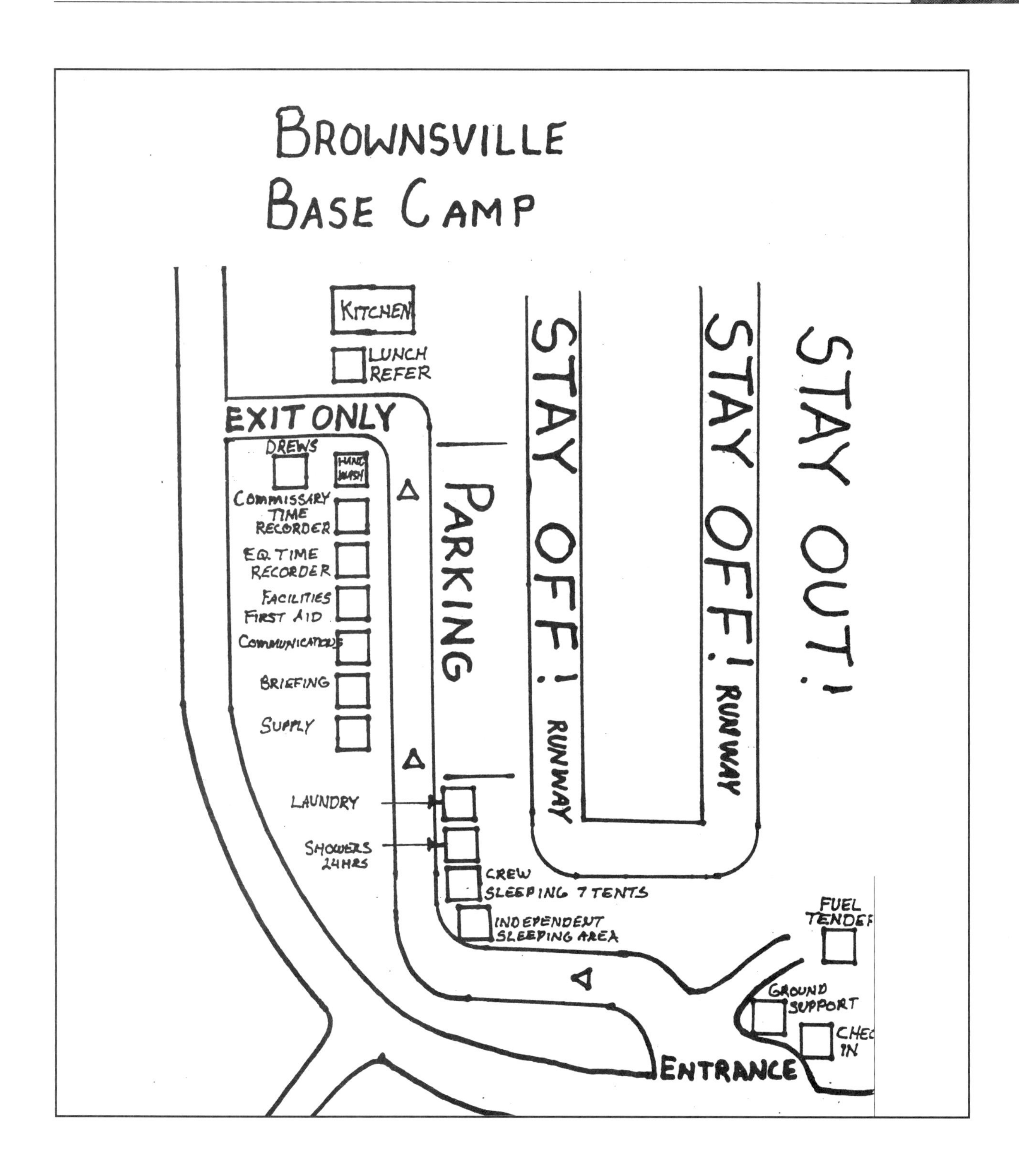
BROWNSVILLE
BASE CAMP
KITCHEN
LUNCH
REFER
EXIT ONLY
DREWS
HAND WASH
COMMISSARY
TIME
RECORDER
EQ. TIME
RECORDER
FACILITIES
FIRST AID
COMMUNICATIONS
BRIEFING
SUPPLY
PARKING
STAY OFF!
STAY OFF!
STAY OUT!
RUNWAY
RUNWAY
LAUNDRY
SHOWERS
24HRS
CREW
SLEEPING 7 TENTS
INDEPENDENT
SLEEPING AREA
FUEL
GROUND
SUPPORT
ENTRANCE

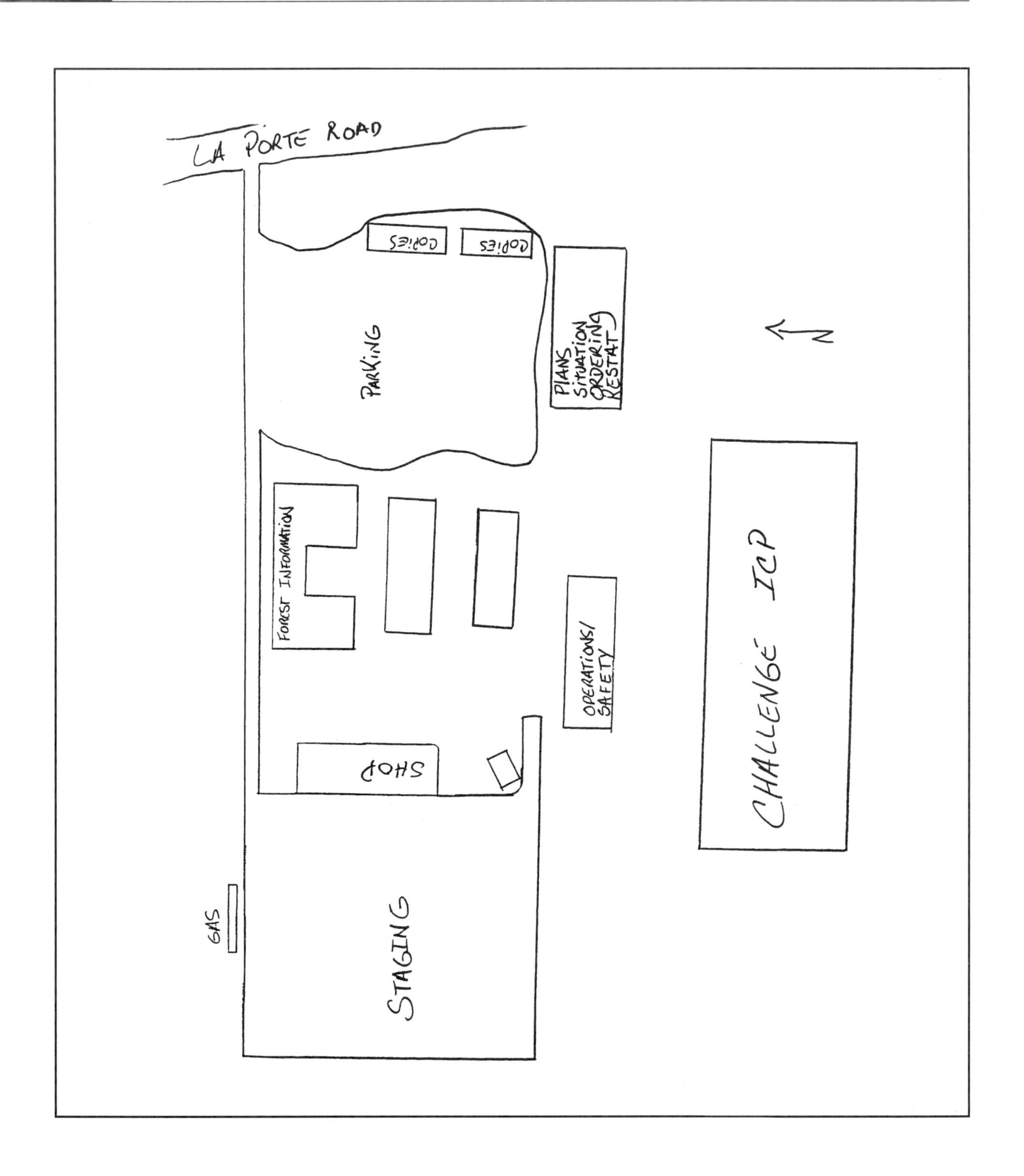
LA PORTE ROAD
COPIES
COPIES
PARKING
PLANS
SITUATION
ORDERING
RESTAT
N
FOREST INFORMATION
OPERATIONS/
SAFETY
SHOP
CHALLENGE ICP
GAS
STAGING

Acronyms

AID	aerial ignition device
AOBD	air operations branch director
AREP	Agency Representative
ASGS	Air Support Group Supervisor
ATGS	Air Tactical Group Supervisor
ATCO	Air Tanker Fixed-Wing Coordinator
ATMU	air transportable modular unit
BATF	Bureau of Alcohol, Tobacco, and Firearms
BLM	Bureau of Land Management
BTU	British thermal unit
CAF	compressed air foam
CCV	crew-carrying vehicle
CDF	California Department of Forestry and Fire Protection
CLMS	Claims Specialist
CMSY	Commissary Manager
DAID	delayed aerial ignition system
DBH	diameter at breast height
DZ	drop zone
EFSA	escaped fire situation analysis
EMC	extinction moisture content
EMT	emergency medical technician
EOC	emergency operations center
ERC	energy release component
ETA	estimated time of arrival
ETD	estimated time of departure
ETE	estimated time en route
FAA	Federal Aeronautics Administration
FAR	federal aviation regulation
FIRESCOPE	Firefighting Resources of California Organized for Potential Emergencies
FLE	fireline explosives
FLI	fire load index
FLIR	forward-looking infrared
FLPMA	Federal Land Policy and Management Act
FS	Forest Service
FWI	fire weather index
FWIS	Forestry Weather Information System
GPS	Global Positioning System
GTS	gum-thickened sulfate
HERO	helibase radio operator
HIGE	hovering in ground effect
HOGE	hovering out of ground effect
IAF	initial fire attack
IAIC	Initial Attack Incident Commander
ICP	Incident Command Post
ICS	Incident Command System
ID	inside diameter
IMET	Incident Meteorologist
IMT	Incident Management Team
IPS	iron pipe standard thread
IPT	iron pipe thread
IR	infrared
KBDI	Keetch-Byram drought index
LAL	lightning activity level
LC	liquid concrete
LCES	lookouts, communications, escape routes, safe zones

LOFR	Liaison Officer
LPG	liquid petroleum gas
LR	lightning risk
MAC	multiagency coordination
MACS	multiagency coordination system
MAFFS	modular airborne firefighting system
MIST	minimum impact suppression techniques
MSL	mean sea level
MOA	military operations area
NFDRS	National Fire Danger Rating System
NFPA	National Fire Protection Association
NH	national hose
NIFC	National Interagency Fire Center
NIFMID	National Interagency Fire Management Integrated Database
NIIMS	National Interagency Incident Management System
NOAA	National Oceanic and Atmospheric Administration
NPSH	national pipe straight mechanical
NPT	national pipe thread
NWCG	National Wildfire Coordinating Group
NWS	National Weather Service
OCC	Operations Coordination Center
OD	outside diameter
PSD	plastic sphere diameter
PVC	polyvinyl chloride
RAWS	remote automatic weather station
RFD	rural fire district
RH	relative humidity
SIMWYE	reversible siamese
TFR	temporary flight restriction
TL	timelag
TPIT	tapered iron pipe thread
USDA	United States Department of Agriculture
USDI	United States Department of the Interior
USGS	United States Geological Survey
UTM	Universal Transverse Mercator
WIMS	Weather Information and Management System

Glossary

Above the ground (AGL) Term frequently used in aviation operations, usually in connection with a stated altitude.

Absolute humidity Total amount of water vapor in the air.

Action Plan Any tactical plan developed by any element of ICS in support of the incident action plan. *See also* Incident Action Plan.

Active crown fire Fire in which a solid flame develops in the crowns of trees, but the surface and crown phases advance as a linked unit dependent on each other.

Active fire Any wildfire on which suppression action has not reached an extensive mop-up stage. Syn.: going fire.

Active resources Resources checked in and assigned work tasks on an incident. Syn.: assigned resources.

Adapter A device for connecting hoses of the same size with nonmatching hose threads or connecting a threaded coupling to a quick-connect coupling.

Adiabatic process Thermodynamic change of state in which no heat is added or subtracted from a system; compression always results in warming, expansion in cooling.

Advancing a line Moving a hose lay toward a specific area from the point where the hose-carrying apparatus has stopped.

Advection The transfer of atmospheric properties by the horizontal movement of air, usually in reference to the transfer of warmer or cooler air, but may also refer to moisture.

Adze Hoe A fire-trenching or digging tool having a sharp, tempered blade, which is useful for heavy grubbing, trenching, and light cutting.

Aerial fuel Standing and supported live and dead combustibles not in direct contact with the ground and consisting mainly of foliage, twigs, branches, stems, cones, bark, and vines.

Aerial ignition Ignition of fuels by dropping incendiary devices or materials from aircraft.

Aerial Ignition Device (AID) Inclusive term applied to equipment designed to ignite wildland fuels from an aircraft. *See also* Delayed Aerial Ignition Devices; Helitorch; Ping-Pong Ball System; Plastic Sphere Dispenser.

Aerial Observer Person specifically assigned to discover, locate, and report wildland fires from an aircraft and to observe and describe conditions at the fire scene.

Aerial Photo Interpreter A person skilled in identification and measurements of natural and cultural features on aerial photographs.

Aerial reconnaissance Use of aircraft for detecting and observing fire behavior, values-at-risk, suppression activity, and other critical factors to facilitate command decisions on strategy and tactics needed for fire suppression.

Affect of slope The affect the slope has on a wildland fire. Fires on steep slopes spread more rapidly because the flames are closer to the fuels and preheat upslope fuels.

Agency A division of government with a specific function, or a nongovernmental organization (e.g., private contractor, business) that offers a particular kind of assistance. In the ICS, agencies are defined as jurisdictional (having statutory responsibility for incident mitigation), or assisting and/or cooperating (providing resources and/or assistance). *See also* Supporting Agency; Cooperating Agency.

Agency/Area Coordination Center Facility that serves as a central point for one or more agencies to use in processing information and resource requests; may also serve as a dispatch center for one of the agencies.

Agency Representative (AREP) Individual assigned to an incident from an assisting or cooperating agency who has been delegated authority to make decisions on matters affecting that agency's participation at the incident. Reports to the incident liaison officer.

Air Attack The deployment of fixed-wing or rotary aircraft on a wildland fire, to drop retardant or extinguishing agents, shuttle and deploy crews and supplies, or perform aerial reconnaissance of the overall fire situation.

Air attack base Permanent facility at which aircraft are stationed for use in air attack operations.

Air cargo All items for transport and delivery by aircraft.

Air guard A common VHF-FM frequency used by natural resource agency aircraft for emergency radio transmissions. Also called *national air safety guard frequency*.

Air mass An extensive body of air having the same properties of temperature and moisture in a horizontal plane.

Air Operations Branch Director (AOBD) The person primarily responsible for management of air operations on an incident.

Air Support Group Supervisor (ASGS) Person responsible for logistical support and management of helibase and helispot operations and temporary fixed-wing bases. The ASGS also maintains liaison with air tanker and fixed-wing aircraft bases.

Air Tactical Group Supervisor (ATGS) Person responsible for directing and coordinating airborne aircraft operations and managing air space for an incident.

Air tanker Fixed-wing aircraft certified by Federal Aeronautic Administration (FAA) as being capable of transport and delivery of fire retardant solution.

Air Tanker/Fixed-wing Coordinator (ATCO) Airborne person responsible for coordinating air tanker/fixed-wing operations on an incident.

Air Transportable Modular Unit (ATMU) A weather data collection and forecasting facility consisting of four modules, weighing a total of 282 pounds and occupying 27.1 cubic feet of space when transported. Used by incident meteorologists on an incident. *See also* Incident meteorologist.

Allocated resources Resources dispatched to an incident that have not yet checked in. *See also* Assigned resources; Available resources.

Anchor point An advantageous location, usually a barrier to fire spread, from which to start constructing a fireline. The anchor point is used to minimize the chance of being flanked by the fire while the line is being constructed.

Apparatus A motor-driven vehicle, or group of vehicles, designed and constructed for the purpose of fighting fires. May be of different types, such as engines, water tenders, or ladder trucks.

Aramid Generic name for a high-strength, flame-resistant, synthetic fabric used in the shirts and jeans of firefighters. Nomex, a brand name for aramid fabric, is the term commonly used by firefighters.

Area command An organization established to (1) oversee the management of multiple incidents that are each being handled by an incident management team (IMT) organization; or (2) to oversee the management of a very large incident that has multiple IMTs assigned to it. Area command has the responsibility to set overall strategy and priorities, allocate critical resources based on priorities, and ensure that incidents are properly managed, objectives met, and strategies followed.

Area ignition Ignition of several individual fires throughout an area, either simultaneously or in rapid succession, and so spaced that they add to and influence the main body of the fire to produce a hot, fast-spreading fire condition. Also called *simultaneous ignition*.

Aspect Cardinal direction toward which a slope faces. *See also* Exposure.

Aspirate (foam) To draw in gases (or other substances); nozzle aspirating systems draw air into the nozzle to mix with the foam solution.

Assigned resources Resources checked in and assigned work tasks on an incident. Syn.: Active resources. *See also* Allocated resources; Available resources.

Assignments Term used as a title for subordinates of the command staff positions. In some cases, assistants are also assigned to unit leader positions in the planning, logistics, and finance/administration sections. Qualifications, technical capability, and responsibility of assistants are normally less than those of the person holding the primary position.

Assisting agency An agency directly contributing tactical or service resources to another agency. *See also* Agency; Cooperating agency; Supporting agency.

Atmospheric inversion (1) Departure from the usual increase or decrease with altitude of the value of an atmospheric property (in fire management usage, nearly always refers to an increase in temperature with increasing height). (2) The layer through which this departure occurs (also called *inversion layer*). The lowest altitude at which the departure is found is called the base of the inversion. *See also* Inversion; Atmospheric stability; Stable layer of air.

Atmospheric pressure Force exerted by the weight of the atmosphere, per unit area.

Atmospheric stability Degree to which vertical motion in the atmosphere is enhanced or suppressed. Vertical motion and pollution dispersion are enhanced in an unstable atmosphere. Thunderstorms and active fire conditions are common in unstable atmospheric conditions. Stability suppresses vertical motion and limits pollution dispersion. *See also* Inversion; Atmospheric inversion; Stable layer of air.

Attack a fire Limit the spread of fire by any appropriate means.

Attack line A line of hose, preconnected to the pump of a fire apparatus and ready for immediate use in attacking a fire. Contrasted to supply lines connecting a water supply with a pump or to feeder lines extended from a pump to various points around the perimeter of a fire.

Auxiliary pump Secondary pump, usually small capacity, on an engine in addition to the main pump.

Available fuel (1) That portion of the total fuel that would actually burn under various environmental conditions. (2) Fuel available for use in a motor vehicle, aircraft, or other motorized equipment.

Available fuel energy Amount of heat released per unit area when the available fuel burns, often expressed in BTUs per square foot.

Available resources Resources assigned to an incident and available for assignment. *See also* Allocated resources; Assigned resources.

Average relative humidity Mathematical average of the maximum and minimum relative humidities measured at a fire weather station from one basic observation time to the next. Part of the National Fire Danger Rating System (NFDRS).

Average temperature Mathematical average of the maximum and minimum dry-bulb temperatures measured at a fire weather station from one basic observation time to the next.

Azimuth Horizontal angle or bearing of a point measured clockwise from true (astronomic) north.

Back azimuth Angle or bearing 180° opposite of azimuth.

Backburn Used in some localities to specify fire set to spread against the wind in prescribed burning.

Backfire Fire set along the inner edge of a fireline to consume the fuel in the path of a wildfire and/or change the direction of force of the fire's convection column. *See also* Burn out.

Backfire torch A flame generating device (e.g., a fount containing diesel oil or kerosene and a wick, or a backpack pump serving a flame-jet). *See also* Drip torch.

Backfiring A tactic associated with indirect attack, intentionally setting fire to fuels inside the control line to slow, knock down, or contain a rapidly spreading fire. Backfiring provides a wide defense perimeter and may be further employed to change the force of the convection column. Backfiring makes possible a strategy of locating control lines at places where the fire can be fought on the firefighter's terms. Except for rare circumstance meeting specified criteria, backfiring is executed on a command decision made through line channels of authority.

Backing fire Fire spreading, or ignited to spread, into (against) the wind or downslope. A fire spreading on level ground in the absence of wind is a backing fire.

Backing wind Wind that changes direction in a counterclockwise motion.

Backpack pump A portable sprayer with hand pump, fed from a liquid-filled container fitted with straps, used mainly in fire and pest control. *See also* Bladder bag.

Baffle A partitioned wall placed in vehicular or aircraft water tanks to reduce shifting of the water load when starting, stopping, or turning.

Ball valve A valve in which fluid flow is controlled by a ball with a hole drilled through it. In one position, fluid flows through the hole. When the valve is turned 90° (1/4 turn), the hole is perpendicular to the flow and the ball stops the flow. Intermediate valve positions can be used to adjust the flow.

Bambi bucket A collapsible bucket slung below a helicopter, used to dip water from a variety of sources for fire suppression.

Banking snags The act of throwing mineral soil about the base of an unlighted snag to prevent its being ignited by a surface fire.

Barrier Any obstruction to the spread of fire, typically an area or strip devoid of combustible fuel.

Base (1) The location at which primary logistics functions for an incident are coordinated and administered. There is only one base per incident. (Incident name or other designator is added to the term "base.") The incident command post may be collocated with the base. (2) The location of initial attack forces. *See also* Camp.

Base manager Person responsible to the facilities unit leader for providing all services, supplies, and nontechnical coordination for all units operating within the incident base.

Base of fire The rear portion of the fire.

Batch mixing Manually adding and mixing a concentrated chemical, such as liquid foam or powdered or liquid retardant with water, or gelling agents with fuel, into solution in a tank or container.

Bearing The horizontal direction to or from any point, usually measured clockwise from true north, or some other reference point through 360°.

Beaufort Wind Scale A system of estimating and reporting wind speeds. In its present form for international meteorological use, it equates (1) Beaufort force (or Beaufort number), (2) wind speed, (3) descriptive term, and (4) visible effects on land objects or sea surface.

Behave A system of interactive computer programs for modeling fuel and fire behavior, comprised of two systems: Burn and Fuel.

Belt weather kit Belt-mounted case with pockets fitted for anemometer, compass, sling psychrometer, slide rule, water bottle, pencils, and book of weather report forms. Used to take weather observations to provide on-site conditions to the fire weather forecaster or fire behavior analyst. Observations include air temperature, wind speed and direction, and relative humidity.

Berm A ridge of soil and debris along the outside edge of a fireline, resulting from line construction. *See also* Throw out.

Blackline Preburning of fuels adjacent to a control line before igniting a prescribed burn. Blacklining is usually done in heavy fuels adjacent to a control line during periods of low fire danger to reduce heat on holding crews and lessen chances for spotting across control line. In fire suppression, a blackline denotes a condition where there is no unburned material between the fireline and the fire edge.

Bladder bag A collapsible backpack portable sprayer made of neoprene or high-strength nylon fabric fitted with a pump. *See also* Backpack pump.

Blowup Sudden increase in fireline intensity or rate of spread of fire sufficient to preclude direct control or to upset existing suppression plans. Often accompanied by violent convection and may have other characteristics of a fire storm. *See also* Extreme fire behavior; Fire storm; Flare-up.

Booster hose Most common type of hose attached and stored on wildland engine booster reels. The hose is made of neoprene and does not appreciably collapse when stored empty.

Booster pump An intermediary pump for supplying additional lift in pumping water uphill past the capacity of the first pump.

Booster reel A reel for the booster hose mounted on a fire engine, often supplied by the auxiliary pump. This reel usually carries a 1-inch (25 millimeter) or ¾-inch (19 millimeter) hose and frequently contains an electric rewind mechanism.

Box canyon Steep-sided, dead-end canyon.

Branch The organizational level having functional or geographical responsibility for major parts of incident operations. The branch level is organizationally between section and division/group in the operations section, and between section and unit in the logistics section. Branches are identified by roman numerals or by functional name (e.g., service, support).

Break left or right Means "turn" left or right. Applies to aircraft in flight, usually on the drop run, and when given as a command to the pilot, implies expectation of prompt compliance.

British Thermal Unit (BTU) Amount of heat required to raise 1 pound of water 1° Fahrenheit (from 59.5° to 60.5° F), measured at standard atmospheric pressure.

Brush A collective term that refers to stands of vegetation dominated by shrubby, woody plants, or low growing trees, usually of a type undesirable for livestock or timber management.

Brush blade Blade attachment with long teeth specially suited to ripping and piling brush with minimum inclusion of soil. Also called *brush rake* or *root rake*.

Brush fire A fire burning in vegetation that is predominantly shrubs, brush, and scrub growth.

Brush hook A heavy cutting tool designed primarily to cut brush at the base of the stem. Used in much the same way as an axe and having a wide blade, generally curved to protect the blade from being dulled by rocks.

Brush patrol unit Any light, mobile vehicular unit with limited pumping and water capacity for off-road operations.

Bubble Building block of foam; bubble characteristics of water content and durability influence foam performance.

Bucket drops Dropping of fire retardants or suppressants from specially designed buckets slung below a helicopter.

Bucking Sawing through the hole of a tree after it has been felled.

Build-up (1) The cumulative effects of long-term drying on current fire danger. (2) The increase in strength of a fire management organization. (3) The accelerated spreading of a fire with time. (4) Towering cumulus clouds, which may lead to thunderstorms later in the day.

Bulk density Weight per unit volume. For fuels, this is usually expressed as pounds per cubic foot; for soils, grams per cubic centimeter.

Bulldozer Any tracked vehicle with a front mounted blade used for exposing mineral soil. *See also* Tractor.

Bulldozer company A resource that includes a bulldozer, its transportation unit, and a standard complement of personnel for its operation.

Bulldozer line Fireline constructed by the front blade of a bulldozer.

Bulldozer tender Any ground vehicle with personnel capable of maintenance, minor repairs, and limited fueling of bulldozers.

Burn Boss Person responsible for supervising a prescribed fire from ignition through mop-up.

Burning conditions The state of the combined factors of the environment that affect fire behavior in a specified fuel type.

Burning index An estimate of the potential difficulty of fire containment as it relates to the flame length at the head of the fire. A relative number related to the contribution that fire behavior makes to the amount or effort needed to contain a fire in a specified fuel type. Doubling the burning index indicates that twice the effort will be required to contain a fire in that fuel type as was previously required, providing all other parameters are held constant.

Burning period That part of each 24-hour period when fires spread most rapidly, typically from 10:00 A.M. to sundown.

Burning torch A flame-generating device (e.g., a fount containing diesel oil or kerosene and a wick, or a backpack pump serving a flame-jet). Syn.: Backfire torch. *See also* Drip torch.

Burn out Setting fire inside a control line to consume fuel between the edge of the fire and the control line. *See also* Backfire.

Burn out time The duration of flaming and smoldering combustion phases at a specified point within a burn or for the whole burn, expressed in convenient units of time.

Burn pattern Characteristic configuration of char left by a fire; can be used to trace a fire's origin. In wildland fires, burn patterns are influenced by topography, wind direction, length of exposure, and type of fuel.

Cache A predetermined complement of tools, equipment, and/or supplies stored in a designated location, available for incident use.

Camp A geographical site(s), within the general incident area, separate from the incident base, equipped and staffed to provide sleeping, food, water, and sanitary services to incident personnel. *See also* Base.

Camp Manager Person responsible to the facilities unit leader for providing all services, supplies, and nontechnical coordination for all units within the camp. There may be one or more camps per incident, each with a camp manager.

Candling The burning of the foliage of a single tree or a small group of trees, from the bottom up. Syn.: Torching.

Canopy The stratum containing the crowns of the tallest vegetation present (living or dead), usually above 20 feet.

Carbon dioxide (CO_2) A colorless, odorless, nonpoisonous gas, which results from fuel combustion and is normally a part of the ambient air.

Carbon monoxide (CO) A colorless, odorless, poisonous gas produced by incomplete fuel combustion.

Cardinal directions North, south, east, west; used for giving directions and information from the ground or air in describing the fire (e.g., the west flank or east flank, not right flank or left flank).

Carrier fuels The fuels that support the flaming front of the moving fire.

Catface Defect on the surface of a tree resulting from a wound where healing has not reestablished the normal cross-section.

Ceiling (1) Height above the Earth's surface of the lowest layer of clouds or obscuring phenomena aloft that is not classified as a thin layer or partial obscuration, that together with all lower clouds or obscuring phenomena covers more than half the sky as detected from the point of observation. (2) Maximum height of a temporary flight restriction (TFR).

Celsius A temperature scale with 0° as the freezing point of water and 100° as the boiling point of water at sea level.

Center firing Method of broadcast burning in which fire is ignited in the center of the area to create a strong draft; additional fires are then ignited progressively nearer the outer control lines (sometimes in one step) as indraft increases so as to draw the flames and smoke toward the center.

Centrifugal pump Pump that expels water by centrifugal force through the ports of a circular impeller rotating at a high speed. With this type of pump, the discharge line may be shut off while the pump is running without damaging the pump or hose.

Chain Unit of measure in land survey, equal to 66 feet (20 meters) (80 chains equal 1 mile). Commonly used to report fire perimeters and other fireline distances, this unit is popular in fire management because of its convenience in calculating acreage (e.g., 10 square chains equal one acre).

Chain of command A series of management positions in order of authority.

Chainsaw A portable powersaw with a continuous chain that carries the cutting teeth.

Char (1) A charred substance or charred remains. (2) In fire simulation, a darkened area within the fire perimeter; usually indicates fire has already passed through; usually created by an opaque material blocking out a selected portion of basic scene illumination.

Charged line Hose filled with water under pressure and ready to use. Also called *live line.*

Check-in Process whereby resources first report to an incident. Check-in locations include: incident command post (resources unit), incident base, camps, staging areas, helibases, helispots, or direct to the line. *See also* Reporting locations.

Check line Temporary fireline constructed at right angles to the control line and used to hold a backfire in check as a means of regulating the heat or intensity of the backfire.

Check valve Valve that permits flow of liquid through a hose or pipe in one direction but prevents a return flow. Uses include the prevention of backflow on uphill hose lays, loss of prime with centrifugal pumps, and chemical contamination in fire chemical mixing systems.

Chevron burn Burning technique in which lines of fire are started simultaneously from the apex of a ridge point and progress downhill, maintaining position along the contour; used in hilly areas to ignite ridge points or ridge ends.

Chief The ICS title for individuals responsible for command functional sections: operations, planning, logistics, and finance/administration.

Cirrus Form of high cloud, composed of ice crystals, that seldom obscures the sun.

Claims Specialist (CLMS) Person responsible to the compensation/claims unit leader for handling all claims-related activities (other than injury) for the incident.

Class A foam Foam intended for use on Class A or woody fuels; made from hydrocarbon-based surfactant, therefore lacking the strong filming properties of Class B foam, but possessing excellent wetting properties.

Class B foam Foam designed for use on Class B or flammable liquid fires; made from fluorocarbon-based surfactants, therefore capable of strong filming action, but incapable of efficient wetting of Class A fuels.

Class of fire Kind of fire for purpose of using proper extinguisher:

Class A—Fires involving ordinary combustible materials (such as wood, cloth, paper, rubber, and many plastics) requiring the heat absorbing (cooling) effects of water, water solutions, or the coating effects of certain dry chemicals, which retard combustion.

Class B—Fires involving flammable or combustible liquids, flammable gases, greases, and similar materials where extinguishment is most readily secured by excluding air (oxygen), inhibiting the release of combustible vapors, or interrupting the combustion chain reaction.

Class C—Fires involving live electrical equipment where safety to the operator requires the use of electrically nonconductive extinguishing agents.

Class D—Fires involving certain combustible metals (such as magnesium, titanium, zirconium, sodium, or potassium) requiring a heat absorbing extinguishing medium not reactive with burning metals. *See also* Size class of fire.

Clean Air Act A federal law enacted to ensure that air quality standards are attained and maintained. Initially passed by Congress in 1963, it has been amended several times.

Clean burn Any fire, whether deliberately set or accidental, that destroys all aboveground vegetation and litter, along with the lighter slash, exposing the mineral soil.

Clear text The use of plain English in radio communications transmissions. No Ten Codes or agency specific codes are used when using clear text.

Climate The prevalent or characteristic meteorological conditions of any place or region, and their extremes.

Cloud A visible cluster of minute water and ice particles in the atmosphere.

Cloudy Adjective class representing the degree to which the sky is obscured by clouds. In weather forecast terminology, expected cloud cover of about 0.7 or more warrants use of this term. In the National Fire Danger Rating System, 0.6 or more cloud cover is termed "cloudy."

Cold front The leading edge of a relatively cold air mass that displaces and may cause warmer air to rise. If the lifted air contains enough moisture, cloudiness, precipitation, and even thunderstorms may result. As fronts move through a region, in the Northern Hemisphere, the winds at a given location experience a marked shift in direction. Ahead of an approaching cold front, winds usually shift gradually from southeast to south, and on to southwest. As a cold front passes, winds shift rapidly to west, then northwest. Typical cold front wind speeds range between 15 and 30 miles per hour but can be much higher.

Cold trailing A method of controlling a partly dead fire edge by carefully inspecting and feeling with the hand for heat to detect any fire, digging out every live spot, and trenching any live edge.

Combination nozzle Also called an "adjustable fog nozzle," this nozzle is designed to provide either a solid stream or a fixed spray pattern suitable for applying water, wet water, or foam solution.

Combination nozzle tip Two attached straight stream nozzle tips of different orifice size used to increase or restrict water flow.

Combustion The rapid oxidation of fuel in which heat and usually flame are produced. Combustion can be divided into four phases: preignition, flaming, smoldering, and glowing.

Combustion efficiency The relative amount of time a fire burns in the flaming phase of combustion, as compared to smoldering combustion. A ratio of the amount of fuel consumed in flaming combustion compared to the amount of fuel consumed during the smoldering phase, in which more of the fuel is emitted as smoke particles because it is not turned into carbon dioxide and water.

Combustion period Total time required for a specified fuel component to be completely consumed.

Combustion rate Rate of heat release per unit of burning area per unit of time. *See also* Reaction intensity.

Command The act of directing, and/or controlling resources by virtue of explicit legal, agency, or delegated authority.

Command Staff The command staff consists of the information officer, safety officer, and liaison officer. They report directly to the incident commander and may have an assistant or assistants, as needed.

Commissary Supply of items such as candy, tobacco products, toilet items, and work clothes that are made available for sale to firefighters working on a wildfire.

Commissary Manager (CMSY) Person responsible to the time unit leader for operating the commissary at an incident base or camp.

Communications Unit An organizational unit in the logistics section responsible for providing and maintaining communication services at an incident. May also be a facility (e.g., a trailer or mobile van) used to provide the major part of an incident communications center.

Compactness Spacing between fuel particles.

Company Any piece of (fire) equipment having a full complement of personnel.

Compass rose A circle, graduated in degrees, printed on some charts or marked on the ground at an airport or heliport, used as a reference to either true or magnetic direction.

Compensation/Claims Unit Functional unit within the finance/administration section responsible for financial concerns resulting from property damage, injuries, or fatalities at the incident.

Compensation-for-Injury Manager Person responsible to the compensation/claims unit leader for administering financial matters arising from serious injuries and deaths occurring at an incident.

Complex Two or more individual incidents located in the same general area assigned to a single incident commander or unified command.

Compressed Air Foam Systems (CAFS) A generic term used to describe foam systems consisting of an air compressor (or air source), a water pump, and foam solution.

Computed gross weight Term used in calculating permissible payload for helicopters. Computed gross weight is the maximum computed gross weight, from performance charts, at which a helicopter is capable of hovering in ground effect or hovering out of ground effect at stated density altitude.

Concentrate A substance that has been concentrated; specifically, a liquid that has been made denser, as by the removal of some of its water.

Condensation The process by which a gas becomes a liquid.

Conduction Heat transfer through a solid material from a region of higher temperature to a region of lower temperature.

Confine a fire The least aggressive wildfire suppression strategy, typically allowing the wildland fire to burn itself out within determined natural or existing boundaries such as rocky ridges, streams, and possibly roads.

Conflagration A raging, destructive fire. Often used to connote such a fire with a moving front as distinguished from a fire storm.

Consistency (foam) Uniformity and size of bubbles.

Consumption The amount of specified fuel type or strata that is removed through the fire process, often expressed as a percentage of the preburn weight.

Contain a fire A moderately aggressive wildfire suppression strategy that can reasonably be expected to keep the fire within established boundaries on constructed firelines under prevailing conditions.

Containment (1) Completion of a control line around a fire and any associated spot fires that can reasonably be expected to stop the fire's spread. (2) The act of controlling hazardous spilled or leaking materials.

Contour Map A map having lines of equal elevation that represent the land surface.

Control a fire To complete the control line around a fire, any spot fire therefrom, and any interior island to be saved; burn out any unburned area adjacent to the fire side of the control lines, and cool down all hot spots that are immediate threats to the control line, until the lines can reasonably be expected to hold under foreseeable conditions. Syn.: Controlled.

Controlled The completion of control line around a fire, any spot fires therefrom, and any interior islands to be saved; burned out any unburned area adjacent to the fire side of the control lines; and cool down all hot spots that are immediate threats to the control line, until the lines can reasonably be expected to hold under the foreseeable conditions. Syn.: Control a fire.

Control line An inclusive term for all constructed or natural barriers and treated fire edges used to control a fire.

Control time The time a fire is declared controlled.

Convection (1) The transfer of heat by the movement of a gas or liquid; convection, conduction, and radiation are the principal means of energy transfer. (2) As specialized in meteorology, atmospheric motions that are predominantly vertical in the absence of wind (which distinguishes this process from advection), resulting in vertical transport and mixing of atmospheric properties.

Convection activity General term for manifestations of convection in the atmosphere, alluding particularly to the development of convective clouds and resulting weather phenomena, such as showers, thunderstorms, squalls, hail, or tornadoes.

Convection column The rising column of gases, smoke, sly ash, particulates, and other debris produced by a fire. The column has a strong vertical component indicating that buoyant forces override the ambient surface wind. *See also* Smoke plume.

Convective-life fire phase Phase of a fire when most of the emissions are entrained into a definite convection column.

Convergence The term for horizontal air currents merging together or approaching a single point, such as at the center of a low pressure area producing a net inflow of air. The excess air is removed by rising air currents. Expansion of the rising air above a convergence zone results in cooling, which in turn often gives condensation (clouds) and sometimes precipitation.

Convergence zone (1) The area of increased flame height and fire intensity produced when two or more fire fronts burn together. (2) In fire weather, that area where two winds come together from opposite directions and are forced upward, often creating clouds and precipitation.

Cooperating agency An agency supplying assistance including but not limited to direct tactical or support functions or resources to the incident control effort (e.g., Red Cross, law enforcement agency, telephone company). *See also* Agency, Supporting agency.

Cooperator Local agency or person who has agreed in advance to perform specified fire control services and has been properly instructed to give such service.

Co-op fire Refers to federal, state, and local cooperative fire progress.

Coordinates The intersection of lines of reference, usually expressed in degrees/minutes/seconds of latitude and longitude, used to determine or report position or location.

Coordination The process of systematically analyzing a situation, developing relevant information, and informing appropriate command authority of viable alternatives for selection of the most effective combination of available resources to meet specific objectives. The coordination process (which can be either intra- or interagency) does not involve dispatch actions. However, personnel responsible for coordination may perform command or dispatch functions within limits established by specific agency delegations, procedures, and legal authority.

Coordination center Term used to describe any facility used for the coordination of agency or jurisdictional resources in support of one or more incidents.

Coriolis force An apparent force due to the rotation of the Earth that causes a deflection of air to the right in the Northern Hemisphere and to the left in the Southern Hemisphere.

Corrosion Result of chemical reaction between a metal and its environment (i.e., air, water, and impurities in same).

Cost-sharing agreements Agreements between agencies or jurisdictions to share designated costs related to incidents. Cooperative fire protection agreements with states, agencies, and jurisdictions outline the procedures for cost sharing.

Cost Unit Functional unit within the finance/administration section responsible for tracking costs, analyzing cost data, making cost estimates, and recommending cost-saving measures.

Council Tool Long-handled combination rake and cutting tool, the blade of which is constructed of a single row of three or four sharpened teeth. Also called *fire rake, council rake. See also* Rich tool.

Counterfire Fire set between main fire and backfire to hasten spread of backfire. Emergency firing to stop, delay, or split a fire front, or to steer a fire. Also called *draft fire.*

Coupling Device that connects the ends of adjacent hoses to other components of hose.

Coverage level Recommended amount of aerially applied retardant keyed to the National Fire Danger Rating System (NFDRS) fuel models and/or fire behavior models. Coverage level 2 represents 2 gallons of retardant per hundred square feet. Levels range from 1 to 6 for most fuel models. A coverage level of greater than 6 is for heavy fuels. The levels can be adjusted for fire behavior.

Coyote tactics A progressive line construction duty involving self-sufficient crews that build fireline until the end of the operational period, remain at or near the point while off duty, and begin building fireline again the next operational period where they left off.

Creeping fire Fire burning with a low flame and spreading slowly.

Crew An organized group of firefighters under the leadership of a crew boss or other designated official.

Crew Boss A person in supervisory charge of usually sixteen to twenty-one firefighters and responsible for their performance, safety, and welfare.

Crew-carrying vehicle A vehicle used to transport handcrews.

Crew shuttle Transportation of fireline personnel to and/or from assigned fireline locations.

Crew transport Any vehicle capable of transporting a specified number of personnel in a specified manner.

Crown consumption Combustion of the twigs and needles or leaves of a tree during a fire.

Crown cover The ground area covered by the crown of a tree as delimited by the vertical projection of its outermost perimeter.

Crown fire A fire that advances from top to top of trees or shrubs more or less independent of a surface fire. Crown fires are sometimes classed as running or dependent to distinguish the degree of independence from the surface fire.

Crowning potential A probability that a crown fire may start, calculated from inputs of foliage moisture content and height of the lowest part of the tree crowns above the surface.

Crown out A fire that rises from ground into the tree crowns and advances from tree top to tree top. To intermittently ignite tree crowns as a surface fire advances.

Crown scorch Browning of needles or leaves in the crown of a tree or shrub caused by heating to lethal temperature during a fire. Crown scorch may not be apparent for several weeks after the fire.

Crown scorch height The height above the surface of the ground to which a tree canopy is scorched.

Cumulonimbus The ultimate growth of a cumulus cloud into an anvil-shaped cloud with considerable vertical development, usually with fibrous ice crystal tops, and usually accompanied by lightning, thunder, hail, and strong winds.

Cumulus A principal low cloud type in the form of individual cauliflower-like cells of sharp nonfibrous outline and less vertical development than cumulonimbus.

Cup trench A fireline trench on the downhill side of fire burning on steep slopes that is supposed to be built deep enough to catch rolling firebrands that could otherwise start fire below the fireline. A high berm on the outermost downhill side of the trench catches material. Also called *gutter trench*.

Curing Drying and browning of herbaceous vegetation or slash.

Dead fuels Fuels with no living tissue in which moisture content is governed almost entirely by absorption or evaporation of atmospheric moisture (relative humidity and precipitation).

Deck The helibase operational area that includes the touchdown pad, safety circle, hover lanes, and external cargo transport area.

Deck Coordinator Person responsible to the helibase manager for coordinating the movement of personnel and cargo at the helibase landing area.

Deep-seated fire A fire burning far below the surface in duff, mulch, peat, or other combustibles as contrasted with a surface fire. A fire that has gained headway and built up heat in a structure so as to require greater cooling for extinguishment.

Delayed Aerial Ignition Device (DAID) Polystyrene balls, 1.25 inches in diameter, containing potassium permanganate. The balls are fed into a dispenser, generally mounted in a helicopter, where they are injected with a water-glycol solution and then dropped through a chute leading out of the helicopter. The chemicals react thermally and ignite in 25–30 seconds. The space between ignition points on the ground is primarily a function of helicopter speed, gear ratio of the dispenser and the number of chutes used (up to four). *See also* Aerial ignition device; Helitorch, Ping-pong ball system; Plastic sphere dispenser.

Delegation of authority A statement provided to the incident commander by the agency executive delegating authority and assigning responsibility. The delegation of authority can include objectives, priorities, expectations, constraints, and other considerations or guidelines as needed. Many agencies require written delegation of authority to be given to incident commanders prior to their assuming command on larger incidents.

Demobilization Release of resources from an incident in strict accordance with a detailed plan approved by the incident commander.

Demobilization unit Functional unit within the planning section responsible for assuring orderly, safe, and efficient demobilization of incident resources.

Density (foam) The ratio of the original volume of the nonaerated foam solution to the resultant volume of foam. The inverse of expansion.

Density altitude Pressure altitude corrected for temperature deviations from standard atmosphere. Used as an index to aircraft performance characteristics such as take-off distance and rate of climb. Density altitude bears the same relation to pressure altitude, as true altitude does to indicated altitude.

Deployment zone Used when fire conditions are such that escape routes and safety zones have been compromised, deployment zones are last ditch areas where fire shelters must be deployed to ensure firefighter survival due to the available space and/or fire behavior conditions at the deployment zone location. *See also* Safe zone.

Deputy A qualified individual who could be delegated the authority to manage a functional operation or perform a specific task. In some cases, a deputy could act as relief for a superior. Deputies can be assigned to the incident commander, general staff, and branch directors.

Desiccant Chemical that, when applied to a living plant, causes or accelerates drying of its aerial parts; used to facilitate burning of living vegetation by substantially lowering fuel moisture content within a few hours.

Dew point The temperature at which the air, if cooled, would reach saturation. At this temperature, dew (or frost) will start to form on an exposed surface.

Direct attack Any treatment applied directly to burning fuel such as wetting, smothering, or chemically quenching the fire or by physically separating the burning from unburned fuel.

Dispatch The implementation of a command decision to move a resource or resources from one place to another.

Dispatch center A facility from which resources are assigned to an incident.

Dispatcher A person who receives reports of discovery and status of fires, confirms their locations, takes action promptly to provide people and equipment likely to be needed for control efforts.

Diurnal Daily, especially pertaining to cyclic actions which are completed within 24 hours, and which recur every 24 hours, such as temperature, relative humidity, and wind.

Divergence The expansion or spreading out of a horizontal wind field, generally associated with high pressure and light winds.

Divert To change aircraft assignment from one target to another or to a new fire.

Division Divisions are used to divide an incident into geographical areas of operation. Divisions are established when the number of resources exceeds the span of control of the operations chief. A division is located within the ICS organization between the branch and the task force/strike team. *See also* Group.

Division Group Supervisor An operation supervisor responsible for all suppression activities on a specific division of a fire.

Division Supervisor An Operations Supervisor responsible for all suppression activities on a specific division of a fire.

Documentation Unit Functional unit within the planning section responsible for collecting, recording, and safeguarding all documents relevant to the incident.

Double-bitted axe A cutting tool used on fireline with cutting edges on both sides; can be used in mop-up operations, to fall snags, and break up stumps or logs.

Double doughnut Two lengths of hose rolled side by side or a single length rolled into two small coils for convenient handling.

Double female coupling A hose-coupling device having two female swivel couplings to permit joining two male hose nipples of the same size and thread type when lines are laid with couplings in opposite or reverse directions.

Double jacket hose Fire hose having two cotton or other fiber jackets outside the rubber lining or tubing.

Double male coupling A hose-coupling device having two male thread nipples for connecting hose and for connecting two female couplings of the same diameter.

Doughnut roll A 50 or 100 foot length of hose or a 50 foot length of hose rolled up for easy handling. There are various ways of forming the doughnut. A convenient one has both couplings close together with the male thread protected by the female coupling.

Downdraft A wind associated with a thunderstorm. In a mature thunderstorm, falling rain within and below the cloud drags air with it and initiates downward-flowing air.

Draft Drawing water from static sources such as a lake, pond, cistern, or river, into a pump that is above the level of the water supply by removing the air from the pump and allowing atmospheric pressure [14.7 pounds per square inch (101 kilopascals) at sea level] to push water through a noncollapsible suction hose into the pump.

Drain time The time in minutes it takes for foam solution to drop out from the foam mass; the time it takes for a specified percent of the total solution contained in the foam to revert to liquid and drain out of the bubble structure.

Drift Effect of wind on smoke, retardant drops, paracargo, smokejumper streamers, and other items.

Drift smoke Smoke that has drifted from its point of origin and is no longer dominated by convective motion. May give false impression of a fire in the general area where the smoke has drifted.

Drip torch Handheld device for igniting fires by dripping flaming liquid fuel on the materials to be burned; consists of a fuel fount, burner arm, and igniter. Fuel used is generally a mixture of diesel and gasoline.

Drizzle Precipitation composed exclusively of water drops smaller than 0.02 inches (0.5 millimeters) in diameter.

Drop That which is dropped in a cargo dropping or retardant dropping operation.

Drop configuration The type of air drop selected to cover the target. Terms that specify drop configuration include: "Salvo drop the entire load," "Trail drop tanks in sequence," "Single or double door drop a partial load."

Drop pass Indicates that the air tanker has the target in sight and will make a retardant drop on this run over the target.

Drop pattern The distribution of an aerially delivered retardant drop on the target area in terms of its length, width, and momentum (velocity × mass) as it approaches the ground. The latter determines the relative coverage level of the fire retardant on fuels within the pattern. Syn.: Pattern.

Drop point An identified area on fireline where supplies and manpower are dropped off.

Drop zone (DZ) Target area for air tankers, helitankers, cargo dropping.

Drought A period of relatively long duration with substantially below normal precipitation, usually occurring over a large area.

Drought index A number representing the net effect of evaporation, transpiration, and precipitation in producing cumulative moisture depletion in deep duff or upper soil layers. *See also* Keetch-Byram Drought Index.

Dry air mass A portion of the atmosphere that has a relatively low dew point temperature and where the formation of clouds, fog, or precipitation is unlikely.

Dry bulb Name given to an ordinary thermometer used to determine the temperature of the air (to distinguish it from the wet bulb).

Dry foam A low-expansion foam with stable bubble structure and slow drain time used primarily for resource and property protection.

Dry hydrant Permanent device with fire engine threads attached to expedite drafting operations in locations where there are water sources suitable for use in fire suppression (e.g., piers, wharves, bridges over streams, highways adjacent to ponds); also permanently installed private fire pumps that depend on suction sources. Also called *suction pipe*.

Dry lightning storm Thunderstorm in which negligible precipitation reaches the ground. Also called *dry storm*.

Dry run A trial pass over the target area by a lead plane and/or an air tanker to pinpoint target areas and warn ground personnel of the impending retardant or extinguishing agent drop.

Dry adiabatic lapse rate The rate of decrease of temperature with height of a parcel of dry air lifted adiabatically through an atmosphere in hydrostatic equilibrium. In the adiabatic lifting process, unsaturated air cools at the fixed rate of approximately 5.5° per 1000 feet increase in altitude.

Dry bulb temperature Temperature of the air.

Dry bulb thermometer In a psychrometer, the thermometer not covered with muslin, which is used to determine air temperature.

Duff The layer of decomposing organic materials lying below the litter layer of freshly fallen twigs, needles, and leaves and immediately above the mineral soil. *See also* Humus; Litter.

Durability (of foam) The effective life span of foam bubbles.

Easting The distance in meters of the position east of the zone line.

Eddy A circularlike flow of a fluid, such as air or water, drawing its energy from a flow of much larger scale and brought about by pressure irregularities as in the downwind (lee) side of a solid obstacle. For example, wind conditions may be erratic on the downwind side of large rock outcroppings or buildings.

Edge (1) The place where plant communities meet or where successional stages or vegetative conditions within plant communities come together. (2) The boundary between two fairly distinct fuel types.

Edge firing Method of burning in which fires are set along the edges of an area and allowed to spread inward.

Eductor A device used to introduce and mix fire chemical into a water stream. An eductor is a fitting with three ports, an inlet for water flow, an outlet for water flow, and an inlet for fire chemical concentrate. The flow of water through the eductor produces a region of lower pressure at the fire chemical inlet, drawing the chemical into the water stream.

Effective windspeed The midflame windspeed adjusted for the affect of slope on fire spread.

Ejector A siphon device used to fill an engine's tank when the water source is below or beyond the engine's drafting capability.

Elapsed time The total time taken to complete any step(s) in fire suppression. Generally divided chronologically into discovery time, report time, getaway time, travel time, attack time, control time, mop-up time, and patrol time.

Elevation loss In hydraulics, the pressure loss caused by raising water through hose or pipe to a higher elevation (roughly equal to 1 pound per square inch for every 2 foot increase in elevation above the pump) (11.3 kilopascals per meter).

Emergency Any incident that requires the response of a fire protection organization's attack units and/or support units.

Emergency Management Coordinator/Director The individual within each political subdivision that has coordination responsibility for jurisdictional emergency management.

Emergency Medical Technician (EMT) A health-care specialist with particular skills and knowledge in prehospital emergency medicine.

Emergency Operations Center (EOC) A predesignated facility established by an agency or jurisdiction to coordinate the overall agency or jurisdictional response and support to an emergency.

Emergency Operations Plan The plan that each jurisdiction maintains for responding to appropriate hazards.

Emission A release of air contaminants such as smoke into the outdoor atmosphere.

Emission factor (EFp) The mass of particulate matter produced per unit mass of fuel consumed (pounds per ton, grams per kilogram).

Emission rate The amount of smoke produced per unit of time (pounds per minute). Emission Rate = Available Fuel × Burning Rate × Emission Factor.

Emission reduction A strategy for controlling smoke from prescribed fires that minimizes the amount of smoke output per unit area treated.

Emission standard Limitation on the release of a contaminant, or multiple contaminants, to the ambient air from a single source.

Energy release component (ERC) The computed total heat release per unit area (British thermal units per square foot) within the flaming front at the head of a moving fire.

Engine Any ground vehicle providing specified levels of pumping, water, and hose capacity but with less than the specified level of personnel.

Engine company A resource that includes an engine of a specific type. The type specifies tank capacity, pump rating, hose capacity, and number of personnel.

Enhancement burn Prescribed fire for recreation and aesthetic purposes (e.g., maintain parklike stands of trees, increase number and visibility of flowering annuals and biennials).

Entrapment A situation where personnel are unexpectedly caught in a fire behavior-related, life-threatening position where planned escape routes or safety zones are absent, inadequate, or compromised. An entrapment may or may not include deployment of a fire shelter for its intended purpose. These situations may or may not result in injury. They include "near misses."

Envelopment Direct attack with multiple anchor points, that allows for multiple points of attack. Generally used as an engine tactic in the wildland/urban interface.

Environment The complex surroundings of an item or area of interest, such as air, water, natural resources, and their physical conditions (temperature, humidity).

Environmental lapse rate The actual rate of decrease of temperature with elevation.

Episode (pollution) Condition of poor contaminant dispersion that may result in concentrations considered potentially harmful to health or welfare. Episodes may also occur during periods of fairly good dispersion if the source of air contaminants is extremely large.

Equilibrium moisture content Moisture content that a fuel particle will attain if exposed for an infinite period in an environment of specified constant temperature and humidity. When a fuel particle reaches equilibrium moisture content, net exchange of moisture between it and its environment is zero.

Equipment Manager Person responsible to the ground support unit leader for servicing, repairing, and fueling all apparatus and equipment on the incident, for obtaining transportation and scheduling its use, and for maintaining records of equipment service and use.

Equipment Time Recorder Person responsible to the time unit leader for assisting all other units at the incident in properly maintaining a daily record of equipment use time, maintaining current records for charges and credits for fuel, parts, services, and commissary items for all equipment, and checking and closing all time record forms before demobilization of equipment.

Escape route A preplanned and understood route firefighters take to move to a safety zone or other low-risk area. When escape routes deviate from a defined physical path, they should be clearly marked (flagged).

Escaped fire Fire that has exceeded or is expected to exceed initial attack capabilities or prescription.

Escaped Fire Situation Analysis (EFSA) A decision-making process that evaluates alternative suppression strategies against selected environmental, social, political, and economic criteria. Provides a record of decisions.

Estimated time en route (ETE) Term used in resource planning/following to estimate time spent between points.

Estimated time of arrival (ETA) Term used in resource planning/following to estimate time of arrival at a point.

Estimated time of departure (ETD) Term used in resource planning/following to estimate time of departure from a point.

Evaporation The transformation of a liquid to its gaseous state; heat is released by the liquid during this process.

Expanded dispatch An organization developed at a dispatch center to support a large-scale incident. The organization expands along with the Incident Command System.

Expansion The ratio of the volume of the foam in its aerated state to the original volume of the nonaerated foam solution.

Exposure (1) Property that may be endangered by a fire burning in another structure or by a wildfire. (2) Direction in which a slope faces, usually with respect to cardinal directions. *See also* Aspect. (3) The general surroundings of a site with special reference to its openness to winds.

Exposure fire Classification for a fire not originating in a building, but which ignites building(s). A fire originating in one building and spreading to another is classified under the original cause of fire.

Extend To drop retardant in such a way that the load slightly overlaps and links a previous drop. "Extend your last drop."

Extended attack incident A wildland fire that has not been contained or controlled by initial attack forces and for which more firefighting resources are arriving, en route, or being ordered by the initial attack incident commander. Extended attack implies that the complexity level of the incident will increase beyond the capabilities of initial attack incident command.

Exterior fire protection The protection of structures from the exterior, with no interior access or activity.

External load A load that is carried or extends outside of the aircraft fuselage.

External payload Maximum external stress load (in pounds) with full fuel and pilot in calm air at standard atmospheric temperature.

Extinguishing agent Substance used to put out a fire by cooling the burning material, blocking the supply of oxygen, or chemically inhibiting combustion.

Extra burning period For any particular fire that is neither contained nor controlled, any 24-hour period following the termination of the first burning period.

Extreme fire behavior "Extreme" implies a level of fire behavior characteristics that ordinarily precludes methods of direct control action. One or more of the following is usually involved: high rate of spread, prolific crowning and/or spotting, presence of firewhirls, strong convection column. Predictability is difficult because such fires often exercise some degree of influence on their environment and behave erratically, sometimes dangerously. *See also* Blowup; Flare-up; Fire storm.

Facilities Unit Functional unit within the logistics section that provides the layout, activation, and management of all incident facilities. These facilities may include the incident base, feeding areas, sleeping areas, and sanitary facilities, among others.

Facility maintenance specialist Person responsible to the base camp manager for general maintenance of the base or camp, including provision of sleeping and sanitation facilities, lighting and electricity, and camp cleanliness.

Fahrenheit A temperature scale on which 32°F denotes the temperature of melting ice, and 212°F the temperature of boiling water, both under standard atmospheric pressure.

Faller A person who fells trees. Also called *sawyer* and *cutter*.

Federal Aviation Regulation (FAR) Refers to the regulations governing all aviation activities of civil aircraft within the United States and its territories.

Federal Land Policy and Management Act (FLPMA) Federal Land Policy and Management Act of 1976 (Public Law 94-570, 90 Stat. 2743, 43 USC 1701).

Female coupling Coupling made to receive a male coupling of the same thread, pitch, and/or diameter.

Field Observer Person responsible to the situation unit leader for collecting and reporting information about an incident obtained from personal observations and interviews.

Filling An increase in the central pressure of a low. Counterclockwise wind flow around the low usually decreases as filling occurs.

Finance/Administration Section The section responsible for all incident costs and financial considerations. Includes the time unit, procurement unit, compensation/claims unit, and cost unit.

Fine fuel moisture The probable moisture content of fast-drying fuels, such as grass, leaves, ferns, tree moss, pine needles, and small twigs (0–1/4 inch diameter) that have a time lag constant of 1 hour or less.

Fine fuels Fast-drying dead fuels, generally characterized by a comparatively high surface area-to-volume ratio, that are less than 1/4 inch in diameter and have a time lag of one hour or less. These fuels (grass, leaves, needles, etc.) ignite readily and are consumed rapidly by fire when dry. *See also* Flash fuels.

Fingers of a fire The long narrow extensions of a fire projecting from the main body.

Fire agency Official group or organization compelled and authorized under statutes or law to control fires within a designated area or on designated lands. *See also* Responsible fire agency.

Fire Analysis Review of fire management actions taken on a specific fire, group of fires, or fire season in order to identify reasons for both effective and ineffective actions, and to recommend or prescribe ways and means of doing a more efficient job. Also called *hot line review*.

Firebase Computerized bibliographic information file that stores and retrieves citations and information digests of fire-related information.

Fire behavior Manner in which a fire reacts to the influences of fuel, weather, and topography.

Fire Behavior Analyst Person responsible to the planning section chief for establishing a weather data collection system and for developing fire behavior predictions based on fire history, fuel, weather, and topography.

Fire behavior forecast Prediction of probable fire behavior, usually prepared by a fire behavior analyst, in support of fire suppression or prescribed burning operations.

Fire behavior prediction system System that uses a set of mathematical equations to predict certain aspects of fire behavior in wildland fuels when provided with data on fuel and environmental conditions.

Fire Blocking Gel A superabsorbent polymer used as a firefighting agent.

Firebrand Any source of heat, natural or human made, capable of igniting wildland fuels. Flaming or glowing fuel particles that can be carried naturally by wind, convention currents, or by gravity into unburned fuels.

Firebreak A natural or constructed barrier used to stop or check fires that may occur, or to provide a control line from which to work.

Fire cache A supply of fire tools and equipment assembled in planned quantities or standard units at a strategic point for exclusive use in fire suppression.

Fire cause For statistical purposes fires are grouped into broad cause classes. The nine general causes used in the United States are: lightning, campfire, smoking, debris burning, incendiary, machine use (equipment), railroad, children, and miscellaneous.

Fire climate Composite pattern of weather elements over time that affect fire behavior in a given region.

Fire concentration (complex) (1) Generally, a situation in which numerous fires are burning in a locality. (2) More specifically, the number of fires per unit area or locality for a given period, generally a year.

Fire crew General term for two or more firefighters organized to work as a unit.

Fire crew work formation Standard crew arrangement used for fireline construction in indirect attack; consists of line locator, line cutters, rakers, torch operators, and mop-up crew.

Fire damage Detrimental fire effects expressed in monetary or other units, including the unfavorable effects of fire-induced changes in the resource base on the attainment of organizational goals.

Fire damage appraisal Method of determining financial or other losses resulting from a fire.

Fire danger Sum of constant and variable danger factors affecting the inception, spread, and resistance to control, and subsequent fire damage; often expressed as an index.

Fire danger index A relative number indicating the severity of wildland fire danger as determined from burning conditions and other variable factors of fire danger.

Fire danger rating A fire management system that integrates the effects of selected fire danger factors into one or more qualitative or numerical indices of current protection needs.

Fire danger rating area Geographical area within which climate, fuel, and topography are relatively homogeneous, hence fire danger can be assumed to be uniform.

Fireday Standard 24-hour period beginning at 1000 hours, during which most wildfires undergo a predictable speeding up and slowing down of intensity, depending primarily on the influence of weather and fuel factors.

Fire death Fire casualty that is fatal or becomes fatal within one year of the fire.

Fire detection Act or system of discovering and locating fires. Syn.: Detection.

Fire discovery The act of determining that a fire exists; does not include determining its location.

Fire district A rural or suburban fire organization, usually tax supported, that maintains fire companies and apparatus. Also called *fire protection district.*

Fire edge The boundary of a fire at a given moment.

Fire effects The physical, biological, and ecological impacts of fire on the environment.

Fire environment The surrounding conditions, influences, and modifying forces of topography, fuel, and weather that determine fire behavior.

Firefinder map A map, generally mounted on a wood or metal base, that is provided with an azimuth circle, at the center of which is pivoted an alidade, and forms part of an Osborne Firefinder.

Fire-flood cycle The greatly increased rate of water runoff and soil movement from steep slopes that may follow removal of the vegetative cover by burning.

Fire frequency The number of fires per unit time in some designated area. The size of the area must be specified (units-number/time/area).

Fire front The part of a fire within which continuous flaming combustion is taking place. Unless otherwise specified, the fire front is assumed to be the leading edge of the fire perimeter. In ground fires, the fire front may be mainly smoldering combustion.

Fire hazard A fuel complex, defined by volume, type condition, arrangement, and location, that determines the degree of ease of ignition and of resistance to control.

Fire hazard index A numerical rating for specific fuel types, indicating the relative probability of fires starting and spreading, and the probable degree of resistance to control; similar to burning index, but without effects of wind speed.

Fire hazardous area Wildland area where the combination of vegetation topography, weather, and the threat of fire to life and property create difficult and dangerous problems.

Fire injury Injury suffered as the result of a fire that requires (or should require) treatment by a practitioner of medicine within one year of the fire, regardless of whether treatment was actually received.

Fire intensity The rate of heat release for an entire fire at a specific point in time.

Fire interval Time (in years) between two successive fires in a designated area (i.e., the interval between two successive fire occurrences); the site of the area must be clearly specified.

Fire investigation Procedure undertaken to determine, at a minimum, when, where, how a fire (or fires) was started, and by whom.

Fireline The part of a control line that is scraped or dug to mineral soil. Also called *fire trail.*

Fireline explosives (FLE) Specially developed coils containing explosive powder detonated to create a fireline through ground fuels.

Fireline intensity The rate of heat energy released during combustion per unit length of fire front. It is usually expressed in BTUs/second/foot.

Fire load The number and size of fires historically experienced on a given unit over a given period (usually one day) at a given index of fire danger.

Fire load index (FLI) Numerical rating of maximum effort required to contain all probable fires occurring within a rating area during the rating period.

Fire management Activities required for the protection of burnable wildland values from fire and the use of prescribed fire to meet land fire management objectives.

Fire management improvements All structures built and used primarily for fire management, for example, lookout towers, lookout cabins, telephone lines, firebreaks, fuelbreaks, and roads to lookouts.

Fire management objective Planned, measurable result desired from fire protection and use based on land management goals and objectives.

Fire management plan Statement, for a specific area, of fire policy, objective, and prescribed action; may include maps, charts, tables, and statistical data.

Firemodel Computer program that, with specific information, predicts an hourly rate of spread from a point of origin.

Fire occurrence The average number of fires in a specified area during a specified time period. Syn.: Fire incidence.

Fire occurrence map A map that shows by symbols the starting points of all fires for a given period.

Fire pack A one-person unit of fire tools, equipment, and supplies prepared in advance for carrying on the back.

Fire perimeter The entire outer edge or boundary of a fire.

Fire planning Systematic technological and administrative management process of designing organization, facilities, and procedures to protect wildland from fire.

Fire plow Heavy duty plowshare or disc plow usually pulled by a tractor to construct a fireline.

Fire prediction behavior model Set of mathematical equations that can be used to predict certain aspects of fire behavior when provided with an assessment of fuel and environmental conditions.

Fire prevention Activities, including education, engineering, enforcement, and administration, that are directed at reducing the number of wildfires, the costs of suppression, and fire-caused damages to resources and property.

Fire progress map A map maintained on a large fire to show at given times the location of the fire, deployment of suppression forces, and progress of suppression.

Fire qualifications Computerized interagency summary of fire suppression qualifications of listed personnel. Available information includes fire training record, fire experience record, and physical fitness testing score for each individual.

Fire regime Periodicity and pattern of naturally occurring fires in a particular area or vegetative type, described in terms of frequency, biological severity, and area extent.

Fire report An official record of a fire, generally including information on cause, location, action taken, damage, costs, and so forth from start of the fire until completion of suppression action. These reports vary in form and detail from agency to agency.

Fire-resistant tree A species with compact, resin-free, thick corky bark and less flammable foliage that has a relatively lower probability of being killed or scarred by a fire than a fire sensitive tree.

Fire resources All personnel and equipment available or potentially available for assignment to incidents.

Fire retardant Any substance except plain water that by chemical or physical action reduces flammability of fuels or slows their rate of combustion.

Fire risk (1) The chance of fire starting, as determined by the presence and activity of causative agents. (2) A causative agent. (3) A number related to the potential number of firebrands to which a given area will be exposed during the rating day (National Fire Danger Rating System).

Fire scar (1) A healing or healed injury or wound to woody vegetation, caused or accentuated by a fire. (2) The mark left on a landscape by fire.

Fire scar analysis Analysis of one or more fire scars to determine individual tree fire frequency or mean fire intervals for specified areas.

FIRESCOPE Firefighting Resources of California Organized for Potential Emergencies. A multiagency coordination system designed to improve the capabilities of California's wildland fire protection agencies. Its purpose is to provide more efficient resource allocation and utilization, particularly in multiple or large fire situations during critical burning conditions.

Fire season (1) Period(s) of the year during which wildland fires are likely to occur, spread, and affect resource values sufficient to warrant organized fire management activities. (2) A legally enacted time during which burning activities are regulated by state or local authority.

Fire sensitive tree A species with thin bark or highly flammable foliage that has a relatively greater probability of being killed or scarred by a fire.

Fire service The organized fire protection service; its members, individually and collectively; allied organizations assisting protection agencies.

Firesetting Igniting of incendiary fires.

Fire severity Degree to which a site has been altered or disrupted by fire; loosely, a product of fire intensity and residence time.

Fire shelter An aluminized tent offering protection by means of reflecting radiant heat and providing a volume of breathable air in a fire entrapment situation. Fire shelters should only be used in life threatening situations as a last resort.

Fire shelter deployment The removing of a fire shelter from its case and using it as protection against fire.

Fire shovel Type of shovel specifically designed for use in constructing a fireline; has a tapered blade with both edges sharpened for scraping, digging, grubbing, cutting, and throwing.

Fire simulator Training device that imposes simulated fire and smoke on a landscape image, for the purpose of instructing fire suppression personnel in different fire situations and fire suppression techniques.

Fire spread model A set of physics and empirical equations that form a mathematical representation of the behavior of fire in uniform wildland fuels.

Fire storm Violent convection caused by a large continuous area of intense fire. Often characterized by destructively violent surface indrafts, near and beyond the perimeter, and sometimes by tornadolike whirls. *See also* Blowup; Extreme fire behavior; Flare-up.

Fire suppressant Any agent used to extinguish the flaming and glowing phases of combustion by direct application to the burning fuel.

Fire suppression All work and activities connected with fire-extinguishing operations, beginning with discovery and continuing until the fire is completely extinguished.

Fire suppression organization (1) The personnel collectively assigned to the suppression of a specific fire or group of fires. (2) The personnel responsible for fire suppression within a specified area. (3) The management structure, usually shown in the form of an organization chart of the persons and groups having specific responsibilities in fire suppression.

Fire swatter A fire tool that consists of a thick, flat piece of rubber on a long handle used to drag over or smother flames of grass fires. *See also* Flapper.

Fire tool cache A supply of fire tools and equipment assembled in planned quantities or standard units at a strategic point for exclusive use in wildland operations.

Fire triangle Instructional aid in which the sides of a triangle are used to represent the three factors (oxygen, heat, fuel) necessary for combustion and flame production; removal of any of the three factors causes flame production to cease.

Fire weather Weather conditions that influence fire ignition, behavior, and suppression.

Fire weather forecast A weather prediction specially prepared for use in wildland fire operations and prescribed fire.

Fire weather index (FWI) A numerical rating in the Canadian fire danger rating system, based on meteorological measurements of fire intensity in a standard fuel type. (The standard fuel type is represented by jack pine and lodgepole pine.) The FWI is comprised of three fuel moisture codes, covering classes of forest fuel of different drying rates, and two indices that represent rate of spread and the amount of available fuel.

Fire weather station A meteorological station specially equipped to measure weather elements that have an important effect on fire behavior.

Fire weather watch A possible critical fire weather pattern. (The National Weather Service has replaced "red flag watch" with "fire weather watch" to avoid the overuse of red flag terminology.) *See also* Red flag warning.

Firewhirl Spinning vortex column of ascending hot air and gases rising from a fire and carrying aloft smoke, debris, and flame. Firewhirls range in size from less than one foot to more than 500 feet in diameter. Large firewhirls have the intensity of a small tornado.

Firing technique Any method of igniting a wildland area to consume the fuel in a prescribed pattern; for example, heading or backing fire, spot fire, strip-head fire, and ring fire.

Fix Geographical position determined by visual reference to the surface, by reference to one or more radio navigational aids, by celestial plotting, or by any other navigational device.

Fixed-point detection Detection of fires from lookout towers or other semipermanent locations as distinguished from roving ground patrols or aerial detection.

Fixed tank A device mounted inside or directly underneath an aircraft that can contain water or retardant for dropping onto a fire.

Flame A mass of gas undergoing rapid combustion, generally accompanied by evolution of sensible heat and incandescence.

Flame angle Angle between the flame at the leading edge of the fire front and the ground surface, expressed in degrees.

Flame depth Depth of the fire front.

Flame height Average maximum vertical extension of flames at the leading edge of the fire front. Occasional flashes that rise above the general level of flames are not considered. This distance is less than the flame length if flames are tilted due to wind or slope.

Flame length Distance between the flame tip and the midpoint of the flame depth at the base of the flame (generally the ground surface), an indicator of fire intensity.

Flaming combustion phase Luminous oxidation of gases evolved from the rapid decomposition of fuel. This phase follows the preignition phase and precedes the smoldering combustion phase, which has a much slower combustion rate. Water vapor, soot, and tar comprise the visible smoke. Relatively efficient combustion produces minimal soot and tar and white smoke; high moisture content also produces white smoke.

Flaming front That zone of a moving fire where the combustion is primarily flaming. Behind this flaming zone combustion is primarily glowing or involves the burning out of larger fuels (greater than about 3 inches in diameter). Light fuels typically have a shallow flaming front, whereas heavy fuels have a deeper front. *See also* Fire front.

Flaming phase That phase of a fire in which the fuel is ignited and consumed by flaming combustion.

Flammability The relative ease with which fuels ignite and burn regardless of the quantity of the fuels. Preferred to "inflammability."

Flank fire A firing technique consisting of treating an area with lines of fire set into the wind that burn outward at right angles to the wind.

Flanking fire suppression Attacking a fire by working along the flanks either simultaneously or successively from a less active or anchor point and endeavoring to connect two lines at the head.

Flanks of a fire Parts of a fire's perimeter that are roughly parallel to the main direction of spread.

Flapper Fire suppression tool, sometimes improvised, used in direct attack for smothering flames along a fire edge; may consist merely of a green pine bough or wet sacking, or be a manufactured tool such as a flap of belting fabric fastened to a long handle. *See also* Fire swatter.

Flare Launcher A handheld flare-launching device that launches a flare 300 feet. When the flare lands, it sprays 4000°F material in a 10 foot arch.

Flare-up Any sudden acceleration in rate of spread or intensification of the fire. Unlike blowup, a flare-up is of relatively short duration and does not radically change existing control plans. *See also* Blowup; Extreme fire behavior; Fire storm.

Flash fuels Fuels such as grass, leaves, draped pine needles, fern, tree moss, and some kinds of slash, that ignite readily and are consumed rapidly when dry. *See also* Fine fuels.

Flashover (1) Rapid combustion and/or explosion of unburned gases trapped at some distance from the main fire front. Usually occurs only in poorly ventilated topography. (2) Stage of a fire at which all surfaces and objects within a space have been heated to their ignition temperature, and flame breaks out almost at once over the surface of all objects within the space.

Flight following The method and process through which an aircraft is tracked from departure point to destination. Flight following is the knowledge of the aircraft location and condition with a reasonable degree of certainty such that, in the event of mishap, those on board may be rescued.

Floating pump Small portable pump that floats in the water source.

Fluid foam A low-expansion foam type with some bubble structure and moderate drain time, exhibiting properties of both wet and dry foam types, which is used for extinguishment, protection, and mop-up.

Foam The aerated solution created by forcing air into, or entraining air in water containing a foam concentrate by means of suitably designed equipment or by cascading it through the air at a high velocity. Foam reduces combustion by cooling, moistening, and excluding oxygen.

Foam blanket A layer of foam that forms an insulating and reflective barrier to heat and is used for fuel protection, suppression, and mop-up.

Foam concentrate The concentrated foaming agent as received from the manufacturer which, when added to water, creates a foam solution; use only those approved for use in wildland fire situations by the authority having jurisdiction.

Foam generation The foam production process of forcing air into or entraining air in foam solution, creating a mass of bubbles.

Foaming agent An additive that reduces the surface tension of water (producing wet water) causing it to spread and penetrate more effectively and which produces foam through mechanical means.

Foam line A body of foam placed along areas to be protected from fire; also used as an anchor for indirect attack in place of hand-made fire trail.

Foam solution A low-expansion foam type with no expansion, therefore lacking bubble structure, which is used for mop-up and flame knockdown.

Foam system The apparatus and technique used to mix concentrate with water to make solution, pump and mix air and solution to make foam, and transport and apply foam. (Systems defined here include compressed air foam and nozzle aspirated.)

Foam type Consistency and viscosity of low-expansion foam as the combination of drain time and expansion.

Foehn wind A warm, dry, and strong general wind that flows down into the valleys when stable, high pressure air is forced across and then down the lee slopes of a mountain range. The descending air is warmed and dried due to adiabatic compression producing critical fire weather conditions. Locally called by various names such as Santa Ana winds, devil winds, north winds, or mono winds.

Folding tank A portable, collapsible water tank with a tubular frame; tank capacities vary in size from 500 to 1500 gallons.

Food Unit Functional unit within the Logistics Section responsible for providing meals for incident personnel.

Foot valve Type of strainer assembly that goes on the end of the suction hose. A strainer is used on the valve to strain out foreign material in the water source. It also contains a spring-loaded assembly to prevent water from running out of the suction hose.

Forecast area Geographical area for which a fire weather forecast is specified.

Forest fire Variously defined for legal purposes (e.g., the State of California Public Resources Code: an uncontrolled fire on lands covered wholly or in part, by timber, brush, grass, grain, or other flammable vegetation). Types of fires are ground, surface, and crown.

Forest floor Organic surface component of the soil supporting forest vegetation; comprised of litter, fermentation, and humus layers.

Forestry Weather Information System (FWIS) A real time system that takes observations and forecasts supplied by the National Weather Service in coded numeric form, reformats that input by computer-based algorithms, and distributes the reformatted information as numeric and worded diagnoses and forecasts for specialized users in localized areas.

Forest Service Generally understood to mean an agency of the U.S. Department of Agriculture. However, some states also use "Forest Service," for example, Colorado State Forest Service.

Forward-looking infrared (FLIR) Handheld or aircraft mounted device designed to detect heat differentials and display their images on a video screen. FLIRs have thermal resolution similar to IR line scanners, but their spatial resolution is substantially less; commonly used to detect hot spots and flare-ups obscured by smoke, evaluate the effectiveness of firing operations, or detect areas needing mop-up.

Forward rate of spread The speed with which a fire moves in a horizontal direction across the landscape, usually expressed in chains per hour or feet per minute.

Free burning The condition of a fire or part of a fire that has not been slowed by natural barriers or by control measures.

Freezing rain Rain that freezes upon contact with objects on the ground.

Friction layer The layer of the atmosphere in which the frictional force of the Earth's surface exercises an appreciable influence on winds.

Friction loss Pressure loss caused by the turbulent movement of water or solution against the interior surface of fire hose, pipe, or fittings; normally measured in pressure loss per length of hose or pipe.

Front In meteorology, the boundary between two air masses of differing atmospheric properties.

Frontal lifting Air being forced up the slope of a warm or cold front. This type of lifting accounts for much of the cloudiness and precipitation in many regions of the country.

Frost Crystals of ice formed and deposited like dew, but at a temperature below freezing.

Fuel Combustible material.

Fuel arrangement A general term referring to the spatial distribution and orientation of fuel particles within a natural setting.

Fuel bed An array of fuels usually constructed with specific loading, depth, and particle size to meet experimental requirements; also, commonly used to describe the fuel composition in natural settings.

Fuel bed depth Average height of surface fuels contained in the combustion zone of a spreading fire front.

Fuelbreak A natural or man-made change in fuel characteristics that affects fire behavior so that fires burning into them can be more readily controlled.

Fuelbreak system A series of modified strips or blocks tied together to form continuous strategically located fuelbreaks around land units.

Fuel characteristics Factors that make up fuels such as compactness, loading, horizontal continuity, vertical arrangement, chemical content, size and shape, and moisture content.

Fuel class Group of fuels possessing common characteristics; part of the National Fire Danger Rating System (NFDRS). Dead fuels are grouped according to 1-, 10-, 100-, and 1,000-hour timelag, and living fuels are grouped as herbaceous (annual or perennial) or woody.

Fuel condition Relative flammability of fuel as determined by fuel type and environmental conditions.

Fuel continuity The degree or extent of continuous or uninterrupted distribution of fuel particles in a fuel bed thus affecting a fire's ability to sustain combustion and spread; applies to aerial fuels and surface fuels.

Fuel depth The average distance from the bottom of the litter layer to the top of the layer of fuel, usually the surface fuel.

Fuel group An identifiable association of fuel elements of distinctive species, form, size, arrangement, or other characteristics. General fuel groups are grass, brush, timber, and slash.

Fuel loading The amount of fuel present expressed quantitatively in terms of weight of fuel per unit area. This amount may be available fuel (consumable fuel) or total fuel and is usually dry weight.

Fuel management Act or practice of controlling flammability and reducing resistance to control of wildland fuels through mechanical, chemical, biological, or manual means, or by fire, in support of land management objectives.

Fuel model Simulated fuel complex for which all fuel descriptors required for the solution of a mathematical rate of spread model have been specified.

Fuel modification Manipulation or removal of fuels to reduce the likelihood of ignition and/or to lessen potential damage and resistance to control (e.g., lopping, chipping, crushing, piling, and burning).

Fuel moisture analog Device that emulates the moisture response of specific classes of dead fuels, constructed from organic or inorganic materials (e.g., half-inch ponderosa pine dowels representing 10-hour timelag fuels).

Fuel moisture content The quantity of moisture in fuel expressed as a percentage of the weight when thoroughly dried at 212°F.

Fuel moisture indicator stick A specially prepared stick or set of sticks of known dry weight continuously exposed to the weather and periodically weighed to determine changes in moisture content as an indication of moisture changes in wildland fuels.

Fuel reduction Manipulation, including combustion, or removal of fuels to reduce the likelihood of ignition and/or to lessen potential damage and resistance to control.

Fuel size class A category used to describe the diameter of down dead woody fuels. Fuels within the same size class are assumed to have similar wetting and drying properties, and to preheat and ignite at similar rates during the combustion process.

Fuel tender Any vehicle capable of supplying engine fuel to ground or airborne equipment.

Fuel type An identifiable association of fuel elements of distinctive species, form, size, arrangement, or other characteristics that will cause a predictable rate of spread or resistance to control under specified weather conditions.

Fuel type classification Division of wildland areas into fire hazard classes.

Fugitive color A coloring agent used in fire retardant that is designed to fade rapidly following retardant application in order to minimize the visual impacts of the retardant.

Function In the incident command system, function refers to the five major activities: command, operations, planning, logistics, and finance/administration.

Fusee A colored flare designed as a railway warning device, widely used to ignite backfires and other prescribed fires.

Gallons per minute (GPM) The measure of water flow in firefighting. It is used to measure the output of wildland and structural fire engines, pumps, hose streams, nozzles, hydrants, and water mains.

Gap Weak or missed area in a retardant drop.

Gate valve A valve with a gatelike disk that moves up and down at right angles to the flow when actuated by a stem screw and hand wheel. Gate valves are best for service that requires infrequent valve operation and where the disk is kept either fully open or closed.

Gear pump Positive displacement pump that uses closely meshed gears to propel water when high pressures and low volumes are desired; can be used safely only with clear water because suspended particles of soil or rocks can quickly wear the gears and reduce pressure and volume of water.

General fire weather forecast A forecast, issued daily during the regular fire season to resource management agencies, that is intended for planning of daily fire management activities, including daily staffing levels, prevention programs, and initial attack on wildfires. Also called *presuppression forecast. See also* Spot weather forecast; Incident weather forecast.

General Staff The group of incident management personnel reporting to the incident commander. Each member may have a deputy, as needed. The General Staff consists of Operations Section Chief, Planning Section Chief, Logistics Section Chief, and Finance/Administration Section Chief.

General winds Large-scale winds caused by high- and low-pressure systems but generally influenced and modified in the lower atmosphere by terrain. *See also* Local winds; Slope winds.

Geographic area A political boundary designated by the wildland fire protection agencies, where these agencies work together in the coordination and effective utilization of resources within their boundaries. The *National Interagency Mobilization Guide* in Chapter 20, section 21.1, identifies the area encompassed by the eleven National Wildfire Coordinating Group (NWCG) geographic areas.

Geographic coordinate system Mapping system that utilizes degrees and minutes.

Global Positioning System (GPS) A system of navigational satellites operated by the U.S. Department of Defense and available for civilian use that can track objects anywhere in the world with an accuracy of approximately 40 feet.

Glowing combustion Oxidation of a solid surface accompanied by incandescence, sometimes evolving flame above it.

Glowing combustion phase Oxidation of solid fuel accompanied by incandescence. All volatiles have already been driven off and there is no visible smoke. This phase follows the smoldering combustion phase and continues until the temperature drops below the combustion threshold value, or until only noncombustible ash remains.

Glowing phase Phase of combustion in which a solid surface of fuel is in direct contact with oxygen, and oxidation occurs, usually accompanied by incandescence, and little smoke production.

Going fire Any wildfire on which suppression action has not reached an extensive mop-up stage.

Gradient wind (1) Wind that flows parallel to pressure isobars or contours and has a velocity such that the pressure gradient, coriolis, and centrifugal force acting in the area are in balance. It does not occur at the Earth's surface due to fractional influence, but occurs at a height of approximately 1,500 feet above mean terrain height. (2) Wind created by differing barometric pressures between high- and low-pressure systems. Velocity is generally 5 to 30 miles per hour, and wind shifts are usually gradual as systems move and shift.

Grass fire Any fire in which the predominant fuel is grass or grasslike.

Gravity tank Water storage tank for fire protection and sometimes community water service that supplies water by gravity pressure.

Greenbelt Irrigated, landscaped, and regularly maintained fuelbreak, usually put to some additional use (e.g., golf course, park, playground).

Greenhouse effect The heating effect exerted upon the Earth because the atmosphere (mainly its water vapor) absorbs and emits infrared radiation.

Grid ignition technique Method of igniting prescribed fires in which ignition points are set individually at predetermined spacing with predetermined timing throughout the area to be burned. Also called *point source ignition technique.*

Ground effect Reaction of a rotor downdraft against the ground surface, forming a "ground cushion" that increases lifting capability of that section of air.

Ground fire Fire that consumes the organic material beneath the surface litter ground, such as a peat fire.

Ground fog Fog that extends vertically to less than 20 feet.

Ground fuel All combustible materials below the surface litter, including duff, tree or shrub roots, punky wood, peat, and sawdust, that normally support a glowing combustion without flame.

Ground Support Unit Functional unit within the logistics section responsible for the fueling, maintaining and repairing of vehicles, and the transportation of personnel and supplies.

Group Established to divide the incident into functional areas of operation and composed of resources assembled to perform a special function not necessarily within a single geographic division. Groups, when activated, are located between branches and resources in the operations section. *See also* Division.

Gum-thickened sulfate (GTS) A dry chemical product that is mixed with water to form a fire retardant slurry.

Gust Rapid fluctuations in wind speed with a variation of 10 knots (11.5 miles per hour) or more between peaks and lulls.

Haines Index An atmospheric index used to indicate the potential for wildfire growth by measuring the stability and dryness of the air over a fire.

Handcrew A number of individuals that have been organized and trained and are supervised principally for operational assignments on an incident.

Handline Fireline constructed with handtools.

Hangup (tree falling) A situation in which a tree is lodged in another, preventing it from falling to the ground.

Hard suction hose Noncollapsible suction hose attached to a pump and used to draft water from a source lower than the pump. Customary hose sizes in wildland fire engines are 2 and 2½ inches in diameter (51 and 64 millimeters).

Hazard A fuel complex defined by kind, arrangement, volume, condition, and location that forms a special threat of ignition and resistance to control.

Hazard map Map of the area of operations that shows all of the known aerial hazards, including but not limited to power lines, military training areas, and hang gliding areas.

Hazardous area Wildland area where the combination of vegetation, topography, weather, and the threat of fire to life and property create difficult and dangerous problems.

Hazard reduction Any treatment of living and dead fuels that reduces the threat of ignition and spread of fire.

Head Pressure due to elevation of water. Equals 0.433 pounds per square inch (PSI) per foot of elevation. Back pressure. (Approximately 0.5 PSI is required to lift water 1 foot in elevation.)

Head fire A fire spreading or set to spread with the wind.

Head (of a fire) The most rapidly spreading portion of a fire's perimeter, usually to the leeward or up slope.

Heading The compass direction in which the longitudinal axis of the aircraft points.

Headlamp Flashlight, ordinarily worn by firefighter on the front of the helmet.

Heat content The net amount of heat that would be given off if fuel burns when it is absolutely dry, noted as BTU per pound of fuel.

Heat low An area of low pressure caused by intense heating of the Earth's surface. High surface temperature causes air to expand and rise, resulting in low atmospheric pressure and induces a weak inflow of air at the surface. Air that rises in a heat low is very dry, so clouds seldom form. Rising air above a heat low produces a warm upper level high and results in a net outflow of air aloft. Heat lows remain practically stationary over areas that produce them.

Heat of combustion The heat energy resulting from the complete combustion of a fuel, expressed as the quantity of heat per unit weight of fuel. The high heat of combustion is the potential available, and the low heat of combustion is the high heat of combustion minus several losses that occur in an open system (primarily heat of vaporization of moisture in the fuel).

Heat per unit area Total amount of heat released per unit area as the flaming front of the fire passes, expressed as BTU/square foot; a measure of the total amount of heat released in flames.

Heat release rate (1) Total amount of heat produced per unit mass of fuel consumed per unit time. (2) Amount of heat released to the atmosphere from the convective-lift fire phase of a fire per unit time.

Heat release rate to the atmosphere The amount of heat released to the atmosphere during the combustion stage of a fire per unit of time.

Heat trough (Heat Low, Thermal Low, Thermal Trough) A heat low that is elongated in shape.

Heat transfer Process by which heat is imparted from one body to another, through conduction, convection, and radiation.

Heat yield The heat of combustion corrected for various heat losses, mainly the presence of moisture in the fuel. To a very close approximation, the quantity of heat per pound of fuel burned that passes through a cross-section of the convection column above a fire that is burning in a neutrally stable atmosphere. Also called *low heat of combustion.*

Heavy equipment transport Any ground vehicle capable of transporting a bulldozer, tractor, or other heavy piece of equipment. Also called *lowboy.*

Heavy fuels Fuels of large diameter, such as snags, logs, or large limbwood, that ignite and are consumed more slowly than flash fuels. Also called *coarse fuels.*

Height The vertical measurement of vegetation from the top of the crown to ground level.

Held line All control line that still contains the fire when mop-up is completed. Excludes lost line, natural barriers not backfired, and unused secondary lines.

Helibase The main location within the general incident area for parking, fueling, maintaining, and loading of helicopters, usually located at or near the incident base.

Helibase crew A crew of individuals who may be assigned to support helicopter operations.

Helibase Manager Person responsible to the air support supervisor for controlling helicopter takeoffs and landings at a helibase, logistically managing helicopters and supplies assigned to the helibase, and managing any fire retardant operations assigned to the helibase.

Helibase Radio Operator (HERO) Person responsible for communicating between incident assigned helicopters, helibase operational units, and the takeoff and landing coordinator.

Helibucket Specially designed bucket carried by a helicopter like a sling load and used for aerial delivery of water or fire retardant.

Helicopter An aircraft that depends principally on the lift generated by one or more rotors for its support in flight.

Helicopter Coordinator Airborne or ground based person responsible to the air attack supervisor for coordinating tactical or logistical helicopter missions on a fire incident. More than one helicopter coordinator may be assigned to an incident, depending on the number and type of missions to be accomplished.

Helicopter tender A ground service vehicle capable of supplying fuel and support equipment to helicopters.

Helicopter Timekeeper Person responsible to the helibase radio operator for keeping time of use of all helicopters assigned to the helibase.

Helipond Small body of water suitable for hover filling a helibucket.

Heliport A permanent facility for the operation of helicopters built to FAA standards and marked on aeronautical charts. Natural resource agencies refer to agency heliports as permanent helibases.

Helispot A natural or improved takeoff and landing area intended for temporary or occasional helicopter use.

Helispot Manager Person supervised by the helibase manager and who is responsible for controlling helicopter takeoffs and landings at a helispot, logistically managing helicopters and supplies at the helispot, and managing any fire retardant operations assigned to the helispot.

Helitack The utilization of helicopters to transport crews, equipment, and fire retardants or suppressants to the fireline during the initial stages of a fire. Also refers to the crew that performs helicopter management and attack activities.

Helitack crew Crew of firefighters specially trained and certified in the tactical and logistical use of helicopters for fire suppression.

Helitack Foreman A supervisory firefighter trained in the tactical use of helicopters for fire suppression.

Helitank Specially designed tank, generally of fabric or metal, fitted closely to the bottom of a helicopter and used for transporting and dropping suppressants or fire retardants.

Helitanker A helicopter equipped with a fixed tank or a suspended bucket-type container used for aerial delivery of water or retardants.

Helitorch An aerial ignition device hung from or mounted on a helicopter to disperse ignited lumps of gelled gasoline. Used for backfires, burnouts, or prescribed burns. *See also* Aerial ignition device; Delayed aerial ignition devices; Ping-pong ball system; Plastic sphere dispenser.

Hidden fire scar Fire scar in a tree resulting from fire injury to the cambium without destruction of the overlying bark and therefore not readily discernible.

Higbee cut Removal of the first (i.e., outside) thread of a female or male coupling to prevent crossing or mutilation of threads. Dimpled rocker on female coupling indicates beginning of Higbee cut.

High expansion foam Foam with an expansion between 201:1 and 1,000:1.

High pressure fog Small capacity spray jet produced at very high pressures (greater than 250 pounds per square inch, the normal maximum pressure for standard ground tankers) and discharged through a small hose with a gun-type nozzle.

Holding forces Resources assigned to do all required fire suppression work following fireline construction but generally not including extensive mop-up.

Holding orbit A predetermined maneuver that keeps an aircraft within a specified airspace while awaiting further orders.

Hose clamp Crimping device for stopping the flow of water in a hose.

Hose lay Arrangement of connected lengths of fire hose and accessories on the ground, beginning at the first pumping unit and ending at the point of water delivery.

Hose reel A rotating drum used for winding booster hose (normally) for storing and dispensing.

Hoseline tee A fitting that may be installed between lengths of hose to provide an independently controlled outlet for a branch line.

Hotshot crew Intensively trained fire crew used primarily in handline construction (Type 1).

Hot spot A particularly active part of a fire.

Hot-spotting Checking the spread of fire at points of more rapid spread or special threat, usually the initial step in prompt control, with emphasis on first priorities.

Hover A stationary in-flight condition for helicopters when no directional flight is achieved.

Hovering in ground effect (HIGE) The situation in which a helicopter is hovering sufficiently close to the ground to achieve added lift due to the effects of "ground cushion." The HIGE ceiling for a given gross weight thus is greater than the HOGE (hovering out of ground effect) ceiling.

Humidity General term referring to the moisture content of the atmosphere.

Humus Layer of decomposed organic matter on the forest floor beneath the fermentation layer and directly above the soil; that part of the duff in which decomposition has rendered vegetation unrecognizable and mixing of soil and organic matter is underway. *See also* Duff; Litter.

Hydrophobicity Resistance to wetting exhibited by some soils, also called *water repellency*. The phenomenon may occur naturally or may be fire induced. It may be determined by water drop penetration time, equilibrium liquid-contact angles, solid-air surface tension indices, or the characterization of dynamic wetting angles during infiltration.

Ignition component Part of the National Fire Danger Rating System (NFDRS), a rating of the probability that a firebrand will cause an actionable fire.

Ignition energy Quantity of heat or electrical energy that must be absorbed by a substance to ignite and burn.

Ignition method The means by which a fire is ignited, such as by handheld drip torch, helitorch, and backpack propane tanks.

Ignition pattern Manner in which a prescribed fire is ignited. The distance between ignition lines or points and the sequence of igniting them is determined by weather, fuel, topography, firing technique, and other factors that influence fire behavior and fire effects.

Ignition probability Chance that a firebrand will cause an ignition when it lands on receptive fuels. Syn.: Ignition index.

Ignition time Time between application of an ignition source and self-sustained combustion of a fuel.

Impulse Term used in weather primarily to describe a weak disturbance that does not necessarily have an associated storm center or surface low. The disturbance usually does not create severe weather and is frequently associated with a marine air push.

Incendiary fire Wildfire willfully ignited by anyone to burn or spread to vegetation or property without consent of the owner or his or her agent.

Incident Human-caused occurrence or natural phenomenon that requires action or support by emergency service personnel to prevent or minimize loss of life or damage to property and/or natural resources.

Incident Action Plan (IAP) Contains objectives reflecting the overall incident strategy and specific tactical actions and supporting information for the next operational period. The plan may be oral or written. When written, the plan may have a number of attachments, including incident objectives, organization assignment list, division assignment, incident radio communication plan, medical plan, traffic plan, safety plan, and incident map. Formerly called *shift plan*.

Incident base Location at the incident where the primary logistics functions are coordinated and administered. (Incident name or other designator will be added to the term "base.") The incident command post may be collocated with the base. There is only one base per incident.

Incident Commander Individual responsible for the management of all incident operations at the incident site.

Incident Command Post (ICP) Location at which primary command functions are executed. The ICP may be collocated with the incident base or other incident facilities.

Incident Command System (ICS) A standardized on-scene emergency management concept specifically designed to allow its user(s) to adopt an integrated organizational structure equal to the complexity and demands of single or multiple incidents, without being hindered by jurisdictional boundaries.

Incident Communications Center The location of the communications unit and the message center.

Incident Communications Manager Person responsible to the communications unit leader for providing dispatch services at the incident and for receiving and transmitting radio and telephone messages to and among incident personnel.

Incident Management Team The incident commander and appropriate general and command staff personnel assigned to an incident.

Incident Meteorologist (IMET) A specially trained meteorologist, responsible to the fire behavior analyst and the planning section chief, who provides site-specific weather forecasts and information to an incident. The IMET is usually accompanied by an ATMU and RAWS. *See also* Air Transportable Modular Unit and Remote Automatic Weather Station.

Incident objectives Statements of guidance and direction necessary for the selection of appropriate strategy(s), and the tactical direction of resources. Incident objectives are based upon agency administrator's direction and constraints. Incident objectives must be achievable and measurable, yet flexible enough to allow for strategic and tactical alternatives.

Incident organization Resources, together with a complement of overhead personnel, calculated to be sufficient to provide fire efficient incident management.

Incident overhead All supervisory positions described in the incident command system.

Incident support organization Includes any off-incident support provided to an incident. Examples include agency dispatch centers, airports, and mobilization centers.

Incident weather forecast A special fire weather forecast for a specific incident prepared by a meteorologist on site at or near the incident area. *See also* General fire weather forecast; Spot weather forecast.

Increment Any resource or grouping of resources on which individual status is maintained.

Incremental drop Air tanker drop in which tank doors are opened in sequence so that fire retardant cascades somewhat continuously.

Independent action Fire suppression activities by other than regular fire suppression organizations or a fire cooperator.

Independent crown fire A fire that advances in the tree crowns alone, not requiring any energy from the surface fire to sustain combustion or movement. Also called *running crown fire*.

Indirect attack A method of suppression in which the control line is located some considerable distance away from the fire's active edge. Generally done in the case of a fast-spreading or high-intensity fire and to utilize natural or constructed firebreaks or fuelbreaks and favorable breaks in the topography. The intervening fuel is usually backfired; but occasionally the main fire is allowed to burn to the line, depending on conditions.

Indraft The process whereby air is being drawn into the fire. Products of combustion are being forced aloft by the superheated air in the convective column. Cool fresh air is drawn in at the fire's base to replace the heated air moving aloft. In the case of a backfire, the larger high-intensity fire draws to it the fire you are lighting because of this concept.

Information Officer A member of the command staff responsible for interfacing with the public and media or with other agencies requiring information directly from the incident. There is only one information officer per incident, but may have assistants.

Infrared (IR) A heat detection system used for fire detection, mapping, and hotspot identification. *See also* Thermal imagery.

Infrared Groundlink A capability through the use of a special mobile ground station to receive air-to-ground IR imagery at an incident.

Initial attack The actions taken by the first resources to arrive at a wildfire to protect lives and property and prevent further extension of the fire.

Initial attack crew Specially trained and equipped fire crew for initial attack on a fire.

Initial attack fire (IAF) Fire that is generally contained by the attack units first dispatched, without a significant augmentation of reinforcements, within 2 hours after initial attack, and full control is expected within the first burning period.

Initial attack incident commander (IAIC) The incident commander at the time the first attack forces commence suppression work on a fire.

Initial response Resources initially committed to an incident.

Inmate crew Any fire crew composed of prison inmates.

Inside diameter (ID) The internal diameter of a tube, conductor, or coupling, as distinguished from its outside diameter (OD). Fire hose sizes are classified by a nominal internal diameter.

In-stand wind Wind speed within a stand at about eye level.

Intermittent smoke Smoke that becomes visible only at intervals.

Internal load Load carried inside the fuselage structure of an aircraft.

Internal payload Allowable aircraft cabin load, in pounds, with full fuel and pilot in calm air at standard atmosphere.

Inversion An increase of temperature with height in the atmosphere. Vertical motion in the atmosphere is inhibited, allowing for pollution buildup. A "normal" atmosphere has temperature decreasing with height. *See also* Atmospheric stability; Stable layer of air; Atmospheric inversion.

Iron pipe thread (IPT) A tapered thread standard used for connecting various sizes of rigid pipe. This standard may be referred to as tapered iron pipe thread (TIPT), national pipe thread (NPT), iron pipe thread (IPT), or iron pipe standard thread (IPS). With tapered thread, the threads and pipe sealant perform the seal at the connection, as opposed to straight thread connectors, which use a gasket to form the seal.

Island An unburned area within a fire perimeter.

Isobar A charted line found on a weather map connecting points of equal atmospheric pressure.

Isotherm A charted line found on a weather map connecting points of equal temperature.

Isothermal layer Layer through which temperature remains constant with elevation.

I-zone The line, area, or zone where structures and other human development meet or intermingle with undeveloped wildland or vegetative fuels. Syn.: Urban interface, Wildland/urban interface.

Jettison Disposing of cargo, fuel, water, or retardant overboard to lighten an aircraft or to improve its stability.

Jump spot Selected landing area for smokejumpers.

Jump suit Approved protection suit worn for smokejumping.

Jurisdiction The range or sphere of authority. Public agencies have jurisdiction at an incident related to their legal responsibilities and authority for incident mitigation. Jurisdictional authority at an incident can be political/geographical (e.g., city, county, state, or federal boundary lines), or functional (e.g., police department, health department). *See also* Multijurisdictional incident.

Jurisdictional agency The agency having land and resource management responsibility for a specific geographical or functional area as provided by federal, state, or local law.

Keetch-Byram Drought Index (KBDI) Commonly used drought index adapted for fire management applications, with a numerical range from 0 (no moisture deficiency) to 800 (maximum drought). *See also* Drought index.

Kindling point Lowest temperature at which sustained combustion can be initiated for a specified substance. Also called *ignition temperature*.

Knock down To reduce the flame or heat on the more vigorously burning parts of a fire edge.

Knot Nautical miles per hour, equal to 1.15 miles per hour.

Ladder fuels Fuel that provides vertical continuity between strata, thereby allowing fire to carry from surface fuels into the crowns of trees or shrubs with relative ease; helps initiate and ensure the continuation of crowning.

Land breeze A convective wind that occurs at night as a result of the land surface cooling more quickly than the water surfaces.

Lapse rate Decrease of an atmospheric variable (temperature unless specified otherwise) with height.

Large aircraft Aircraft in which maximum certified gross weight at takeoff exceeds 12,500 pounds.

Large fire (1) For statistical purposes, a fire burning more than a specified area of land, for example, 300 acres. (2) A fire burning with a size and intensity such that its behavior is determined by interaction between its own convection column and weather conditions above the surface.

Latitude Angular distance, measured in degrees, creating imaginary lines circling the earth's globe. The lines extend in an easterly and westerly direction, parallel with the equator, which is at 0 degrees latitude. The degrees of latitude increase as one proceeds away from the equator toward either the north or south poles where the latitude is 90 degrees.

Lead plane Aircraft with pilot used to make trial runs over the target area to check wind, smoke conditions, and topography and to lead air tankers to targets and supervise their drops.

Liaison Officer (LOFR) A member of the command staff responsible for coordinating with agency representatives from assisting and cooperating agencies.

Life-safety Refers to the joint consideration of both the life and physical well-being of individuals.

Lifting processes Any of the processes that lead to upward vertical motion in the atmosphere. These processes may include low level convergence, heating or thermal convection, orographic lifting over the mountains, and frontal lifting.

Light burn Degree of burn that leaves the soil covered with partially charred organic material; heavy fuels are not deeply charred.

Light (fine) fuels Fast-drying fuels, generally with a comparatively high surface area-to-volume ratio, which are less than 1/4 inch in diameter and have a timelag of 1 hour or less. These fuels ignite readily and are rapidly consumed by fire when dry.

Lightning activity level (LAL) A number, on a scale of 1 to 6, that reflects frequency and character of cloud-to-ground lightning (forecasted or observed). The scale for 1 to 5 is exponential, based on powers of 2 (i.e. , LAL 3 indicates twice the lightning of LAL 2). LAL 6 is a special category for dry lightning and is closely equivalent to LAL 3 in strike frequency. Part of the National Fire Danger Rating System (NFDRS).

Lightning fire Wildfire caused directly or indirectly by lightning.

Lightning fire occurrence index Numerical rating of the potential occurrence of lightning-caused fires. Part of the National Fire Danger Rating System (NFDRS).

Lightning risk (LR) A number related to the expected number of cloud-to-ground lightning strokes to which a protection unit is expected to be exposed during the rating period; the LR value used in the occurrence index includes an adjustment for lightning activity experienced during the previous day to account for possible holdover fires. Part of the National Fire Danger Rating System (NFDRS).

Lightning risk scaling factor Factor derived from local thunderstorm and lightning-caused fire records that adjusts predictions of the basic lightning fire occurrence model to local experience, accounting for factors not addressed directly by the model (e.g., susceptibility of local fuels to ignition by lightning, fuel continuity, topography, regional characteristics of thunderstorms). Part of the National Fire Danger Rating System (NFDRS).

Light wind Wind speed less than 7 miles per hour (6 knots) measured at 20 feet above ground. At eye level, light winds are less than 3 miles per hour (3 knots).

Limbing Removing branches from a felled or standing tree, or from brush.

Limited containment Halting of fire spread at the head, or that portion of the flanks of a prescribed fire that is threatening to exceed prescription criteria, and ensuring that this spread rate will not be encountered again; does not indicate mop-up.

Line cutter Fire crew member in the progressive method of line construction who cuts and clears away brush, small saplings, vines, and other obstructions in the path of the fireline; usually equipped with ax or brush hook.

Lined fire hose Fire hose with a smooth inner coating of rubber or plastic to reduce friction loss.

Line firing Setting fire to only the border fuel immediately adjacent to the control line.

Line holding Ensuring that the established fireline has completely stopped fire progress.

Line ignition Setting a line of fire (e.g., backing fire) as opposed to individual spots.

Line officer Agency administrator.

Line scout A firefighter who determines the location of a fireline.

Liquid concentrate (LC) Liquid phosphate fertilizers used as fire retardants, usually diluted three to five times prior to application.

Litter The top layer of forest floor, composed of loose debris of dead sticks, branches, twigs, and recently fallen leaves or needles; little altered in structure by decomposition. *See also* Duff; Humus.

Live line Hose line filled with water under pressure and ready to use. *See also* Charged line.

Live burning Burning of green slash progressively as it is cut.

Live fuel moisture content Ratio of the amount of water to the amount of dry plant material in living plants.

Live fuels Living plants, such as trees, grasses, and shrubs, in which the seasonal moisture content cycle is controlled largely by internal physiological mechanisms rather than by external weather influences.

Live herbaceous moisture content Ratio of the amount of water to the amount of dry plant material in herbaceous plants, that is, grasses and forbs.

Live woody moisture content Ratio of the amount of water to the amount of dry plant material in shrubs.

Living fuels Naturally occurring fuels in which moisture content is physiologically controlled within the living plant.

Local winds Winds generated over a comparatively small area by local terrain and weather. They differ from those which would be appropriate to the general pressure pattern. *See also* General winds; Slope winds.

Logging debris Unwanted tree parts (crowns, logs, uprooted stumps) remaining after harvest.

Logistics Section Section responsible for providing facilities, services, and materials for the incident.

Logistics Section Chief A member of the General Staff who is responsible for providing facilities, services, and material in support of the incident.

Longitude Angular distance, measured in degrees, creating imaginary lines extending from the north pole to the south pole that identify geographical positions on the Earth's globe. The lines are based from the prime meridian of 0 degrees longitude that runs through Greenwich, England, extending 180 degrees westward and eastward.

Longline Line or set of lines, usually in 50 foot increments, used in external load operations that allow the helicopter to place loads in areas in which the helicopter could not safely land.

Long-range forecast Fire weather forecast for a period greater than 5 days in advance.

Long-range spotting Large glowing firebrands are carried high into the convection column and then fall out downwind beyond the main fire starting new fires. Such spotting can easily occur 1/4 mile or more from the firebrand's source.

Long-term fire retardant Chemical that inhibits combustion primarily through chemical reactions between products of combustion and the applied chemicals, even after the water component has evaporated. Other chemical effects also may be achieved, such as film-forming and intumescence.

Lookout (1) Person designated to detect and report fires from a vantage point. (2) Location from which fires can be detected and reported. (3) Fire crew member assigned to observe the fire and warn the crew when there is danger of becoming trapped.

Lookout firefighter Person combining the functions of lookout and firefighters

Lookout(s), Communication(s), Escape Route(s), and Safety Zone(s) (LCES) Elements of a safety system used by fire fighters to routinely assess their current situation with respect to wildland firefighting hazards.

Lookout tower Structure that elevates a person above nearby obstructions to sight for fires; generally capped by some sort of house or cupola.

Lopping After felling, cutting branches, tops, and unwanted boles into lengths such that resultant logging debris will lie close to the ground.

Lopping and scattering Lopping logging debris and spreading it more or less evenly over the ground.

Lost Line Any part of a fireline rendered useless by a breakover of the fire.

Low An area of relatively low atmospheric pressure in which winds tend to move in a counterclockwise direction, spiraling in toward the low's center.

Low expansion foam Foam with an expansion between 1:1 and 20:1.

Macroclimate General large-scale climate of a large area or country as distinguished from smaller scale mesoclimate and microclimate.

Main ridge Prominent ridgeline separating river or creek drainages. Usually has numerous smaller ridges (spur) extending outward from both sides.

Male coupling An externally threaded hose nipple that fits in the thread of a female (internally threaded) swivel coupling of the same pitch and diameter. Nozzles attach to this type of coupling.

Management by objectives In ICS, a top-down management activity that involves a three-step process to achieve the incident goal. The steps are: establishing the incident objectives, selection of appropriate strategy(s) to achieve the objectives; and the tactical direction associated with the selected strategy. Tactical direction includes selection of tactics, selection of resources, resource assignments, and performance monitoring.

Managers Individuals within ICS organizational units assigned specific managerial responsibilities, for example, staging area manager or camp manager.

Manicured line Fireline built exactly to standards.

Manually regulated A proportioning method or device that requires a manual adjustment to maintain a desired mix ratio over a changing range of water flows and pressures.

Map's legend A key accompanying a map that shows information needed to interpret that map. Each type of map has information represented in a different way relating to its subject matter. The legend can explain map scales, symbols, and color.

Marine air Air that has a high moisture content and the temperature characteristics of an ocean surface due to extensive exposure to that surface. An intrusion of marine air moderates fire conditions; absence of marine air in coastal areas may lead to more severe fire danger.

Marine climate Regional climate under the predominant influence of the sea, that is, a climate characterized by marine air; the opposite of a continental climate.

Maritime air Air that has assumed high moisture content and the temperature characteristics of a water surface due to extensive exposure to that surface.

Mass fire A fire resulting from many simultaneous ignitions that generates a high level of energy output.

Mass transport Heat carried ahead of the fire in the form of firebrands.

Mattock Hand tool with a narrow hoeing surface at one end of the blade and a pick or cutting blade at the other end; used for digging and grubbing.

McLeod A combination hoe or cutting tool and rake, with or without removable blades.

Mean fire interval Arithmetic average of all fire intervals determined, in years, in a designated area during a specified time period; size of the area and the time period must be specified.

Mean sea level (MSL) Average height of the surface of the sea for all stages of the tide over a 19-year period. *Note:* when the abbreviation MSL is used in conjunction with a number of feet, it implies altitude above sea level (e.g., 1,000 feet MSL).

Medical Unit Functional unit within the logistics section responsible for the development of the medical emergency plan and for providing emergency medical treatment of incident personnel.

Medium expansion foam Foam with an expansion between 21:1 and 200:1.

Medevac Mobile medical treatment and transportation.

Meter A basic unit of length in the metric system equal to 39.37 inches.

Midflame windspeed The speed of the wind measured at the midpoint of the flames, considered to be most representative of the speed of the wind that affects fire behavior.

Military operations area (MOA) Military operations area found on aeronautical charts.

Military time The 24-hour clock system where midnight is 2400, one minute after midnight is 0001 and progresses to 2400 daily.

Millibar A unit of pressure equal to a force of 1,000 dynes per square centimeter. (A dyne is the force that would give a free mass of one gram an acceleration of one centimeter per second per second.)

Mineral soil Soil layers below the predominantly organic horizons; soil with little combustible material.

Minimum Impact Suppression Techniques (MIST) The application of strategy and tactics that effectively meet suppression and resource objectives with the least environmental, cultural, and social impacts.

Mineral ash The residue of mineral matter left after complete combustion of wood (wood ash) or other organic material; consists largely of oxides, carbonates, and phosphates of Ca, K, and Mg, together with other compounds [formerly used as a source of potash (K_2CO_3)].

Mixing A random exchange of air parcels on any scale from the molecular to the largest eddy.

Mixing chamber A tube, constructed with deflectors or baffles, that mixes foam solution and air to produce tiny, uniform bubbles.

Mixing height Measured from the surface upward, the height to which relatively vigorous mixing occurs due to convection. Also called *mixing depth.*

Mixing layer That portion of the atmosphere from the surface up to the mixing height; The layer of air, usually a subinversion layer, within which pollutants are mixed by turbulence and diffusion. Also called *mixed layer.*

Mixmaster The person in charge of fire retardant mixing operations with responsibility for quantity and quality of the slurry and for the loading of aircraft in land based air tanker operations.

Mix ratio The ratio of liquid foam concentrate to water, usually expressed as a percent.

Mobile pump A method whereby a fire engine pumps a hoseline while moving.

Mobile weather unit forecast A special fire weather forecast for a specific fire prepared by a meteorologist on site at or near the fire area.

Mobilization The process and procedures used by all organizations, federal, state, and local, for activating, assembling, and transporting all resources that have been requested to respond to or support an incident.

Moderate burn Degree of burn in which all organic material is burned away from the surface of the soil, which is not discolored by heat; any remaining fuel is deeply charred. Organic matter remains in the soil immediately below the surface.

Modified suppression Suppression action dictated by one or more management constraints that affect strategy and/or tactics.

Modular Airborne Firefighting System (MAFFS) A manufactured unit consisting of five interconnecting tanks, a control pallet, and a nozzle pallet, with a capacity of 3,000 gallons, designed to be rapidly mounted inside an unmodified C-130 (Hercules) cargo aircraft for use in cascading retardant chemicals on wildfires.

Moist adiabatic lapse rate Rate of decrease of temperature with increasing height of an air parcel lifted at saturation via adiabatic process through an atmosphere in hydrostatic equilibrium. Rate varies according to the amount of water vapor in the parcel and is usually between 2° and 5° F per 1,000 feet (3.6° and 9.2°C per 1000 meters)

Moisture of extinction The fuel moisture content, weighed over all the fuel classes, at which the fire will not spread. Also called *extinction moisture content* (EMC).

Mop-up Extinguishing or removing burning material near control lines, felling snags, and trenching logs to prevent rolling after an area has burned, to make a fire safe, or to reduce residual smoke.

Mop-up crew A portion or all of a regular fire crew assigned to mop-up work after the fire or a portion of the fire has been controlled.

Mop-up time Elapsed time from control of a fire until organized mop-up is complete.

Move-up System of redistributing remaining personnel and equipment following dispatch of other forces among a network of fire stations to provide the best possible response within the fire department's direct protection area in the event of additional calls for emergency assistance.

Multiagency Coordination (MAC) A generalized term that describes the functions and activities of representatives of involved agencies and/or jurisdictions who come together to make decisions regarding the prioritizing of incidents, and the sharing and use of critical resources. The MAC organization is not a part of the on-scene ICS and is not involved in developing incident strategy or tactics.

Multijurisdictional incident An incident requiring action from multiple agencies that have a statutory responsibility for incident mitigation. In ICS, these incidents are managed under unified command. *See also* Jurisdiction.

Mutual aid A system wherein two or more fire departments, by prior agreement, operate essentially as a single agency to respond routinely across jurisdictional boundaries to render mutual assistance in combating fire emergencies.

Mutual aid agreement Written agreement between agencies and/or jurisdictions in which they agree to assist one another upon request, by furnishing personnel and equipment.

Mutual threat zone A geographical area between two or more jurisdictions into which those agencies would respond on initial attack. Also called *mutual response zone* or *initial action zone*.

National Fire Danger Rating System (NFDRS) A uniform fire danger rating system that focuses on the environmental factors that control the moisture content of fuels.

National Interagency Fire Center (NIFC) Facility located in Boise, Idaho, jointly operated by several federal agencies, dedicated to coordination, logistical support, and improved weather services in support of fire management operations throughout the United States.

National pipe straight hose thread (NPSH) Straight (nontapered) thread standard with the same threads per inch as the appropriate size iron pipe thread; requires a gasket to seal and is the thread standard used by most U.S. industry. Also known as National Pipe Straight Mechanical (NPSM) thread.

National Standard Thread (NH) Specifically defined screw thread used on fire hose couplings. Abbreviated (NH) for national hose.

National Wildfire Coordinating Group (NWCG) Group formed under the direction of the secretaries of the Interior and Agriculture to improve the coordination and effectiveness of wildland fire activities and provide a forum to discuss, recommend appropriate action, or resolve issues and problems of substantive nature.

Native species A species which is a part of the original fauna or flora of a given area.

Natural barrier Any area where lack of flammable material obstructs the spread of wildfires.

Near miss Any potential accident which, through prevention, education, hazard reduction, or luck, did not occur.

Neutral atmosphere Condition in which temperature decrease with increasing altitude is equal to the dry adiabatic lapse rate (i.e., the atmosphere neither aids nor hinders large-scale vertical motion).

Nomex Trade name for a fire-resistant synthetic material used in the manufacturing of flight suits and pants and shirts used by firefighters. Aramid is the generic name.

Nonconvective-life fire phase Phase of a fire when most emissions are not entrained into a definite convection column.

Normal fire season (1) A season when weather, fire danger, and number of distribution of fires are about average. (2) Period of the year that normally comprises the fire season.

Normal fire year The year with the third greatest number of fires in the past ten.

Northing The distance in meters of the position north of the equator, measured along a zone line.

Nozzle-aspirated foam system A foam-generating device that mixes air at atmospheric pressure with foam solution in a nozzle chamber.

Nozzle operator A person assigned to operate a fire hose nozzle, usually on a handline.

One-hour timelag fuels Fuels consisting of dead herbaceous plants and roundwood less than about one-fourth inch (6.4 millimeters) in diameter. Also included is the uppermost layer of needles or leaves on the forest floor.

One-hour timelag fuel moisture (1-h TL FM) Moisture content of one-hour timelag fuels.

One-hundred hour timelag fuels Dead fuels consisting of roundwood in the size range of 1 to 3 inches (2.5 to 7.6 centimeters) in diameter and very roughly the layer of litter extending from approximately three-fourths of an inch (1.9 centimeters) to 4 inches (10 centimeters) below the surface.

One-hundred hour timelag fuel moisture (100-h TL FM) The moisture content of the 100-hour timelag fuels.

One-lick method A progressive system of building a fireline on a wildfire without changing relative positions in the line. Each worker does one to several "licks," or strokes, with a given tool and then moves forward a specified distance to make room for the worker behind.

One-thousand hour timelag fuels Dead fuels consisting of roundwood 3–8 inches in diameter and the layer of the forest floor more than about 4 inches below the surface.

One-thousand hour timelag fuel moisture (1,000-h TL FM) The moisture content of the 1,000-hour timelag fuels.

Onshore flow Wind blowing from water to land.

Operational period The period of time scheduled for execution of a given set of tactical actions as specified in the incident action plan; can be of various lengths, although usually not more than 24 hours.

Operations Branch Director Person under the direction of the operations section chief who is responsible for implementing that portion of the incident action plan appropriate to the branch.

Operations Coordination Center (OCC) Primary facility of the multiagency coordination system (MACS); houses staff and equipment necessary to perform the MACS function.

Operations Section The section responsible for all tactical operations at the incident; includes branches, divisions and/or groups, task forces, strike teams, single resources, and staging areas.

Operations Section Chief A member of the General Staff responsible for the management of all tactical operations as set forth in the Incident Action Plan, who reports directly to the Incident Commander.

Ordering Manager Person responsible to the supply unit leader for ordering all supplies and equipment needed at an incident.

Organic Material That fraction of the soil that includes plant and animal residues at various stages of decomposition, cells and tissues of soil organisms, and substances synthesized by the soil population.

Organic soil Any soil or soil horizon containing at least 30% organic matter (e.g., muck, peat).

Orographic Pertaining to, or caused by, mountains.

Orographic lifting Air being forced aloft by a slope, hill, or mountain range. The air is forced upward on the windward side and is cooled. It is an important process in producing clouds and precipitation.

Orthophoto Photograph obtained from the orthogonal (i.e., horizontal) projection of a correctly oriented stereoscopic model formed by two overlapping aerial photographs; an orthophoto is free of tilt and relief displacements.

Orthophoto maps Aerial photographs corrected to scale such that geographic measurements may be taken directly from prints. They may contain graphically emphasized geographic features and may be provided with overlays of such features as water systems or facility location.

Out-of-service resources Resources assigned to an incident but unable to respond for mechanical, rest, or personal reasons.

Outside aid Firefighting assistance given to adjacent areas and nearby communities by contract or other agreement that covers conditions and payment for assistance rendered and services performed. Contrasted to mutual aid, in which neighboring firefighting organizations assist each other without charge.

Overhead Personnel assigned to supervisory positions, including incident commander, command staff, general staff, branch directors, supervisors, unit leaders, managers, and staff.

Overwintering fire A fire that persists through the winter months until the beginning of fire season.

Packing ratio The fraction of a fuel bed occupied by fuels, or the fuel volume divided by bed volume.

Pack test Used to determine the aerobic capacity of fire suppression support personnel and assign physical fitness scores. The test consists of walking a specified distance, with or without a weighted pack, in a predetermined period of time, with altitude corrections.

Parallel attack Method of fire suppression in which fireline is constructed approximately parallel to, and just far enough from the fire edge to enable workers and equipment to work effectively, although the fireline may be shortened by cutting across unburned fingers. The intervening strip of unburned fuel is normally burned out as the control line proceeds but may be allowed to burn out unassisted where this occurs without undue delay or threat to the fireline.

Parallel burning (1) A type of suppression fire. Igniting a narrow strip of fuel adjacent to a control line and then burning successively wider adjacent strips as the preceding strip burns out. (2) Burning only a relatively narrow strip or strips through an area of slash, leaving the remainder. (3) Burning slash in strips generally 100–300 feet wide along roads or barriers to subdivide the slash area into blocks.

Parallel pumping Procedure by which the flow from two fire pumps is combined into one hose line.

Parallel tandem pumping Procedure by which the flow from two fire pumps is combined into a third pump.

Particle size The size of a piece of fuel, often expressed in terms of size classes.

Particulate matter Any liquid or solid particles. "Total suspended particulates" as used in air quality are those particles suspended in or falling through the atmosphere. They generally range in size from 0.1 to 100 microns.

Particulates Fine liquid or solid particles such as dust, smoke, mist, fumes, or smog found in air or emissions.

Parts of a fire On typical free-burning fires the spread is uneven with the main spread moving with the wind or upslope. The most rapidly moving portion is designated the head of the fire, the adjoining portions of the perimeter at right angles to the head are known as the flanks, and the slowest moving portion is known as the rear or the base of the fire.

Passive crown fire A fire in the crowns of' trees in which trees or groups of trees torch, ignited by the passing front of the fire. The torching trees reinforce the spread rate, but these fires are not basically different from surface fires.

Patch burning Burning in patches to prepare sites for group planting or sowing or to form a barrier to subsequent fires.

Patrol (1) To travel over a given route to prevent, detect, and suppress fires. (2) To go back and forth vigilantly over a length of control line during and/or after construction to prevent breakovers, suppress spot fires, and extinguish overlooked hot spots. (3) A person or group of persons who carry out patrol actions.

Patrol unit Any light, mobile unit with limited pumping and water capacity.

Pattern The distribution of an aerially delivered retardant drop on the target area in terms of its length, width, and momentum (velocity × mass) as it approaches the ground. The latter determines the relative coverage level of the fire retardant on fuel within the pattern. Syn.: Drop pattern.

Payload Weight of passengers and/or cargo being carried by an aircraft.

Peak wind The greatest 5-second average wind speed during the previous hour that exceeded 25 knots.

Perennial plant A plant that continues to grow year after year.

Perimeter access Fireline suitable for vehicle travel.

Permafrost A short term for "permanently frozen ground"; any part of the Earth's crust, bedrock, or soil mantle that remains below 32°F (0°C) continuously for a number of years.

Personal Protective Equipment (PPE) That equipment and clothing required to mitigate the risk of injury from or exposure to hazardous conditions encountered during the performance of duty. PPE includes, but is not limited to, fire-resistant clothing, hard hat, flight helmets, shroud, goggles, gloves, respirators, hearing protection, chainsaw chaps, and shelter.

Personnel Time Recorder Person responsible to the time unit leader for daily recording of the time of all personnel at the incident, posting commissary charges to the time reports, closing out all records prior to personnel leaving the incident, and distributing completed time documents according to each agency's policy.

Pincer action Direct attack around a fire in opposite directions by two or more attack units.

Ping-Pong Ball System Mechanized method of dispensing delayed aerial ignition devices (DAIDs) at a selected rate. *See also* Aerial Ignition Device; Delayed Aerial Ignition Devices; Helitorch; Plastic Sphere Dispenser.

Planning interval Period of time between scheduled planning meetings.

Planning meeting A meeting held regularly throughout the duration of an incident to select specific strategies and tactics for incident control operations and to plan for needed service and support. On larger incidents, the planning meeting is a major element in the development of the incident action plan.

Planning Section Responsible for the collection, evaluation, and dissemination of tactical information related to the incident, and for the preparation and documentation of incident action plans. The section also maintains information on the current and forecasted situation, and on the status of resources assigned to the incident. Includes the situation, resource, documentation, and demobilization units, as well as technical specialists.

Plan of attack The selected course of action and organization of personnel and equipment in fire suppression, as applied to a particular fire or to all fires of a specific type.

Plastic Sphere Dispenser (PSD) Device installed, but jettisonable, in a helicopter, which injects glycol into a plastic sphere containing potassium permanganate, which is then expelled from the machine and aircraft. This produces an exothermic reaction resulting in ignition of fuels on the ground for prescribed or wildland fire applications. *See also* Aerial ignition device; Delayed aerial ignition devices; Helitorch; Ping-pong ball system.

Plow line Fireline constructed by a fire plow, usually drawn by a tractor or other motorized equipment. *See also* Bulldozer; Tractor; Tractor plow.

Plume The segment of the atmosphere occupied by the emissions from a single source or a grouping of sources close together. A convection column, if one exists, forms a specific part of the plume.

Plume-dominated wildfire A wildland fire whose activity is determined by the convection column.

Plume rise How high above the level of release an emission plume rises.

Pockets of a fire Unburned indentations in the fire edge formed by fingers or slow burning areas.

Point of attack That part of the fire on which work is started when suppression crews arrive.

Point of origin Point of original ignition of a fire.

Portable pump Small gasoline-driven pump that can be carried to a water source by one or two firefighters or other conveyance over difficult terrain.

Portatank Container, either with rigid frame or self-supporting, which can be filled with water or fire chemical mixture from which fire suppression resources can be filled. It can also be a source for charging hose lays from portable pumps or stationary engines.

Positive displacement pump A pump that moves a specified quantity of water through the pump chamber with each stroke or cycle; it is capable of pumping air, and therefore is self-priming, but must have pressure relief provisions if plumbing or hoses have shut-off nozzles or valves. Gear pumps and piston pumps are common examples of this type.

Preattack A planned, systematic procedure for collecting, recording, and evaluating prefire and fire management intelligence data for a given planning unit or preattack block. The planning phase is usually followed by a construction and development program integrated with other resources and activities.

Preattack planning Within designated blocks of land, planning the locations of firelines, fire camps, water sources, and helispots; planning transportation systems, probable rates of travel, and constraints of travel on various types of attack units; and determining what types of attack units likely would be needed to construct particular firelines, their probable rate of fireline construction, and topographic constraints on fireline construction.

Precipitation Any or all forms of water particles, liquid or solid, that fall from the atmosphere and reach the ground.

Precipitation duration Time, in hours and fraction of hours, that a precipitation event lasts. More precisely, for fire danger rating purposes, the length of time that fuels are subjected to liquid water.

Preconnected Hard suction hose or discharge hose carried connected to pump, eliminating delay occasioned when hose and nozzles must be connected and attached at fire.

Premo Mark III This aerial firing device dispenses polystyrene spheres partially filled with potassium permanganate crystals and injected with ethylene glycol. The spheres are dispensed by an internally mounted sphere dispensing machine on a helicopter. Also known as a *Ping-Pong Ball System*.

Prescribed burning Controlled application of fire to wildland fuels in either their natural or modified state, under specified environmental conditions, which allows the fire to be confined to a predetermined area and produce the fire behavior and fire characteristics required to attain planned fire treatment and resource management objectives.

Prescribed fire A management ignited wildland fire that burns under specified conditions where the fire is confined to a predetermined area and produces the fire behavior and fire characteristics required to attain planned fire treatment and resource management objectives.

Prescribed natural fire Naturally ignited wildland fire that burns under specified conditions where the fire is confined to a predetermined area and produces the fire behavior and fire characteristics to attain planned fire treatment and resource management objectives.

Prescription A written statement defining the objectives to be attained as well as the conditions of temperature, humidity, wind direction and speed, fuel moisture, and soil moisture under which a fire will be allowed to burn. A prescription is generally expressed as acceptable ranges of the prescription elements and the limit of the geographic area to be covered.

Pressure gradient The difference in atmospheric pressure between two points on a weather map. That is, the magnitude of pressure difference between two points at sea level or at constant elevation above sea level. Wind speed is directly related to pressure gradient. If distance between constant pressure lines is reduced by one-half, wind speed will be doubled. Conversely, if distance between lines is doubled, wind speed will be reduced by one-half.

Pressure loss Reduction in water pressure between a pump or hydrant and a nozzle due to expenditure of pressure energy required to move water through a hose; includes losses due to back pressure, friction loss, elevation loss, and/or losses in fittings.

Pressure pattern The distribution of surface atmospheric pressure features over an area of the Earth as shown on a weather map. Surface pressure features include lines of constant pressure (isobars), highs, lows, and pressure gradient. The pressure pattern is directly related to wind speeds and directions at specific locations.

Presuppression Activities in advance of fire occurrence to ensure effective suppression action. Includes planning the organization, recruiting and training, procuring equipment and supplies, maintaining fire equipment and fire control improvements, and negotiating cooperative and/or mutual aid agreements.

Pretreat The use of water, foam or retardant along a control line in advance of the fire, often used where ground cover or terrain is considered best for control action.

Prevention Activities directed at reducing the incidence of fires, including public education, law enforcement, personal contact, and reduction of fuel hazards (fuels management).

Prime meridian An imaginary line on the ground running north and south that is accurately laid out to serve as the reference meridian in land survey. The prime meridian is 0 degrees longitude.

Priming Filling pump with water when pump is taking water not under a pressure head. Necessary for centrifugal pumps.

Probability A number representing the chance that a given event will occur. The range is from 0% for an impossible event, to 100% for an inevitable event.

Probability forecast A forecast of the probability of occurrence of one or more of a mutually exclusive set of weather contingencies as distinguished from a series of categorical statements.

Probability of ignition The chance that a firebrand will cause an ignition when it lands on receptive fuels.

Procurement unit Functional unit within the finance administration section responsible for financial matters involving vendor contracts.

Progressive hose lay A hose lay in which double shutoff wye (Y) valves or Tees are inserted in the main line at intervals and lateral lines are run from the wyes or tees to the fire edge, thus permitting continuous application of water during extension of the lay.

Progressive method of line construction A system of organizing workers to build fireline in which they advance without changing relative positions in line.

Proportioner A device that adds a predetermined amount of foam concentrate to water to form foam solution.

Psychrometer The general name for instruments designed to determine the moisture content of air. A psychrometer consists of dry- and wet-bulb thermometers that give the dry- and wet-bulb temperatures, which in turn are used to determine relative humidity and dew point.

Public Information Officer A member of the Command Staff who acts as the central point of contact for the media.

Pulaski A combination chopping and trenching tool widely used in fireline construction, which combines a single-bitted axe blade with a narrow adzelike trenching blade fitted to a straight handle.

Punk Partly decayed material, such as old wood, in which fire can smolder unless it is carefully mopped up and extinguished. A good receptor for firebrands when dry.

Pyrolysis The thermal or chemical decomposition of fuel at an elevated temperature; the preignition combustion phase of burning during which heat energy is absorbed by the fuel which, in turn, gives off flammable tars, pitches, and gases.

Quadrangle Mapping unit that defines an area in terms of longitude and latitude distance. Two common scales are 1:24,000 quadrangles, which are 7.5′ longitude × 7.5′ latitude, and 1:62,500 quadrangles, which are 15′ longitude × 15′ latitude.

Range fire Any wildfire on rangeland.

Rate of spread The relative activity of a fire in extending its horizontal dimensions, expressed as rate of increase of the total perimeter of the fire, as rate of forward spread of the fire front, or as rate of increase in area, depending on the intended use of the information. Usually it is expressed in chains or acres per hour for a specific period in the fire's history.

Rate of spread factor A factor usually on a scale of 1 to 100 that represents a relative rate of forward spread for a specific fuel condition and fixed weather conditions (or fuel model). Factors can be used as multipliers, arguments for entering tables, or to provide a ratio of values between two fuels.

Reaction intensity The rate of heat release, per unit area of the flaming fire front, expressed as heat energy/area/time, such as BTU/square foot/minute, or Kcal/square meter/second. *See also* Combustion rate.

Rear (of a fire) (1) That portion of a fire spreading directly into the wind or downslope. (2) That portion of a fire edge opposite the head. (3) Slowest spreading portion of a fire edge. Also called *heel of a fire*.

Reburn (1) Repeat burning of an area over which a fire has previously passed, but left fuel that later ignites when burning conditions are more favorable. (2) An area that has reburned.

Recorders Individuals within ICS organizational units who are responsible for recording information. Recorders may be found in planning, logistics, and finance/administration units.

Red card Fire qualification card issued to fire-rated persons showing their training needs and their qualifications to fill specified fire suppression positions in a large fire suppression or incident organization.

Red flag warning Term used by fire weather forecasters to alert forecast users to an ongoing or imminent critical fire weather pattern. *See also* Fire weather watch.

Reel A frame on which hose is wound, now chiefly used for "booster" or small hose (3/4 or 1 inch hose) (19 or 25 millimeters) supplied by a water tank on the apparatus; also, a hand-drawn two-wheel frame for 2½ inch (64 millimeters) hose used in industrial plants.

Rehabilitation The activities necessary to repair damage or disturbance caused by wildfire or the wildfire suppression activity.

Rekindle Reignition due to latent heat, sparks, or embers or due to presence of smoke or steam.

Relative humidity (RH) The ratio of the amount of moisture in the air to the maximum amount of moisture that air would contain if it were saturated; the ratio of the actual vapor pressure to the saturated vapor pressure.

Relay Use of two or more fire pumps to move water a distance that would require excessive pressures in order to overcome friction loss if only one pump were employed at the source.

Relay tank A tank, usually collapsible, used as a reservoir in the relay of water from one fire pump to another.

Relief valve A pressure-controlled device that bypasses water at a fire pump to prevent excessive pressures when a nozzle is shut down.

Remote Automatic Weather Station (RAWS) An apparatus that automatically acquires, processes, and stores local weather data for subsequent transmission to the GOES Satellite, from which they are retransmitted to an Earth receiving station for use in the National Fire Danger Rating System.

Reportable fire Any wildfire that requires fire suppression to protect natural resources or values associated with natural resources or is destructive to natural resources.

Reporting locations Location or facilities where incoming resources can check in at the incident. Check-in locations include incident command post (resources unit), incident base, camps, staging areas, helibases, helispots, and direct to the line. *See also* Check-in.

Report time Elapsed time from fire discovery until the first personnel charged with initiating action for fire suppression are notified of its existence and location.

Residual smoke Smoke produced by smoldering material after the initial fire front has passed through the fuel.

Resistance to control The relative difficulty of constructing and holding a control line as affected by resistance to line construction and by fire behavior. Also called *difficulty of control*.

Resistance to line construction The relative difficulty of constructing control line as determined by the fuel, topography, and soil.

Resource order Form used by dispatchers, service personnel, and logistics coordinators to document the request, ordering, or release of resources, and the tracking of those resources on an incident.

Resources (1) Personnel, equipment, services, and supplies available, or potentially available, for assignment to incidents. Personnel and equipment are described by kind and type, for example, ground, water, air, and may be used in tactical, support, or overhead capacities at an incident. (2) The natural resources of an area, such as timber, grass, watershed values, recreation values, and wildlife habitat. *See also* Values-at-risk.

Resource Status Board Visual aid containing pertinent information regarding fire organization, current operational period resources, previous operational period resources, and next operational period resources being prepared; placed at a convenient location in fire camp for review by fireline overhead personnel on large fires.

Resource Unit Functional unit within the planning section responsible for recording the status of resources committed to the incident. The unit also evaluates resources currently committed to the incident, the impact that additional responding resources will have on the incident, and anticipated resource needs.

Resource Unit Leader The person responsible for establishing all check-on activities; preparation and processing of resource status change information; the preparation and maintenance of displays, charts, and lists that reflect the current status and location of suppression resources, transportation and support vehicles; and for maintaining a master check-in list of resources assigned to the incident.

Resource Use Specialist Person responsible to the planning section chief for determining capabilities and limitations of resources at an incident.

Resource value-at-risk Fire suppression planning tool providing a relative expression (in five classes) of fire effects on all resources (not the value of the resources themselves).

Response Movement of an individual firefighting resource from its assigned standby location to another location or to an incident in reaction to dispatch orders or to a reported alarm.

Responsible fire agency Agency with primary responsibility for fire suppression on any particular land area. *See also* Fire agency.

Retardant A substance or chemical agent that reduces the flammability of combustibles.

Retardant coverage Area of fuel covered and degree of coverage on the fuel by a fire retardant, usually expressed in terms of gallons per hundred square feet (liters per square meter).

Retardant drop Fire retardant cascaded from an air tanker or helitanker.

Reversible siamese (SIMWYE) Hose fitting that performs the functions of a siamese or a wye. *See* Siamese; Wye.

Rich tool A long-handled combination rake and cutting tool, the blade of which is constructed of a single row of mowing machine cutter teeth fastened to a piece of angle iron. Also called *fire rake* or *council rake*. *See also* Council tool.

Ridge An elongated area of relatively high atmospheric pressure extending from the center of a high-pressure region. *See also* Surface high.

Ring firing A technique generally used as an indirect attack and backfire operation. It involves circling the perimeter of an area with a control line and then firing the entire perimeter. Ring firing is often used to burn out around structures, preserve historic and archeological sites, or protect endangered species.

Risk (1) The chance of fire starting as determined by the presence and activity of causative agents. (2) A causative agent. (3) In the NFDRS, a number related to the potential of firebrands to which a given area will be exposed during the rating day.

Risk index A number related to the probability of a firebrand igniting a fire.

Roll cloud A turbulent altocumulus cloud formation found in the lee of some large mountain barriers. The air in the cloud rotates around an axis parallel to the range. Also sometimes refers to part of the cloud base along the leading edge of a cumulonimbus cloud; it is formed by rolling action in the wind shear region between cool downdrafts within the cloud and warm updrafts outside the cloud. Also called *rotor cloud*.

Rough The accumulation of living and dead ground and understory vegetation, especially grasses, forest litter, and draped dead needles, sometimes with addition of underbrush such as palmetto, gallberry, and wax myrtle. Most often used for southern pine.

Run (of a fire) Rapid advance of the head of a fire, characterized by a marked transition in fireline intensity and rate of spread with respect to that noted before and after the advance.

Running fire Behavior of a fire spreading rapidly with a well-defined head.

Rural Any area wherein residences and other developments are scattered and intermingled with forest, range, or farm land and native vegetation or cultivated crops.

Rural fire district (RFD) An organization established to provide fire protection to a designated geographic area outside of areas under municipal fire protection. Usually has some taxing authority and officials may be appointed or elected.

Rural fire protection Fire protection and firefighting problems that are outside of areas under municipal fire prevention and building regulations and that are usually remote from public water supplies.

Safe refuge area An area that can be used to shelter firefighters or residents in place until the flaming front passes. In the case of a house, this may provide only temporary shelter as it may later burn as the flame front passes. Once the flame front passes occupants of the house can exit into the burned area created by the passage of the flame front.

Safety circle An obstruction-free circle around the (helicopter) landing pad.

Safety officer A member of the command staff responsible to the incident commander for monitoring and assessing hazardous and unsafe situations and developing measures for assessing personnel safety.

Safe zone An area cleared of flammable materials used for escape in the event the line is outflanked or in case a spot fire causes fuels outside the control line to render the line unsafe. In firing operations, crews progress so as to maintain a safety zone close at hand allowing the fuels inside the control line to be consumed before going ahead. Safety zones may also be constructed as integral parts of fuelbreaks; they are greatly enlarged areas that can be used with relative safety by firefighters and their equipment in the event of blowup in the vicinity. *See also* Deployment zone.

Salvo Dropping by an air tanker of its entire load of fire retardant at one time.

Scorch height Average heights of foliage browning or bole, blackening caused by a fire.

Scratch line An unfinished preliminary control line hastily established or constructed as an emergency measure to check the spread of fire.

Scrubbing The process of agitating foam solution and air within a confined space (usually a hose) that produces tiny, uniform bubbles. The length and type of hose determine the amount of scrubbing and, therefore, foam quality.

Sea breeze A convective wind that occurs during the day as a result of the land mass becoming warmer than the adjacent water surface.

Sea-level pressure Pressure value obtained by the theoretical reduction or increase of station pressure to sea level. The average atmospheric pressure at sea level is 14.7 pounds per square inch.

Secondary line Any fireline constructed at a distance from the fire perimeter concurrently with or after a line already constructed on or near to the perimeter of the fire. Generally constructed as an insurance measure in case the fire escapes control by the primary line.

Secondary lookout (1) A lookout point intermittently used to supplement the visible area coverage of the primary lookout system when required by fire danger, poor visibility, or other factors. (2) The person who occupies such a station.

Section That organizational level with responsibility for a major functional area of the incident, such as operations, planning, logistics, finance/administration. The section is organizationally between branch and incident commander.

Security Manager Person responsible to the facilities unit leader for safeguarding property from loss or damage, and also personnel and their personal goods while at the incident.

Segment A geographical area in which a task/strike team leader or supervisor of a single resource is assigned authority and responsibility for the coordination of resources and implementation of planned tactics. A segment may be a portion of a division or an area inside or outside the perimeter of an incident. Segments are identified with arabic numbers (e.g., A-1) and are not to be used as radio designators.

Service branch A branch within the logistics section responsible for service activities at the incident. Includes the communications, medical, and food units.

Severe burn Degree of burn in which all organic material is removed from the soil surface and soil surface is discolored (usually red) by heat; organic material below the surface is consumed or charred.

Severity funding Funds provided to increase wildland fire suppression response capability necessitated by abnormal weather patterns, extended drought, or other events causing abnormal increase in the fire potential and/or danger.

Severity index A number that indicates the relative net effects of daily fire danger on the fire load for an area during a specified period, such as a fire season.

Short-term fire retardant Fire retardant that inhibits primarily by the cooling and smothering action of water. Chemicals may be added to the water to alter its viscosity or retard its evaporation, thereby increasing its effectiveness.

Shoulder carry Method of carrying hose on the shoulders.

Shrub A woody perennial plant differing from a perennial herb by its persistent and woody stem and from a tree by its low stature and habit of branching from the base.

Shutoff nozzle Common type of fire hose nozzle permitting stream flow to be controlled by the firefighter at the nozzle rather than only at the source of supply.

Shutoff pressure Maximum pressure a centrifugal pump will attain when water flow is clamped or shut off.

Siamese Hose fitting (preferably gated) for combining flow from two or more lines of hose into a single stream; one male coupling to two female couplings. *See also* Reverse siamese; Wye.

Simms bucket Self-leveling helibucket slung under a helicopter that can be filled by hovering over a water source.

Simple hose lay A hose lay consisting of consecutively coupled lengths of hose without laterals. The lay is extended by inserting additional lengths of hose in the line between pumps and nozzle. Also called *single hose lay*.

Single resource An individual, a piece of equipment and its personnel complement, or a crew or team of individuals with an identified work supervisor that can be used on an incident.

Situation analysis Analysis of factors that influence suppression of an escaped fire from which a plan of attack will be developed; includes development of alternative strategies of fire suppression and net effect of each.

Situation Unit Functional unit within the planning section responsible for the collection, organization, and analysis of incident status information, and for analysis of the situation as it progresses. Reports to the planning section chief.

Situation Unit Leader The person responsible for the collection and organization of incident status and situation information as well as the evaluation, analysis, and display of that information for use by ICS personnel and agency dispatchers.

Size class of fire (As to size of wildfire):

Class A—one-fourth acre or less

Class B—more than one-fourth acre, but less than 10 acres

Class C—10 acres or more, but less than 100 acres

Class D—100 acres or more, but less than 300 acres

Class E—300 acres or more, but less than 1,000 acres

Class F—1,000 acres or more, but less than 5,000 acres

Class G—5,000 acres or more.

See also Class of fire.

Size-up The evaluation of the fire to determine a course of action for suppression.

Skidder unit A self-contained unit consisting of a water tank, fire pump, and hose specially designed to be carried on a logging skidder for use in forest fire suppression.

Sky cover Amount of clouds and/or other obscuring phenomena detectable from the point of observation.

Slash Debris resulting from such natural events as wind, fire, or snow breakage; or such human activities as road construction, logging, pruning, thinning, or brush cutting; includes logs, chunks, bark, branches, stumps, and broken understory trees or brush.

Sling A net attached by a lanyard to a helicopter cargo hook and used to haul supplies.

Sling load Any cargo carried beneath a helicopter and attached by a lead line and swivel.

Sling psychrometer A hand-operated instrument for obtaining wet- and dry-bulb temperature readings and, subsequently, relative humidity.

Slope class Code that designates the most common slope encountered in the primary fire problem area on a protection unit. Slope class 1 is 0–20%, slope class 2 is 21–40%, slope class 3 is 41–55%, slope class 4 is 56–74%, and slope class 5 is 75% or greater. Part of the National Fire Danger Rating System (NFDPS).

Slope percent The ratio between the amount of vertical rise of a slope and horizontal distance as expressed in a percent. One hundred feet of rise to 100 feet of horizontal distance equals 1.00 percent.

Slope winds Small-scale convective winds that occur due to local heating and cooling of a natural incline of the ground. *See also* General winds; Local winds.

Slopover A fire edge that crosses a control line or natural barrier intended to confine the fire. Also called *breakover.*

Slug flow The discharge of distinct pockets of water and air due to the insufficient mixing of foam concentrate, water, and air in a compressed air foam system.

Smoke A term used when reporting a fire or probable fire in its initial stages. In fire control, the following types of smokes are recognized: legitimate smoke, false smoke, drift smoke, intermittent smoke, smoke haze, and smoke column.

Smokejumper A specifically trained and certified firefighter who travels to wildland fires by aircraft and parachutes to the fire.

Smoke management Application of fire intensities and meteorological processes to minimize degradation of air quality during prescribed fires.

Smoke plume The gases, smoke, and debris that rise slowly from a fire while being carried along the ground because the buoyant forces are exceeded by those of the ambient surface wind. *See also* Convection column.

Smoke vent height Level, in the vicinity of the fire, at which the smoke ceases to rise and moves horizontally with the wind at that level.

Smoldering A fire burning without flame and barely spreading.

Smoldering combustion Combustion of a solid fuel, generally with incandescence and smoke but without flame.

Smoldering combustion phase Combined processes of dehydration, pyrolysis, solid oxidation, and scattered flaming combustion and glowing combustion that occur after the flaming combustion phase of a fire; often characterized by large amounts of smoke consisting mainly of tars. Emissions are at twice that of the flaming combustion phase.

Smoldering phase The overall reaction rate of the fire has diminished to a point at which concentrations of combustible gases above the fuel are too low to support a persistent flame envelope. Consequently, the temperature drops and gases condense. The smoke evolved during this phase is virtually soot-free, consisting mostly of tar droplets less than a micrometer in size.

Smudge Spot in a fire or along a fire edge that has not been extinguished, and that is producing smoke; term is commonly used during the mop-up stage of a fire.

Snag A standing dead tree or part of a dead tree from which at least the leaves and smaller branches have fallen. Often called a stub if less than 20 feet tall.

Snorkel tank A fixed tank attached to the belly of the helicopter that has a pump-driven snorkel attached. The helicopter hovers over the water source with the end of the snorkel immersed, and the pump then fills the tank.

Soft suction Commonly accepted term for short length of large diameter soft hose used to connect a structural or wildland engine with a hydrant. No vacuum is involved because the hose is useful only when the engine receives water at a rate of flow in excess of the demand of the pump.

Soot Carbon dust formed by incomplete combustion.

Spanner Metal wrench used to tighten and free hose connections.

Span of control The supervisory ratio of from three to seven individuals, with five-to-one being established as optimum.

Specific heat The heat required to raise the temperature of 1 kilogram of a substance one degree kelvin. The heat capacity of a system per unit mass; that is, the ratio of the heat absorbed (or released) to the corresponding temperature rise (or fall).

Spike-out Standby crew in an area of expected high fire occurrence, generally on a day of critical fire weather.

Split drop Retardant drop made from one compartment at a time from an air tanker with a multicompartment tank.

Spot burning A modified form of broadcast slash burning in which the greater accumulations of slash are fired and the fire is confined to these spots. Sometimes called *jackpot burning* or *jackpotting.*

Spot fire Fire ignited outside the perimeter of the main fire by a firebrand.

Spot fire technique A method of lighting prescribed fires where ignition points are set individually at a predetermined spacing and with predetermined timing throughout the area to be burned.

Spotter In smokejumping, rappelling, and paracargo operations, the individual responsible for selecting drop target and supervising all aspects of dropping smokejumpers, rappellers, or cargo.

Spotting Behavior of a fire producing sparks or embers that are carried by the wind and which start new fires beyond the zone of direct ignition by the main fire.

Spot weather forecast A special forecast issued to fit the weather of each specific fire. These forecasts are issued upon request of the user agency and are more detailed, timely, and specific than zone forecasts. (Usually special on-site weather observations are required for the forecasting office.) *See also* General fire weather forecast; Incident weather forecast.

Spread component A rating of the forward rate of spread of the head of a fire, part of the National Fire Danger Rating System (NFDRS).

Spread index A number used to indicate relative (not actual) rate of spread.

Spread index meter Device for combining measured ratings of various fire danger factors into numerical classes or rates of spread.

Spur ridge A small ridge that extends fingerlike from a main ridge.

Squall Sudden increase in wind speed to at least 17 mph (15 knots) that is sustained for at least 1 minute but not more than 5 minutes.

Squall line Any nonfrontal line or narrow band of active thunderstorms extending across the horizon. It is of importance to fire behavior due to accompanying strong gusty winds and the possibility of such a line passing between regular weather observation stations without being reported. Also called *line squall.*

Stable Condition of the atmosphere in which the temperature decrease with increasing altitude is less than the dry adiabatic lapse rate. In this condition, the atmosphere tends to suppress large-scale vertical motion. Also called *stable air.*

Stable layer of air A layer of air having a temperature change (lapse rate) of less than dry adiabatic (approximately –5.4°F per 1,000 feet) thereby retarding either upward or downward mixing of smoke. *See also* Inversion; Atmospheric stability; Atmospheric inversion.

Staging area Location set up at an incident where resources can be placed while awaiting a tactical assignment on a 3-minute available basis. Staging areas are managed by the operations section.

Stagnant conditions Atmospheric conditions under which pollutants build up faster than the atmosphere can disperse them.

Standard coupling Fire hose coupling with American National Standard (NH) threads.

State of weather A code that expresses the amount of cloud cover, kind of precipitation, and/or restriction to visibility being observed at the fire danger station at basic observation time.

Static pressure Water pressure head available at a specific location when no water is being used so that no friction loss is being encountered. Static pressure is that pressure observed on the engine inlet gauge before any water is taken from the hydrant.

Static water supply Supply of water at rest that does not provide a pressure head for fire suppression but that may be employed as a suction source for fire pumps (e.g., water in a reservoir, pond, or cistern).

Status Check-In Recorder Person responsible to the resources unit leader for checking-in all resources arriving at an incident. There is at least one check-in recorder at each check-in location.

Step test Five-minute test used to predict a person's ability to take in, transport, and use oxygen (aerobic capacity), the most important factor limiting the ability to perform arduous work.

Straight stream Water or fire retardant projected directly from the nozzle (as contrasted with a fog or spray cone), provided by a solid stream orifice or by adjusting a fog jet into a straight stream pattern.

Straight stream nozzle A hose tip spout designed to provide the maximum reach of water without feathering.

Strainer A wire or metal guard used to keep debris from clogging pipe or other openings made for removing water; used in pumps and on suction hose to keep foreign material from clogging or damaging pumps.

Strategy The general plan or direction selected to accomplish incident objectives.

Strength of attack Number of resources used to attack a fire.

Strike team Specified combinations of the same kind and type of resources, with common communications, and a leader.

Stringer A narrow finger or band of fuel that connects two or more patches or areas of wildland fuel.

Strip burning (1) Burning by means of strip firing. (2) In hazard reduction, burning narrow strips of fuel and leaving the rest of the area untreated by fire.

Strip firing Setting fire to more than one strip of fuel and providing for the strips to burn together. Frequently done in burning out against a wind where inner strips are fired first to create drafts that pull flames and sparks away from the control line.

Strip-head fire A series of lines of fire ignited near and upwind (or downslope) of a firebreak or backing fire so they burn with the wind (or upslope) toward the firebreak or backing fire.

Subsidence Downward or sinking motion of air in the atmosphere. Subsiding air warms due to compression. Increasing temperature and decreasing humidities are present in subsiding air. Subsidence results in a stable atmosphere inhibiting dispersion. Subsidence is generally associated with high atmospheric pressure.

Subsidence inversion An inversion caused by subsiding air often resulting in very limited atmospheric mixing conditions.

Suction life In fire service, the number of feet (meters) of vertical lift from the surface of the water to the center of the pump impeller. In testing, fire department pumpers are required to discharge their rated capacity at 150 pounds (1034 kilopascals) net pump pressure at a 10 foot (3 meter) lift. The suction gauge would indicate the vertical suction lift in inches of mercury when the pump was primed with no appreciable water flowing.

Supervisor The ICS title for individuals responsible for command of a division or group.

Supplies Minor items of equipment and all expendable items assigned to an incident.

Supply Unit Functional unit within the support branch of the logistics section responsible for ordering equipment and supplies required for incident operations.

Support Branch A branch within the logistics section responsible for providing personnel, equipment, and supplies to support incident operations. Includes the supply, facilities, and ground support units.

Supporting agency An agency providing suppression or other support and resource assistance to a protecting agency. *See also* Agency; Assisting agency; Cooperating agency.

Suppress a fire The most aggressive wildfire suppression strategy leading to the total extinguishment of a wildfire.

Suppressant An agent that extinguishes the flaming and glowing phases of combustion by direct application to the burning fuel.

Suppression crew Two or more firefighters stationed at a strategic location for initial action on fires. Duties are essentially the same as those of individual firefighters.

Suppression firing Intentional application of fire to speed up or strengthen fire suppression action on wildfires. Types of suppression firing include burning out, counterfiring, and strip burning.

Surface area to volume ratio The ratio between the surface area of an object, such as a fuel particle, to its volume. The smaller the particle, the more quickly it can become wet, dry out, or become heated to combustion temperature during a fire.

Surface fire Fire that burns loose debris on the surface, which includes dead branches, leaves, and low vegetation.

Surface fuel Fuels lying on or near the surface of the ground, consisting of leaf and needle litter, dead branch material, downed logs, bark, tree cones, and low stature living plants.

Surface high An area on the Earth's surface where atmospheric pressure is at a relative maximum. Winds blow clockwise around highs in the Northern Hemisphere but, due to friction with the Earth's face, tend to cross constant pressure lines away from the high center. Air is usually subsiding within a surface high. This causes warming due to air compression, which results in stable atmospheric conditions and light surface winds. *See also* Ridge.

Surface low An area on the earth's surface where atmospheric pressure is at a relative minimum. Winds blow counterclockwise around lows in the Northern Hemisphere but, due to friction with the Earth's surface, tend to cross constant pressure lines toward the low center. Upon converging at the low center, air currents are forced to rise. As air rises it cools due to expansion. Cooling reduces its capacity to hold moisture; so cloudiness and precipitation are common in lows. If a low center intensifies sufficiently, it will take on the characteristics of a storm center with precipitation and strong winds.

Surface tension The elasticlike force at the surface of a liquid, tending to minimize the surface area and causing drops to form, expressed as Newtons per meter or dynes per centimeter.

Surface wind Wind measured at a surface observing station, customarily at some distance (usually 20 feet) above the average vegetative surface to minimize the distorting effects of local obstacles and terrain.

Surfactant A surface active agent; any wetting agent. A formulation which, when added to water in proper amounts, materially reduces the surface tension of the water and increases penetration and spreading abilities of the water.

Surge Rapid increase in flow, which may result in an attendant pressure rise.

Survival zone A natural or cleared area of sufficient size and location to protect fire personnel from known hazards while inside a fire shelter. Examples include rock slides, road beds, clearings, knobs, wide ridges, benches, bulldozer lines, wet areas, cleared areas in light fuels, and previously burned areas. These are all areas where you expect no flame contact or prolonged heat and smoke. *See also* Deployment zone; Safe zone.

Sustained attack Continuing fire suppression action until fire is under control.

Swamper (1) A worker who assists fallers and/or sawyers by clearing away brush, limbs and small trees. Carries fuel, oil, and tools and watches for dangerous situations. (2) A worker on a bulldozer crew who pulls winch line, helps maintain equipment, and other jobs to speed suppression work on a fire.

Tactical direction Direction given by the operations section chief that includes the tactics appropriate for the selected strategy, the selection and assignment of resources, and performance monitoring for each operational period.

Tactics Deploying and directing resources on an incident to accomplish the objectives designated by strategy.

Tag-on Connecting a (air tanker) drop to the forward part of a previous drop.

Tailgate safety session Brief meetings held at the beginning, or end, or during an operational period to discuss new work assignments, new work methods, changes in plans, use of tools and equipment, and recognition and protection against work hazards that may be encountered.

Tandem Two or more units of any one type working one in front of the other to accomplish a specific fire suppression job; the term can be used in connection with crews of firefighters, power pumps, bulldozers, and so forth.

Task force Any combination of single resources assembled for a particular tactical need, with common communications and a leader. A task force may be preestablished and sent to an incident, or formed at an incident.

Technical Advisory Unit Consists of advisors with special skills, for example, in the areas of water resources, environmental concerns, resource use, and training, who are activated only when needed.

Technical Specialists Personnel with special skills that can be used anywhere within the ICS organization. These personnel may perform the same duties during an incident that they perform in their everyday job.

Ten-hour timelag fuel moisture (10-h TL FM) The moisture content of the 10-hour timelag roundwood fuels.

Ten-hour timelag fuels Dead fuels consisting of roundwood 1/4 to 1 inch (0.6 to 2.5 cm) in diameter and, very roughly, the layer of litter extending from immediately below the surface to 3/4 inch (1.9 cm) below the surface.

Termination point A planned point in the fireline where a firing operation is terminated.

Terra Torch Device for throwing a stream of flaming liquid, used to facilitate rapid ignition during burnout operations on a wildfire or during a prescribed fire operation.

Test fire A prescribed fire set to evaluate such things as fire behavior, detection performance, control measures.

Thermal belt An area of mountainous slope (characteristically the middle third), where the top of the radiation inversion intersects the slope. It typically experiences the least variation in diurnal temperatures and has the highest average temperatures and, thus, the lowest relative humidity. Its presence is most evident during clear weather with light wind.

Thermal imagery The display or printout of an infrared scanner operating over a fire. Also called *infrared imagery*. *See also* Infrared.

Thermal lifting Air being lifted aloft by local heating of the land mass. As heated surface air becomes buoyant, it is forced aloft and cools.

Thin layer Layer of clouds whose ratio of dense sky cover to total sky cover is 1/2 or less.

Thin sky cover Sky cover through which higher clouds or the sky can be detected.

Thread The specific dimensions of screw thread employed to coupled fire hose and equipment. American National Standards (NH) have been adopted for fire hose couplings threads in $\frac{3}{4}$, $1\frac{1}{2}$, $2\frac{1}{2}$, $3\frac{1}{2}$, 4, 5, and 6 inch sizes.

Threat fire Any uncontrolled fire near to or heading toward an area under organized fire protection.

Throw out Soil pushed over the edge of the fireline by the fire plow. *See also* Berm.

Thunderstorm Localized storm characterized by one or more electrical discharges.

Tie-in Act of connecting a control line to another line or an intended firebreak.

Timelag (TL) Time needed under specified conditions for a fuel particle to lose about 63% of the difference between its initial moisture content and its equilibrium moisture content. If conditions remain unchanged, a fuel will reach 95% of its equilibrium moisture content after four timelag periods.

Time-temperature curve Graph showing the increase in temperature at a specified point in a fire as a function of time, beginning with ignition and ending with burnout.

Time unit Functional unit within the finance/administration section responsible for recording time for incident personnel and hired equipment.

Tips Nozzle tips used to change orifice size of a hose stream.

Torch Ignition and subsequent envelopment in flames, usually from bottom to top, of a tree or small group of trees.

Tractor A rubber tired or tracked rider-controlled automotive vehicle used in wildland fire management for pulling a disk or a plow to construct fireline by exposing mineral soil. *See also* bulldozer; Plow line; Tractor plow.

Tractor plow Any tractor with a plow for constructing fireline by exposing mineral soil. Also as a resource for typing purposes, a tractor plow includes the transportation and personnel for its operation. *See also* Bulldozer; Plow line; Tractor.

Trail drop A drop from an aircraft in which fire retardant is dropped sequentially in order to extend the length of the drop.

Training Specialist Person responsible to the Planning Section Chief for coordinating the use of trainees on the incident and for ensuring that the trainees meet their training objectives and receive performance evaluation reports.

Trench A small ditch often constructed below a fire on sloping ground (undercut or underslung line) to catch rolling material.

Trigger point A point in the fire's path that is used to alert firecrews of the arrival of the impending fire.

Trough An elongated area of relatively low atmospheric pressure, usually extending from the center of a low pressure system.

Truck trail Substantial transportation route for fire suppression motor vehicles, built prior to a fire. Also called *fire road.*

True bearing Bearing by true north rather than magnetic north.

Turnaround time Time used by an air tanker or helitanker to reload and return to the fire.

Turn the corner Contain a fire along a flank of the fire and begin containing it across the head. Refers to ground or air attack.

Type Refers to resource capability. A Type 1 resource provides a greater overall capability due to power, size, capacity, etc., than would be found in a Type 2 resource. Resource typing provides managers with additional information in selecting the best resource for the task.

Uncontrolled fire Any fire that threatens to destroy life, property, or natural resources, and (a) is not burning within the confines of firebreaks, or (b) is burning with such intensity that it could not be readily extinguished with ordinary tools commonly available. *See also* Wildfire.

Underburn A fire that consumes surface fuels but not trees and shrubs.

Undercut line A fireline below a fire on a slope. Should be trenched to catch rolling material. Also called *underslung line. See also* Cup trench.

Understory burning Prescribed burning under a forest canopy.

Unified area command A unified area command is established when incidents under an area command are multijurisdictional.

Unified command In ICS, unified command is a unified team effort that allows all agencies with jurisdictional responsibility for the incident, either geographical or functional, to manage an incident by establishing a common set of incident objectives and strategies. This is accomplished without losing or abdicating authority, responsibility, or accountability.

Uniform fuels Fuels distributed continuously, thereby providing a continuous path for fire to spread.

Unit The organizational element of an incident having functional responsibility for a specific activity in the planning, logistics, or finance/administration activity.

Unity of command The concept by which each person within an organization reports to one and only one designated person.

Universal Transverse Mercator A rectangular coordinate system of determining location on the Earth's surface, similar to latitude and longitude, but defined in meters rather than degrees. UTM is commonly used in Global Positioning System (GPS) and Geographic Information System (GIS) mapping, as it allows for a high degree of precision.

Unlined fire hose Hose commonly of cotton, linen, or synthetic fiber construction without rubber tube or lining, often used for wildfires because of its light weight and self-protecting (weeping) characteristics; such hose is attached to first-aid standpipes in buildings. At a specified flow, friction loss in unlined hose of a stated diameter is about twice that of lined fire hose.

Upper level (cold) low (Upper Level Disturbance, Cold Low Aloft) A circulation feature of the upper atmosphere where pressure, at a constant altitude, is lowest. Winds blow counterclockwise around the center in an approximately circular pattern. Upper level lows are usually quite small. The mechanics of these upper lows is such that a pool of cool moist air always accompanies their development. There is often no evidence of low pressure at the Earth's surface. An upper low may exist above a surface high pressure system.

Upper level (cold) trough (Trough, Trough Aloft, Upper Level [Cold] Low) An elongated area of relatively low pressure, at constant altitude, in the atmosphere. The opposite of an upper level ridge. Upper level troughs are usually oriented north-south with the north end open. That is, air currents moving from west to east around the Earth flow around three sides of the trough then turn eastward rather than toward the west, as in the case of a closed circulation. A large upper level trough may have one or more small upper level closed low circulation systems within it.

Upper level high (Upper High, High Aloft, Upper Level Ridge) A circulation feature of the upper atmosphere where pressure, at a constant altitude, is higher than in the surrounding region. Winds blow clockwise around an upper level high. Air in an upper level high is usually subsiding. This results in comparatively warm dry air with light winds over a large area. An upper level high may exist without there being high pressure at the Earth's surface.

Upper level ridge (Upper Level High, Ridge Aloft) An elongated area of relatively high pressure, at a constant altitude, in the atmosphere. The opposite of an upper level trough. Upper level ridges are often oriented north-south, alternating between upper level troughs, however, during summer they may assume random orientations and vast dimensions.

Urban Area in which residences and other human developments form an essentially solid covering of the landscape, including most areas within cities and towns, subdivisions, commercial and industrial parks, and similar developments whether inside city limits or not.

Urban interface The line, area, or zone where structures and other human development meet or intermingle with undeveloped wildland or vegetative fuels. Syn.: I-zone, wildland urban interface.

Values-at-risk Natural resources, improvements, or other values that may be jeopardized if a fire occurs; estimated damages and benefits that may result from fires in a particular presuppression or suppression situation. *See also* Resources.

Variable wind direction Wind direction that varies by 60° or more during the period of time the wind direction is being determined.

Vectors Directions of fire spread as related to rate of spread calculations (in degrees from upslope).

Vee pattern To make two separate drops in an overlapping configuration, usually to stop the head.

Vertical fuel arrangement Fuels above ground and their vertical continuity, which influences fire reaching various levels or vegetation strata.

Virga Precipitation falling out of a cloud but evaporating before reaching the ground.

Viscosity An indication in the ability of the foam to spread and cling, as well as to cling to itself, upon delivery.

Vortex turbulence Miniature whirlwinds trailing from the wingtips of any aircraft in flight. Vortex will be in the form of a horizontal whirlwind with velocities up to 25 miles per hour (40 kilometers) per hour or more. Also created by action of rotor blades on helicopters; these whirlwinds tend to move downward toward the ground. If an aircraft flies low over a fire, vortices may reach the ground and suddenly cause violent and erratic fire behavior.

Warm front The leading edge of a relatively warm air mass that moves in such a way that warm air displaces colder air. Winds associated with warm frontal activity are usually light and mixing is limited. The atmosphere is relatively stable when compared to cold front activity.

Water bar A shallow channel or raised barrier, for example, a ridge of packed earth or a thin pole laid diagonally across the surface of a road or trail so as to lead off water, particularly storm water. Frequently installed in firelines on steep slopes to prevent erosion.

Water hammer A force created by the rapid acceleration or deceleration of water, commonly created by opening or closing a valve too quickly. Pressures developed in a water hammer, proportional to the mass multiplied by the square of the velocity, can damage a pipe or hose.

Water Resources Specialist Person responsible to the planning section chief for collecting information of water resources in and adjacent to the incident area, determining water use requirements of firefighting resources, and providing input to the incident action plan about available water resources and/or anticipated shortages.

Water source Any strategically located supply of water that is readily available for pumps, tanks, trucks, helicopters, or fire camp use.

Water supply map A map showing location of supplies of water readily available for pumps, tanks, trucks, or camp use.

Water tender Any ground vehicle capable of transporting specified quantities of water.

Water thief A type of bleeder valve designed for installation at convenient points in hose lines to permit drawing off water for filling backpack pumps or other use without interfering with pump or nozzle operation.

Weather Information and Management System (WIMS) An interactive computer system designed to accommodate the weather information needs of all federal and state natural resource management agencies. Provides timely access to weather forecasts, current and historical weather data, the National Fire Danger Rating System (NFDRS), and the National Interagency Fire Management Integrated Database (NIFMID).

Weather Observer Person responsible to the situation unit leader for collecting current weather data and information at the incident and providing them to an assigned meteorologist, fire behavior specialist, or the situation unit leader.

Wet-bulb thermometer In a psychrometer, the thermometer with its bulb covered with a jacket of clean muslin that is saturated with distilled water before an observation.

Wet foam A low expansion foam type with few and varied bubbles and rapid drain time used for rapid penetration and fire extinguishment.

Wet line A line of water, or water and chemical retardant, sprayed along the ground, which serves as a temporary control line from which to ignite or stop a low-intensity fire.

Wetting agent A chemical that when added to water reduces the surface tension of the solution and causes it to spread and penetrate exposed objects more effectively than the untreated water.

Widow maker A loose limb or top or piece of bark lodged in a tree that may fall on anyone working beneath it.

Wilderness An area established by the federal government and administered either by the Forest Service, USDA or National Park Service, Fish and Wildlife Service, or Bureau of Land Management, in order to conserve its primeval character and influence for public enjoyment, under primitive conditions, in perpetuity.

Wildfire A fire occurring on wildland that is not meeting management objectives and thus requires a suppression response. *See also* Uncontrolled fire.

Wildland An area in which development is essentially nonexistent, except for roads, railroads, powerlines, and similar transportation facilities. Structures, if any, are widely scattered.

Wildland fire Any fire occurring on the wildlands, regardless of ignition source, damages or benefits.

Wildland urban interface The line, area, or zone where structures and other human development meet or intermingle with undeveloped wildland or vegetative fuels. Syn.: I-zone, urban interface.

Wind The horizontal movement of air relative to the surface of the Earth.

Wind direction Compass direction from which wind is blowing.

Wind-driven wildland fire A wildland fire controlled by a strong consistent wind.

Windfall Tree that has been uprooted or broken off by wind.

Wind profile A chart of wind speed in relation to height, most commonly determined by a pilot balloon observation.

Windrow burning Burning slash that has been piled into long continuous rows. Also includes wildfire in vegetation planted to protect improvements or agriculture.

Winds aloft Generally, wind speeds and wind directions at various levels in the atmosphere above the domain of surface weather observations.

Wind shift A change in the average wind direction of 45° or more that takes place in less than 15 minutes if the wind speed during this period is 6 knots or greater.

Wind speed (1) Rate of horizontal motion of air past a given point. (2) (NFDRS) Wind, in miles per hour, measured at 20 feet above ground, or above the average height of vegetation, and averaged over at least a 10-minute period. Also called *wind velocity.*

Wind vectors Wind directions used to calculate fire behavior.

Woody vegetation condition A code reflecting the moisture content of the foliage and small twigs (less than ¼ inch (0.6 centimeters) of living woody plants. Part of the National Fire Danger Rating System (NFDRS).

Woven jacket fire hose Fire hose of conventional construction, woven on looms from fibers of cotton or synthetic fibers. Most fire department hose is double jacketed (i.e., it has an outer jacket protecting the inner one against wear and abrasion).

Wye A hose connection with two outlets permitting two connections of the same coupling diameter to be taken from a single supply line. *See also* Reversible siamese; Siamese.

Index